T0296817

Protecting Transportation

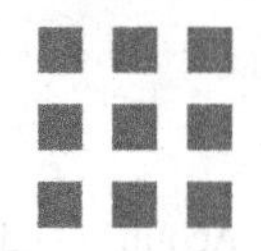

Protecting Transportation

Implementing Security Policies and Programs

R. William Johnstone

AMSTERDAM • BOSTON • HEIDELBERG • LONDON
NEW YORK • OXFORD • PARIS • SAN DIEGO
SAN FRANCISCO • SINGAPORE • SYDNEY • TOKYO
Butterworth-Heinemann is an imprint of Elsevier

Acquiring Editor: Sara Scott
Editorial Project Manager: Marisa LaFleur
Project Manager: Punithavathy Govindaradjane
Designer: Matthew Limbert

Butterworth-Heinemann is an imprint of Elsevier
The Boulevard, Langford Lane, Kidlington, Oxford OX5 1GB, UK
225 Wyman Street, Waltham, MA 02451, USA

ISBN: 978-0-12-408101-7

British Library Cataloguing-in-Publication Data
A catalogue record for this book is available from the British Library

Library of Congress Cataloging-in-Publication Data
A catalog record for this book is available from the Library of Congress

For information on all Butterworth-Heinemann publications
visit our website at http://store.elsevier.com/

Table of Contents

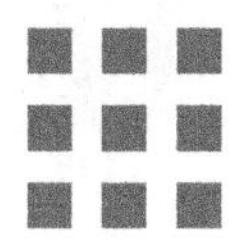

Digital Assets

Thank you for selecting Butterworth Heinemann's *Protecting Transportation.* To complement the learning experience, we have provided a number of online tools for instructors to accompany this edition.

Please consult your local sales representative with any additional questions.

For the Instructor

Qualified adopters and instructors need to register at the this link for access: http://textbooks.elsevier.com/web/manuals.aspx?isbn=9780124081017

- **Test Bank** Compose, customize, and deliver exams using an online assessment package in a free Windows-based authoring tool that makes it easy to build tests using the unique multiple choice and true or false questions created for *Protecting Transportation.* What's more, this authoring tool allows you to export customized exams directly to Blackboard, WebCT, eCollege, Angel, and other leading systems. All test bank files are also conveniently offered in Word format.
- **PowerPoint Lecture Slides** Reinforce key topics with focused PowerPoints, which provide a perfect visual outline with which to augment your lecture. Each individual book chapter has its own dedicated slideshow.
- **Instructor's Manual** Design your course around customized learning objectives, critical thinking questions, and key terms.

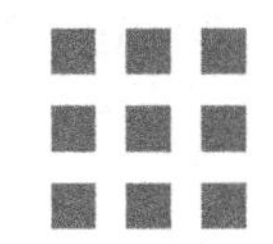

Introduction

Transportation security as a distinct discipline scarcely existed before the 1970s. From that period forward, security systems have been developed and evolved largely in response to the occurrence (or absence) of high-profile incidents, which increasingly involved acts of terrorism. Thus, by 2001, there were internationally recognized standards in place for maritime and aviation security (with enforcement left largely to national governments and thus of a highly variable quality), as well as a limited number of significant localized land transportation security efforts (including the security system developed to protect London's passenger rail network in response to decades-long attacks by the Irish Republican Army). National aviation security programs in the United States, the United Kingdom, and certain other countries had been significantly increased after terrorist bombings of passenger aircraft in the 1980s. However, the 9/11 aircraft hijackings in the United States produced the largest expansion in transportation security, resulting in its current state in which tens of billions of dollars and tens of thousands of workers are assigned to provide such security in the United States alone.

Protecting transportation systems is an extremely difficult undertaking. The front-line airport screeners, customs agents, transit operators, security inspectors, law enforcement personnel, intelligence officers, and others involved in implementing transportation security measures are confronted with the daunting mission of securing a globally connected, largely open network of airports, seaports, rail tracks, roads, tunnels, bridges, and stations, as well as the passengers, cargo, and transportation workers within those systems. This must be accomplished so as to minimize disruptions to commerce; inconveniences to passengers; and costs to shippers, customers, and taxpayers. Much of their effort is directed at continuously defeating a terrorist enemy who may choose the time, place, and method of attack. Furthermore, these defenders must cope with the fact that the nature of the terrorist threat means that security measures must be maintained over time, although actual incidents—which help to galvanize attention and secure resources—are likely to be limited in number and infrequent in occurrence.

Many works have been written about the achievements and shortcomings of these efforts to implement transportation security, with much of the focus directed to the most visible, and most expensive, component: screening of passengers and luggage at commercial airports. These are important and useful documents, and gaining an understanding of the details of the technologies, systems, and methods used in carrying out specific security measures is an important part of learning about transportation security.

Although this volume also seeks to describe the major programs that define the security measures being deployed for maritime, land, and aviation transportation systems, its primary concern is with another key component of the transportation security picture: policymaking and the strategies, plans, international agreements, laws, appropriations, and regulations that comprise it. It is transportation security policies that define security standards, authorize and fund programs, and develop and enforce regulations. These policies determine whether, and how, a particular threat is to be addressed or a program is to be reconciled with privacy and cost concerns. If the front-line implementers are confronted with substantial challenges in carrying out their responsibilities, so, too, are those who set the policies being implemented.

- Because of the impossibility of financing and carrying out efforts to protect all potential targets, governments seek to use risk management principles to focus security activities on the most vulnerable of these targets and those whose destruction or impairment would produce the most harmful consequences. Yet calculations of the threat, vulnerability, and consequence components required to inform proper risk-based decision making are all fraught with significant uncertainties and other limitations.
- Attempts to evaluate the effectiveness of security measures—of determining what works and what does not—are impeded by recurring difficulties in developing appropriate performance measurements, and much of the information that would be useful in this regard is classified (to prevent its disclosure to those who pose threats to transportation systems) and thus unavailable to many transportation security stakeholders.
- Most policy is either made directly by, or strongly influenced by, elected officials and thus is inevitably subject to political factors. In the United States, this influence has posed a number of challenges in making coherent transportation security policy, including, among others, partisan divisions that have produced uncertainty and delay in the funding process, parochial allocations of grant money, and fragmented congressional oversight of security programs.
- As has historically been the case, transportation security policies remain subject to singular events (the 9/11 hijackings being the most severe example) that can produce rapid and major changes in policy priorities.

Part of the purpose of this text is to promote awareness and understanding of these and other policymaking challenges, as well as the means developed by policymakers in coping with them and the policies that have emerged from this process. Another central aim stems from this author's belief, as expressed in a 2006 work, that shortcomings in *policy* are responsible for many of the current problems in transportation security:

> *In the pre-9/11 world, [efforts to significantly boost transportation security in the U.S. were] doomed to fail, with neither the White House, nor the Congress, nor the American public prepared to accept the financial costs and inconveniences [such actions] would have entailed. However, 9/11 was a watershed event, and in its aftermath there was a sea change in attitudes toward the terrorist threat and the priority to be attached to homeland security. And the national leadership has responded with a multi-billion dollar increase in federal expenditures and a raft*

> *of policy initiatives.... It has clearly become possible to do much more to bolster transportation security than was ever the case prior to 2001. If significant systemic problems persist in aviation and transportation security, as the available evidence indicates, the post-9/11 failure is, then, one of policy and national policy makers.*
>
> *Johnstone, 2006*

Before 9/11, transportation security policies in the United States and elsewhere were much more limited in terms of objectives, authorities, and resources than became the case after that catastrophe, when the scope of policymaking became much wider. Thus, it is the opinion here that the opportunities for the greatest improvements in transportation security, in terms of performance and cost effectiveness, lie in the policy arena and that learning more about that aspect should be a higher priority in transportation security coursework.

To both provide as up-to-date information as possible in the rapidly evolving field of transportation security and expose readers to the "world" of policymakers, multiple "primary" documents from official governmental sources (including agency websites and reports, the *Federal Register*, Government Accountability Office [GAO] and Department of Homeland Security Office of Inspector General [DHS OIG] reports, and others) are used whenever possible, supplemented as appropriate by independent analyses ("think tanks," nongovernmental stakeholders, and so on). This is meant to provide useful information not only to those who are, or aspire to be, transportation security professionals but also to policymakers themselves and to citizens seeking to understand and evaluate transportation security policies.

The focus of the book is on the U.S. transportation security system, which is, in many respects, the most elaborate such system in the world while also exerting a strong influence on what has been done at the international level. However, to place the U.S. system into greater context, some attention is given to security efforts in the European Union and other nations.

Protecting Transportation is organized into 10 chapters.

The opening chapters provide an historical overview of the evolution of threats to transportation systems and the security response to those threats. Chapter 1 covers the period before September 2001 and considers maritime piracy, the terrorist threat to each transportation mode, and the key incidents (including the 1985 seizure of the cruise ship *Achille Lauro*, the 1988 bombing of Pan Am Flight 103, and the 1995 nerve gas attack on the Tokyo subway system) that manifested these threats and provoked international and national security responses (including the international Convention for the Suppression of Unlawful Acts against the Safety of Maritime Navigation and the U.S. Aviation Security Act of 1990).

Chapter 2 addresses the events of 9/11, as well as the security measures in place on that day and how they were circumvented. In addition, the chapter describes the immediate policy reaction to the hijackings in the United States and elsewhere, as represented by new laws in the U.S. (the Aviation and Transportation Security Act of 2001, the Homeland

Security Act and Maritime Transportation Security Act of 2002, and the 2004 statute implementing many of the 9/11 Commission's recommendations) and new international security protocols (including the International Civil Aviation Organization's Aviation Security Plan of Action and the International Maritime Organization's International Ship and Port Facility Security Code).

In Chapter 3, major characteristics of transportation systems in the various modes (maritime, aviation, highways and motor carriers, mass transit and passenger rail, freight rail, and pipelines) are described, and the historical narrative of significant transportation security incidents is completed with coverage of post-9/11 events (including the terrorist attacks on Madrid commuter rail trains in 2004, against various targets in Mumbai in 2008, and on Northwest Flight 253 in 2009). Last, the concept of risk management—which has been adopted by the U.S. DHS and many other homeland security agencies as the preferred means for making policy decisions—and its key components (assessments of threat, vulnerability, and consequences) are outlined.

Chapter 4 turns to the roles and responsibilities of the various governmental and nongovernmental entities involved in transportation security at the international, national, state and local levels, and provides brief descriptions of the organizations involved as well as the instruments utilized in defining their roles (including DHS's Transportation Systems Sector-Specific Plan). Though the primary focus is on the U.S., the organizational structures in Canada, India and the United Kingdom are also discussed.

Chapter 5 explores the policymaking process in general, and how this has been applied to transportation security in particular. The American system (involving authorizing and appropriations legislation, the budget process, Presidential directives and nominations, and regulations) is again the focal point (with a brief description of the quite different process used in parliamentary systems also provided). The chapter includes actions through Fiscal Year 2013, as well as the President's FY 2014 budget proposal.

The following three chapters describe the policies and programs (including the international codes that serve as the basis for many of those policies) that implement transportation security in the United States. Independent assessments of the programs (typically performed by the GAO or the DHS OIG) are also provided where available. Chapter 6 deals with maritime security and its three major sub-components: port and vessel security (for which the Coast Guard has principal responsibility); supply chain security (where Customs and Border Protection has the lead role); and maritime domain awareness (with the Coast Guard again in the lead). The two key international frameworks for maritime security—the International Ship and Port Facility Security Code for port and vessel security and the SAFE Framework of Standards to Secure and Facilitate Global Trade for supply chain security—are also considered.

The various land (or surface) modes of transportation are the subject of Chapter 7, including programs for protecting mass transit and passenger rail, freight rail, highways and motor carriers, and pipeline systems. Though the Transportation Security Administration (TSA) has the primary responsibility at the federal level (with the various U.S. Department of Transportation modal administrations continuing to have important and sometimes overlapping roles), local authorities, as well as private owners and operators of these systems, play a significant part in policy implementation in these sectors.

Chapter 8 covers aviation security, including commercial (passenger) aviation (which has received, by far, the most policy attention and funding), air cargo, and general aviation (all other civil aviation aircraft and facilities). TSA is the lead federal agency for all of these. Chapter 8 details the programs involved in providing each of the security "layers" for commercial aviation: airport security; passenger pre-screening; passenger and carry-on baggage screening; checked bag screening; and aircraft and onboard security.

In Chapter 9, several different perspectives for evaluating the effectiveness of U.S. transportation security policy are presented, including: performance assessments conducted by the Department of Homeland Security, GAO and DHS OIG; assessments of Congressional oversight of homeland security efforts; measures of DHS workforce morale; and public opinion.

The volume closes, in Chapter 10, with consideration of efforts to balance transportation security with economic efficiency (through benefit-cost analyses), civil liberties (through governmental and private privacy advocates, laws, and judicial proceedings) and budgetary constraints (through expanded targeting of programs, increased usage of user fees, and/or elimination of programs). As part of the latter, final action on FY 2013 and FY 2014 appropriations for transportation security is provided, along with the President's FY 2015 budget proposal. Last, the emerging priority of cybersecurity—and its application to transportation systems—is introduced.

Although efforts have been made to make the material herein as timely as possible, technological, political and other external changes—not to mention the occurrence of significant security events and/or the evolution of the terrorist threat—may lead to substantial modifications in transportation security policies over relatively short time spans. Readers are encouraged to seek out information on the latest developments by consulting key websites and updated versions of major source documents cited throughout this book, including the following:

For maritime security
International Maritime Organization: www.imo.org
United States Coast Guard: www.uscg.mil
United States Customs and Border Protection: www.cbp.gov
World Customs Organization: www.wcoomd.org

For land transportation security
International Union of Railways: www.uic.org
Transportation Security Administration: www.tsa.gov
United States Department of Transportation: www.dot.gov

For aviation security
International Civil Aviation Organization: www.icao.int
Transportation Security Administration: www.tsa.gov

For assessments and evaluations

U.S. Department of Homeland Security, Office of Inspector General: www.oig.dhs.gov.

U.S. Government Accountability Office: www.gao.gov.

For funding actions (U.S.)

U.S. Congress. Congress.gov: https://www.congress.gov/

U.S. Department of Homeland Security: www.dhs.gov

Reference

Johnstone, R.W., 2006. 9/11 and the Future of Transportation Security. Praeger, Westport, CT, p. 106.

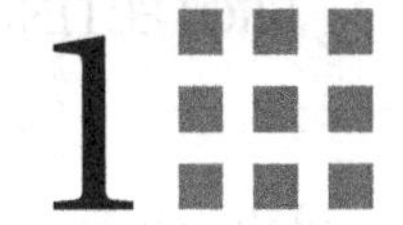

Transportation Security Before 9/11/01

CHAPTER OBJECTIVES:

In this chapter, you will learn about transportation security before 2001, including:

- The evolution of transportation and transportation security systems
- Trends in terrorist and other attacks on transportation systems
- The content and effectiveness of transportation security measures
- The content, reliability, and comparability of data on terrorist and other attacks on transportation systems

Introduction

Throughout human history, transportation systems and their passengers and freight have encountered natural disasters, accidents, and intentional acts of violence, including robberies on highways, piracy on the high seas, sabotage and hijackings in the air, and bombings and assaults on the rails. In recent decades, one of the major sources of such attacks has been terrorism, defined most recently in the U.S. Code as "premeditated, politically motivated violence perpetrated against noncombatant targets by subnational groups or clandestine agents" (22 USC 2656).[1] Nonterrorist criminal acts accounted for a large majority of intentional attacks on all modes of transportation and in all time periods, but terrorist actions (with their generally higher visibility and greater consequences) have served as the principal motivating force for the development and elaboration of security measures, especially after 1970.

Pre-9/11 Maritime Security

Before the Industrial Revolution of the latter half of the 18th century, there were no motorized forms of transportation, and systems on land were limited in speed, efficiency, and extent. Instead, it was waterways—first the great river systems of Eurasia and northeast Africa (including the Tigris and Euphrates, Nile, Indus, Ganges, and Huang He Rivers) where early civilizations and trading systems were centered. By the Middle Ages, large maritime transportation networks had developed using the Mediterranean, Baltic, and North Seas

[1]This is the definition used by the U.S. government's National Counterterrorism Center in compiling its database of terrorist incidents, called the Worldwide Incidents Tracking System in its unclassified form (National Counterterrorism Center, 2011).

and the coastal waters, navigable rivers, and canals in Europe and China. In the early 15th century in China and later that century in Europe, voyages of discovery opened up the Pacific, Indian, and Atlantic Oceans, which led to a further expansion of maritime trade (Rodrigue et al., 2006, pp. 14–18).

The advent of steam-powered ships in the late 18th century was followed by the establishment of the first truly worldwide maritime routes in the early 19th century, with sail-driven vessels gradually being supplanted by steamships. Ongoing improvements in propulsion systems, fuels, construction materials, and ship designs produced not only explosive growth in maritime trade but also required massive investments in port infrastructure to accommodate the larger vessels and increased volume of cargo. Harbors became major industrial centers, which supported both production and transshipment of goods. And beginning in the 1880s, ships provided the world's first international passenger transport (Rodrigue et al., 2006, pp. 20–22).

Two major developments during the 20th century were the introduction of increasingly large tankers in midcentury (especially oil tankers to carry petroleum from the Middle East) and the development of standardized, multimodal containers as the central element of modern freight transportation. The first container ship was launched in 1956, the first specialized container terminal (in Port Elizabeth, NJ) was built in 1960, and the first regular maritime container route (between North America and Western Europe) was established in 1966. By the early 1980s, container shipping had become the dominant form of international transportation (Rodrigue et al., 2006, pp. 23–24).

Piracy

Since its beginnings, maritime trade was accompanied by the use of force aimed at seizing vessels, crews, or cargo. These acts of piracy[2] are recorded at least as far back as the first millennium BCE in the Persian Gulf and Mediterranean Sea. The development of transoceanic trade routes starting in the 15th century led to more far-ranging pirates. These latter included privateers, which were chartered by European monarchies and authorized to seize ships and cargo on the high seas (especially the gold and silver shipments to Spain from its New World colonies). This officially sanctioned piracy stood in contrast to the limited success governments had in suppressing such acts during this era. The takeover of colonial responsibilities from private companies by European governments, the recognition of more distinct colonial boundaries, the development of more aggressive and effective antipiracy patrols (especially by the Royal Navy of Great Britain), and the establishment of both national (e.g., Britain's Piracy Act of 1721, which extended penalties for piracy to

[2]For purposes of this work, piracy is defined as an act of boarding or attempting to board any ship with the apparent intent to commit theft or any other crime and with the apparent intent or capability to use force in furtherance of that act. This is the terminology adopted by the International Maritime Bureau and is broader than the official definition under the 1982 United Nations Convention on the Law of the Sea, which focuses only on attacks beyond the territorial waters of any state, thus excluding acts in ports or coastal waters (Chalk, 2008, p. 3).

include all those who traded with pirates) and international (including the 1856 Declaration of Paris, which renounced the use of privateers) sanctions produced sharp declines in piracy by the mid-19th century (Bennett, 2008, pp. 150–151).

In more recent times, the International Maritime Bureau (IMB), established by the International Chamber of Commerce in 1979 primarily to combat maritime fraud, found a resurgence of piracy dating as far back as 1970. However, it was not until 1983 that the international community felt sufficient need to respond significantly with the adoption by the United Nations' (UN's) International Maritime Organization (IMO) of a resolution that

- Expressed great concern about the rising number of incidents involving piracy.
- Recognized the grave risks to lives, navigation, and the environment posed by such acts.
- Urged governments "to take, as a matter of highest priority, all measures necessary to prevent and suppress acts of piracy and armed robbery from ships in or adjacent to their waters, including strengthening of security measures."

The IMO subsequently indicated that between 1984 and the end of 2000, there had been 1700 reported pirate attacks on ships around the globe, although it estimated that the actual number of incidents was likely double that figure (International Maritime Organization, 2000, pp. 1–2).

Bennett (2008) cites several factors that contributed to the reemergence of piracy as a significant threat.

- More potential targets produced by the rapid growth in maritime trade and vessels
- Reduced governmental naval deployments after the end of the Cold War
- Inadequate national and international antipiracy laws and limited jurisdiction over attacks in international waters
- Increased shipboard automation, leading to smaller crews, which in turn offer fewer defenders against pirate attacks
- Higher fuel costs that necessitated cost-driven reductions in ship speeds, making pirate pursuits easier to accomplish
- Reduced costs for private arms and technology purchases, allowing pirates to procure better tracking equipment, assault vessels, and weapons (p. 152)

The distribution and nature of pirate attacks shifted during the period from the early 1980s through 2000, with assaults in port in west Africa (especially Nigeria) most common at the beginning, attacks within the Malacca Strait in Southeast Asia particularly prominent in the late 1980s and early 1990s, attacks in international waters within the South China Sea the biggest problem area in the mid-1990s, and incidents in port or in territorial waters in both the Malacca Strait and South China Sea the leading trouble spots at the end of the period (International Maritime Organization, 2000, pp. 2–4).

The costs of 20th century piracy, in both economic and human terms, are harder to assess. One estimate of the financial effect on the shipping industry put the figure at somewhere between $450 million and $1 billion in losses in 2000, but found that, even at the upper end of the range, the impact was relatively small, amounting to an additional cost of less than 40 cents on each $10,000 worth of shipping (Gottschalk and Flanagan, 2000, p. 106.) With regard to casualties, Chalk (2008) cited figures from the IMB indicating that between 1995 and 2000, 256 maritime crew and passengers were killed during pirate attacks, another 203 were injured and 1780 were taken hostage (p. 9).

Maritime Terrorism

Terrorism "within the maritime environment, using or against vessels or fixed platforms at sea or in port, or against any one of their passengers or personnel, [or] against coastal facilities or settlements, including tourist resorts, port areas and port towns or cities"[3] has been much less frequent than terrorist acts against land or air transportation. Chalk (2008) explains:

> *Part of the reason for this relative paucity has to do with the fact that many terrorist organizations have neither been located near coastal regions nor possessed the means to extend their physical reach beyond purely land theaters. There are also several problems associated with carrying out waterborne strikes. . . . Most intrinsically, operating at sea requires terrorists to have mariner skills, access to appropriate assault and transport vehicles, the ability to mount and sustain operations from a non-land-based environment, and certain specialist capabilities (for example, surface and underwater demolition techniques). Limited resources have traditionally prevented groups from accessing these options.*
>
> *Chalk, 2008, p. 19*

He also cites the costs and unpredictability of maritime attacks and the fact that "an attack on a ship is less likely to elicit the same publicity—either in scope or immediacy—as land-based targets" as additional factors in the relative rarity of maritime terrorism (p. 20).

■ ■ Critical Thinking ■

The explanations cited by Bennett as contributing to increased acts of piracy after 1970 would appear to apply to maritime terrorism as well. Referring to Chalk's list of constraints on would-be maritime terrorists, as well as other reasons you might think of, what one or two factors were most important in inhibiting a similar rise in maritime terrorism during this period?

One of the first attempts to quantify the terrorist threat to the maritime sector was a September 1983 report by the Rand Corporation. While acknowledging worries by some security experts that growing international terrorism, combined with the increasing importance of sea-borne trade and offshore oil platforms, might lead to an upsurge in maritime terrorism, the authors noted many of the aforementioned constraints and concluded

[3]This is the definition of maritime terrorism adopted by Chalk (2008) and derived from the Council for Security Cooperation in the Asia Pacific Working Group on Maritime Terrorism.

FIGURE 1.1 Italian cruise ship *Achille Lauro. (Courtesy of D. R. Walker.)*

that "increased terrorist attacks on maritime targets are not inevitable." The Rand study reported that "although terrorist groups have not operated extensively in the maritime environment," in the period between 1960 and 1983, they attacked 47 ships, hijacked 11 ships, and sank or destroyed 12 sea-going vessels. The main perpetrators in this era were anti-Castro Cuban exile groups, with the Irish Republican Army (IRA), Islamic separatists in the southern Philippines, Christian extremists in Lebanon, northern African Polisario guerrillas, Portuguese dissidents, Angolan rebels, and members of the Maltese National Front also involved in multiple incidents (Jenkins, 1983, pp. 1–3).

Two years after the Rand report, on October 7, 1985, the Italian cruise liner *Achille Lauro* (Figure 1.1) was seized by four Palestinian terrorists as it left the Egyptian port of Alexandria. The perpetrators, members of the Palestine Liberation Organization (PLO), demanded the release of 50 Palestinians held in Israeli jails in exchange for the approximately 400 passengers and crew then onboard. They directed the ship to sail to Syria, where they were denied permission to enter port. At that point, the terrorists murdered a disabled American passenger (Leon Klinghoffer) and threw his body overboard. Egyptian and PLO officials negotiated an end to the incident in which the perpetrators were to be given safe passage out of Egypt in return for the release of all of the hostages. President Ronald Reagan ordered U.S. Navy aircraft to intercept the Egyptian airliner carrying the perpetrators and PLO leader Abu Abbas, the reported mastermind of the operation. The Egyptian plane was forced to land at a U.S.–Italian military base in Sicily, where a jurisdictional dispute arose between the American and Italian governments. Ultimately, the Italians asserted authority and arrested the four perpetrators while allowing Abu Abbas to leave the country (Simon, 1986, pp. 2–3).

Even though there was only one casualty and the ship was recovered intact, the *Achille Lauro* incident produced a significant reaction in the United States and elsewhere and carried with it a number of implications for maritime security.

- The potential for maritime terrorism to yield both substantial media attention and major economic damage (with deep declines in Mediterranean cruise ship bookings following the attack)
- Confirmation of the eastern Mediterranean—and the Arab–Israeli conflict—as a new focal point for maritime terrorism
- The inadequacy of security measures undertaken by the maritime industry in general and cruise ships in particular, with one official noting that such measures on cruise ships consisted almost solely of limiting access to the vessel to ticketed passengers and to guests who register when they come onboard
- The need for timely and accurate intelligence gathering and sharing, as well as for comprehensive risk assessments for the maritime sector
- The critical importance of international cooperation, which was complicated in the maritime case by the great potential for jurisdictional confusion and disputes, especially when incidents occur beyond territorial waters (Simon, 1986, pp. 1, 7–10)

Despite the concerns that the *Achille Lauro* attack might trigger an increase in maritime terrorism, no clear trend emerged. The Global Terrorism Database (GTD) maintained by the National Consortium for the Study of Terrorism and Responses to Terrorism (START)[4] lists a total of 145 acts of maritime terrorism between 1970 and 2000 (just one of which was in U.S. territory: the September 16, 1976 bombing by an anti-Castro group of a Soviet cargo ship docked in Port Elizabeth, NJ, producing no casualties; Figure 1.2). Of these, 95 (66%) occurred after the *Achille Lauro* incident. However, many of the events recorded by START between 1985 and 2000 were part of ongoing domestic conflicts in Morocco, Sudan, Sri Lanka, the Philippines, and elsewhere. Indeed, in figures compiled for the Rand Database of Worldwide Terrorism Incidents,[5] of the 123 documented acts of *international* maritime terrorism committed between the beginning of 1968 and the end of 2000, just 28 (23%) occurred after 1985, and of the 115 fatalities in that same time frame, only 47 happened after 1985 (with 17 of those resulting from the attack on the *USS Cole* in October of 2000).

The GTD reports 643 fatalities in maritime terrorism incidents between 1970 and 2000, with 300 of those stemming from a February 2, 1984, ship bombing in the Sudan.

[4]The Global Terrorism Database was established at the University of Maryland in 2005 as a U.S. Department of Homeland Security Center for Excellence. Its data are drawn from three major sources: the Pinkerton Global Intelligence Services database, a privately created source covering acts of domestic and international terrorism from 1970 to 1997; a retrospective compilation by START covering 1998 to 2007; and START's real-time compilation beginning in 2007. As of November 2014, the database contained no data for 1993. See http://www.start.umd.edu/gtd.

[5]This database was begun as a result of a 1972 request from the U.S. Government's Cabinet Committee to Combat Terrorism that Rand examine recent trends in terrorism, including the attacks at the 1972 Munich Olympics and the Red Army attack on Israel's Lod Airport. As a result, a group of Rand analysts developed a database of terrorist incidents, originally termed the *Rand terrorism chronology*. See http://www.rand.org/nsrd/projects/terrorism-incidents/about.html.

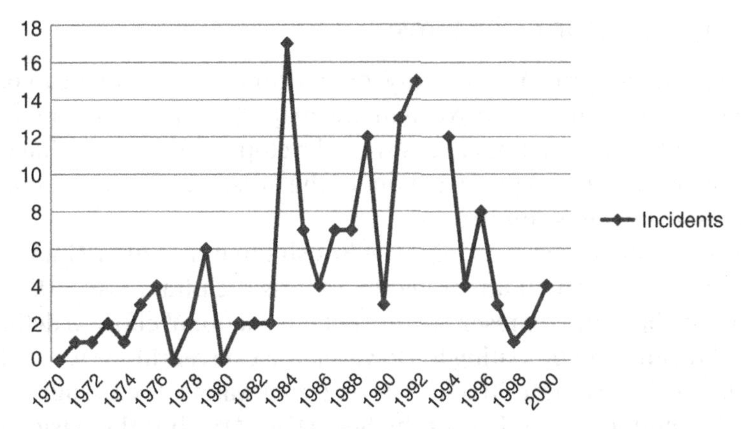

FIGURE 1.2 Maritime terrorism incidents, 1970 to 2000. *(Source: START Global Terrorism Database.)*

ATTACK ON THE *USS COLE*

Prelude to 9/11

On October 12, 2000, a small boat filled with explosives pulled alongside the *Cole*, a U.S. Navy destroyer on a refueling stop in the harbor of Aden, Yemen. Those on the boat reportedly made "friendly gestures" to the *Cole's* crew but then detonated the explosives. Seventeen crew members were killed, at least 40 more were wounded, and the vessel was severely damaged. The Navy, the Department of Defense (DOD), and the FBI all undertook investigations to determine responsibility for the attack and to review existing security procedures. The Navy and DOD findings were released in January 2001 and reached similar conclusions that the commanding officer of the *Cole* acted "reasonably" in adjusting his security procedures based on the information available to him at the time, but to quote Chief of Naval Operations Admiral Vern Clark, "the investigation clearly shows that the commanding officer of the *Cole* did not have the specific intelligence, focused training, appropriate equipment or on-scene security support to effectively prevent or deter such a determined, preplanned assault on his ship" (U.S. Department of Defense, 2001).

A December 21, 2000, CIA presentation indicated that, based on strong circumstantial evidence, its "preliminary judgment" was al Qaeda "supported the attack," but the CIA had "no definitive answer on [the] crucial question of outside direction of the attack—how and by whom" because the intelligence was ambiguous. The 9/11 Commission reported in 2004, "The plot, we now know, was a full-fledged al Qaeda operation, supervised directly by bin Laden. He chose the target and location of the attack, selected the suicide operatives, and provided the money needed to purchase explosives and equipment" (The 9/11 Commission report, 2004, pp. 190–197).

The *Cole* attack produced a number of consequences. It emboldened al Qaeda and energized its recruitment efforts. It exposed shortcomings in American intelligence and security measures. It provoked investigations in the Congress into intelligence operations with respect to terrorist attacks and U.S. terrorism policy. It also led to a reassessment by the CIA and National Security Council of U.S. counterterrorism policy in the closing weeks of the Clinton Presidency, although these efforts were largely superseded by the change in administrations.

Pre-9/11 Maritime Security Systems

Before 2001, measures for providing security for maritime vessels, crew, cargo, passengers, and support facilities were not extensive. With the major threat throughout this period being piracy, rudimentary international laws were developed to address piracy on the high seas, while security incidents in port were under the jurisdiction of the criminal laws and other regulations of the port's country.

Whereas certain earlier measures (e.g., the British Piracy Act of 1721 and the multinational 1856 Declaration of Paris) attempted to address the piracy issue, the 1958 Geneva Convention on the High Seas represented the first international effort to define piracy and establish a legal regime for combating it. However, the Convention was very limited, covering only certain offenses committed in international waters. It was slightly modified by the 1982 UN Convention on the Law of the Sea (UNCLOS), but the basic provisions remained in force. Bennett (2008) writes:

> *Although this legal regime grants universal jurisdiction over piracy, maritime piracy is limited to depredations (1) committed on the high seas (including, for this purpose, exclusive economic zones), (2) for private ends, (3) by persons from another ship or aircraft. Excluded from the definition are acts (1) occurring in a nation's territorial sea, or its internal or archipelagic waters; (2) motivated by a political purpose, rather than economic gain, although occurring on the high seas; or (3) undertaken by corrupt military units (unless the crew has actually mutinied). In these cases, jurisdiction normally resides in, and enforcement is up to, (1) the coastal state in whose waters the depredation occurs; (2) the flag state of the vessel on which the "political statement" has taken place; or (3) the country to whom the military belongs.*

Thus, the pre-9/11 international legal framework did not encompass either attacks in national waters (where most acts of piracy occur) or terrorist actions (which, by definition, are political in nature) (pp. 158–160).

As was (and is) often the case with respect to transportation security, it took a major incident to provoke further action. In this instance, the 1985 seizure of the *Achille Lauro* led to calls for improvements in international maritime security laws, culminating in the 1988 Convention for the Suppression of Unlawful Acts against the Safety of Maritime Navigation. Under this agreement,

- All unlawful acts against a ship, its crew, or passengers are covered, regardless of the location, source, or motive of the attack.
- All nations party to the convention are obligated to criminalize such acts and may prosecute any alleged offender found in its territory regardless of nationality or location of the event.
- In instances when a state is unable or unwilling to prosecute an alleged offender, it is required to transfer that person to another state party with jurisdiction that is willing to do so (Bennett, 2008, pp. 160–161).

Consistent with the limited international legal response to piracy and maritime terrorism, the IMO, the UN agency responsible for the safety and security of shipping, did little before 2001 beyond seeking to develop improved statistics and investigatory techniques on global incidents of piracy and armed robbery against ships, and calling for enhanced security measures by nations, ship owners, and seafarers. Furthermore, the IMO developed a circular providing "guidance to ship owners, ship operators, shipmasters and crews on preventing and suppressing acts of piracy and armed robbery against ships" in 1993 (revised in 1999). Among the recommended measures were reducing temptation for attackers by eliminating the need to carry large sums of cash onboard; exercising caution in transmitting information via radio about cargo or valuables onboard; enhancing security watches; providing appropriate surveillance and detection equipment; and developing a ship security plan that addresses such matters as surveillance and lighting procedures, appropriate crew responses to attacks, radio alarm procedures, and proper reporting procedures after an attack (International Maritime Organization, 2000, pp. 2–7).

At the national level, where most responsibility for maritime security in ports and territorial waters resided under both international law and customary practice, the security systems varied, although all were generally limited in scope and in resources. In the United Kingdom, the chief instrument for pre-9/11 maritime (as well as aviation) security was the Aviation and Maritime Security Act of 1990, which consolidated a number of previous statutes and incorporated certain features of common law (e.g., making ship hijacking illegal). It also was designed to give effect to the Convention for the Suppression of Unlawful Acts against the Safety of Maritime Navigation and the related Protocol for the Suppression of Unlawful Acts against the Safety of Fixed Platforms Located on the Continental Shelf. The law

- Criminalized the hijacking of ships, seizure of fixed platforms, destruction of ships or fixed platforms or endangering their safety, commission of other acts endangering or likely to endanger safe navigation, and the making of threats pursuant to such offenses
- Criminalized the making of false statements with respect to baggage, cargo, supplies, and identity documents
- Gave the Secretary of State authority to require certain information from British ship owners and those operating within British harbors to designate restricted zones in harbor areas, within which persons or property seeking to enter a ship must be searched by appropriate authorities before being permitted to do so, and ships seeking to leave port must also be searched; to authorize inspections of ships within British harbors and any other part of harbor areas; and to regulate "sea cargo agents" involved in handling cargo (Government of the United Kingdom, 1990).

In the United States, the Coast Guard was given responsibility for the security of U.S. ports and waterways, starting with the Espionage Act of 1917, which, among many other provisions, provided it with the power to issue regulations to prevent damage to harbors and ships in harbor during wartime. The 1950 Magnuson Act considerably expanded this authority by assigning the Coast Guard a permanent mission to safeguard U.S. ports, harbors, vessels, and waterfront facilities from accidents, sabotage, or other subversive acts. Included was the authority to search ships in U.S. territorial waters and to control the movement of foreign vessels in American ports (U.S. Department of Homeland Security, United States Coast Guard, n.d.)

One of the few attempts to provide a comprehensive assessment of maritime security undertaken before 2001—albeit limited to one facet in one country—was the "Report of the Interagency Commission on Crime and Security in U.S. Seaports." The Commission was established by President Bill Clinton in 1999 and charged with providing "a comprehensive review of the nature and extent of seaport crime and the overall state of security in seaports, as well as the ways in which governments in all levels are responding to the problem."

The commission report indicated that although it "was not able to determine the full extent of serious crime at seaports" because of inadequacies in data collection and reporting, it did find "significant criminal activity is taking place at most of the 12 seaports" it surveyed, with drug smuggling the prevalent crime at most locations, and stowaways and alien smuggling, trade fraud, cargo theft, and stolen vehicles also reported at a majority of the seaports. With regard to terrorism, the report stated: "Although seaports represent an important component of the nation's transportation infrastructure, there is no indication that U.S. seaports are currently being targeted by terrorists. The FBI considers the present threat of terrorism to be low, even though their vulnerability to attack is high. The Commission believes that such an attack has the potential to cause significant damage" (Report of the Interagency Commission on Crime and Security in U.S. Seaports, 2000, pp. iii–v).

The commission rendered a mixed evaluation of existing security measures.

> *The state of security in U.S. seaports generally ranges from poor to fair, and, in a few cases, good. There are no widely accepted standards or guidelines for physical, procedural, and personnel security for seaports, although some ports are making outstanding efforts to improve security. Control of access to the seaport or sensitive areas within the seaports is often lacking. Practices to restrict or control the access of vehicles to vessels, cargo receipt and delivery operations, and passenger processing operations at seaports are either not present or not consistently enforced, increasing the risk that violators could quickly remove cargo or contraband. Many ports do not have identification cards issued to personnel to restrict access to vessels, cargo receipt and delivery operations, and passenger processing operations...Many seaports rely on private security personnel who lack the crime prevention and law enforcement training and capability of regular police officers.*
>
> *Report of the Interagency Commission on Crime and Security in U.S. Seaports, 2000, pp. v–vi*

■ ■ Critical Thinking ■

Maritime security was not afforded a high priority by the United States and other governments before 9/11. *Based on the evidence available at the time,* do you believe this was an appropriate decision? Why or why not?

Pre-9/11 Land Transportation Security[6]

Although land transportation systems in the form of paths and roadways go back to the dawn of human civilization, until the past 250 years, constraints on their efficiency generally limited them to highly localized impact. A notable exception was the Silk Road, described by Rodrigue et al. (2006) as "the most enduring trade route of human history, being used for about 1,500 years [from around 139 BCE]. . . . The Silk Road consisted of a succession of trails followed by caravans through Central Asia, about 6,400 km in length. . . . Economies of scale, harsh conditions and *security considerations* [emphasis added] required the organization of trade into caravans slowly trekking from one stage (town and/or oasis) to the other" (pp. 14–15).

In addition to being restricted in scope, the vast majority of inland transportation systems in this era were low in capacity and slow in speed, with, for example, certain trade routes conveying an annual amount of goods that would not fill a modern freight train at an average speed of 2 miles an hour (Rodrigue et al., 2006, p. 18). Although the entwinement of land transport networks throughout societies around the globe guaranteed that they would be the scene of countless criminal and violent acts, the implications of these were usually local, rarely involving the intervention of national authorities.

The beginning of the Industrial Revolution around 1750 exposed shortcomings in the capacity of the road networks, which were predominantly unpaved but were the basis of most land transportation at the time. In England, this led to the development of turnpike trusts, which were authorized to construct and maintain a specific road segment through the imposition of tolls on travelers and freight. These roads significantly improved the speed and efficiency of road travel. About 25,000 km of turnpikes had been built in England by 1770 and 32,500 by 1836, but by then, rail transportation had begun to emerge as the dominant form of surface transport (Rodrigue et al., 2006, pp. 18–19).

The first steam-powered railway was established in 1814 in England to haul coal, with the initial commercial rail line (linking Manchester and Liverpool) built in 1830. From that point, the new technology spread rapidly throughout England, the rest of Europe, and North America. By 1845, the United States possessed the world's largest rail network, of 5458 km, followed by Great Britain (3083 km), Germany (2956 km), France (817 km), and Belgium (508 km). Railroads continued to expand in size and importance throughout the 19th century, and by 1929, they carried 74% of all freight moved in the United States (Rodrigue et al., 2006, p. 20; Sweet, 2006, pp. 21–23).

The growth of urban areas in the latter part of the 19th century spurred the development of public urban mass transportation systems, including both underground rail systems (with London being the first, in 1863) and above-ground, electricity-powered streetcars in

[6]"Land transportation" is the preferred term in this work because it is more descriptively accurate than the alternative—used by the Transportation Security Administration in the United States as well as by other security authorities—of "surface transportation." In both cases, what is generally being referred to includes highways, railways, mass transit systems (including underground systems), and associated bridges and tunnels, as well as pipelines.

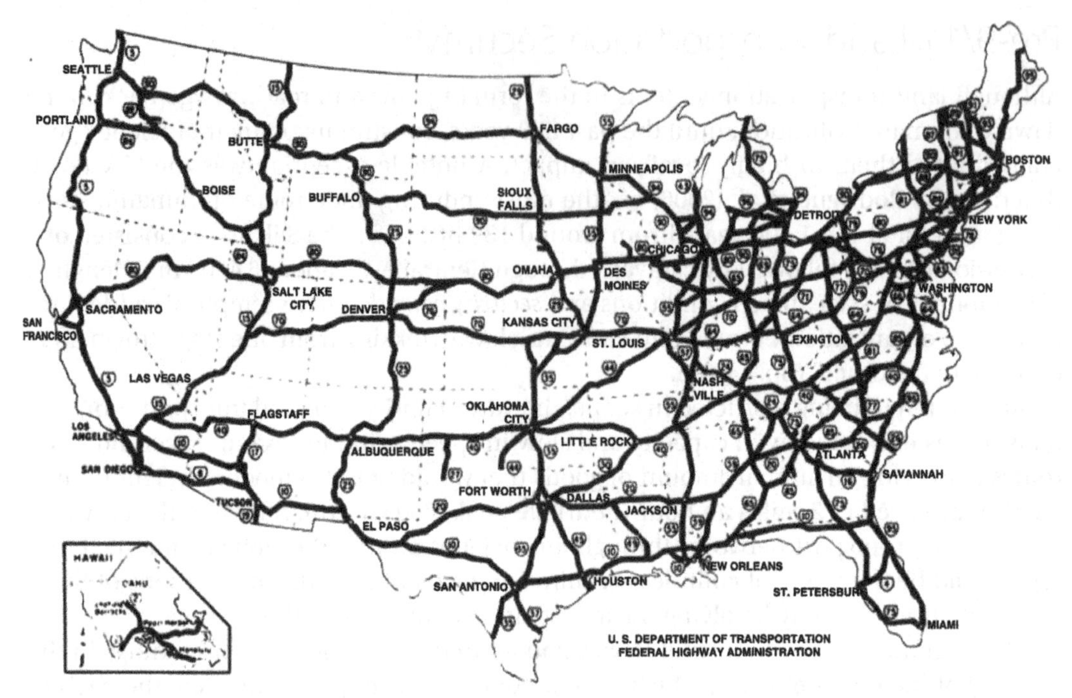

FIGURE 1.3 U.S. Interstate highway system.

Western Europe and the United States (beginning in the 1880s). Introduction of the pneumatic tire (1885) and the internal combustion engine (1889) produced more capable road transportation vehicles, including automobiles, buses, and trucks. Combined with the development of improved road surfaces and the construction of longer range road systems (e.g., the Interstate Highway System in the United States [Figure 1.3]), this made land transportation modes increasingly important in the movement of passengers and goods as the 20th century progressed (Rodrigue et al., 2006, pp. 22–23).

Pre-9/11 Attacks on Land Transportation

The first systematic analysis and reporting of terrorist and other violent attacks on land transportation systems was even more recent than was the case in the maritime sector. The most thorough of the early efforts was a 1997 report by Brian Jenkins for the Mineta International Institute for Surface Transportation, which provided a chronology of worldwide terrorist attacks and other significant criminal incidents involving land transportation systems (Jenkins, 1997).

The chronology was derived from three major sources: (1) Peter Semmens' *Railway Disasters of the World: Principal Passenger Train Accidents of the 20th Century*, from which 14 major attacks on passenger trains were included, ranging from 1920 to 1966; (2) the beginnings of what has become the Rand Database of Worldwide Terrorism Incidents, starting with 1968 events; and (3) the Information Services of the Kroll-O'Gara Company computerized chronology of armed conflicts, significant acts of terrorism, and other major crimes. The result was a compilation of 631 entries involving guerilla and terrorist attacks and other incidents of serious crime directed against trains, subways, train and subway stations and rails, buses, bus terminals, bridges, tunnels, and other land transportation targets, as well as passengers on these systems. It excluded conventional warfare and acts of wartime sabotage (Jenkins, 1997, pp. 104–105). A 2001 update of the chronology added more than 200 incidents from where the first version left off through the end of 2000.

- Terrorist attacks on surface transportation systems had increased over the preceding 30 years, with bombing the most common tactic (accounting for 60% of all attacks).
- Buses (32% of attacks) were the most frequent target followed by subways and trains (26%), subway and train stations (12%), rails (8%), bus terminals (7%), tourist buses (7%), bridges and tunnels (5%), school buses (1%), and other (2%).
- India (69 incidents), Pakistan (41), Algeria (17), Egypt (12), and Russia (11) were the sites of the most incidents with fatalities. A large portion of the attacks involved ongoing conflicts, including civil and guerilla wars and organized terrorist campaigns. Only nine incidents occurred within the United States during the entire period.
- Terrorist attacks on public transportation were particularly likely to produce casualties. Whereas approximately 20% of all incidents of international terrorism involved fatalities, almost twice as many (37%) of such attacks on public transportation did so (Jenkins and Gersten, 2001, pp. 67–74).

■ ■ Critical Thinking ■

Why do you think so few acts of terrorism and other major crimes were reported for highways and bridges compared with other land transportation systems?

The GTD compilation of domestic and international terrorist incidents involving land transportation lists 3746 such events between 1970 and 2000, with 7375 reported fatalities. Dividing the period into three 10-year intervals (with no data for 1993), just 251 attacks (7% of the total) producing 265 casualties (4% of the total) occurred from 1970 to 1979; 1753 incidents (47%) and 3372 fatalities (46%) were recorded from 1980 to 1989; and similar levels of 1742 attacks (46%) and 3738 fatalities (50%) were reported from 1990 to 2000. However, a noticeable decline in both incidents and fatalities began in 1998 and continued for several years beyond 2000. As was true with respect to maritime terrorism, very few of these many attacks on land transportation took place within the United States, with just 14 events and two fatalities during the 1970 to 2000 time frame (Figure 1.4).

The 2001 Rand report offered an explanation of the frequency of attacks on land transportation systems:

> *Open to relatively easy penetration, trains, buses, and light rail systems offer an array of vulnerable targets to terrorists who seek publicity, political disruption, or high*

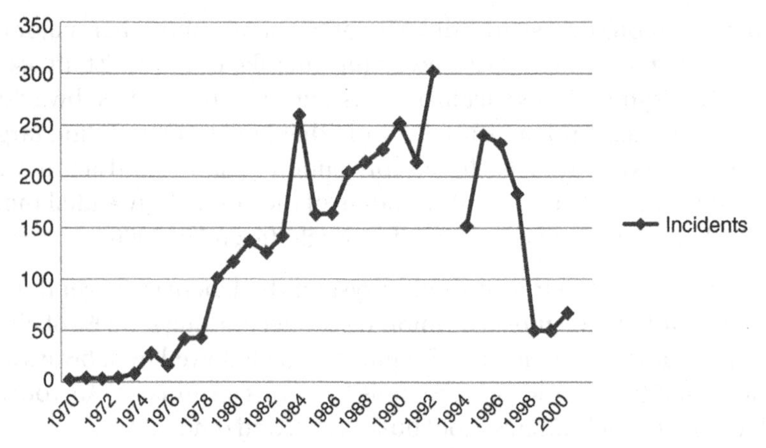

FIGURE 1.4 Land transportation terrorism incidents, 1970 to 2000. *(Source: START Global Terrorism Database.)*

> *body counts. High concentrations of people in relatively crowded quarters are inviting fodder for those who would cause mayhem and death. The massive amounts of explosives needed for truck bombs are unnecessary in crowded train stations, bus depots, carriages, or coaches. Even without large numbers of casualties, disruptions to transit can seriously impact a region's economy and the public's faith in the government's ability to provide basic protections to its citizens.*
>
> *Jenkins and Gersten, 2001, pp. 1–2*

ATTACK ON THE TOKYO SUBWAY SYSTEM

Chemical Weapon Attack by the Aum Shinrikyo Cult

Five members of the Japanese religious cult Aum Shinrikyo released the nerve gas sarin into the Tokyo subway system on March 20, 1995. The cult had been established in 1987 and eventually grew to 10,000 members, with assets estimated at more than $2 billion. In the words of Jenkins and Gersten (2001), "Aum Shinrikyo was noteworthy in combining its apocalyptic world view, a fascination with weapons of mass destruction, an ability to recruit scientists, and access to immense financial resources." Part of the group's apocalyptic vision was that it would take over Japan as a prelude to the end of the world in 1997 (pp. 50–51).

Before 1995, the cult had undertaken a number of attempts to use exotic weapons against various targets in Japan, including the spraying of botulinum toxin in central Tokyo (which went unnoticed and produced no fatalities); the spraying of anthrax spores from a tall building in Tokyo (which failed because the spores were incubated improperly); a 1994 attack on the town of Matsumoto using sarin dispersed from a refrigerated delivery truck (which resulted in seven deaths and more than 200 injuries); and earlier March 1995 attacks using the botulinum toxin, first on a Yokohama commuter train (which resulted in the hospitalization of 11 passengers) and then on the Tokyo subway system (which failed because the toxin was not loaded into the containers used in the attack) (Jenkins and Gersten, 2001, pp. 51–52).

The lack of impact of the earlier attempts, as well as the decentralized nature of Japanese police forces and a reluctance to interfere with religious groups delayed the authorities' recognition of the threat posed by Aum. By November 1994, local police had identified sarin as the agent used in the Matsumoto attacks, but Aum was not initially suspected. However, over the next few months, police began to piece together evidence from the various incidents Aum had instigated, as well as related complaints from citizens and records of Aum's unusual chemical purchases, and planned a nationwide search of all known Aum facilities for the spring of 1995. The March 1995 attacks were the result of the cult's perception that the Japanese authorities were about to move against them (Jenkins and Gersten, 2001, p. 52).

The March 20 event involved five Aum members boarding trains on three Tokyo subway lines at around 8:00 AM in order to coincide with morning rush hour. They carried a total of 11 plastic bags containing sarin in the hope of killing hundreds of Tokyo police on their way to work (with the thousands of collateral civilian casualties considered an acceptable consequence). The attackers succeeded in releasing the gas by puncturing the bags and escaped into the crowd by 8:10 AM (Jenkins and Gersten, 2001, pp. 52–55).

The first emergency calls to police came at 8:20 AM, but there was great uncertainty as to the cause of the incident, with sarin not identified as the source until 10:30 AM. Because police were aware of Aum's possession of the nerve gas, they had trained and equipped units to deal with it. By 1:30 PM, these units had identified and recovered the plastic bags that had contained the sarin, and assisted by the Japanese military, had begun decontaminating the stations and trains. By March 22, police had conducted searches of Aum's facilities in Tokyo and on Mount Fuji and by mid-April had arrested more than 100 of the cult's members but none tied directly to the sarin attack. However, the group was able to carry out additional attacks (largely unsuccessfully) up until July, when ongoing arrests and seizure of Aum's assets finally brought the assaults to an end (Jenkins and Gersten, 2001, pp. 55–59).

Twelve individuals died from the sarin attack, with 3398 injured. The number of fatalities would likely have been much higher had the gas not contained a significant amount of impurities. The incident produced a large increase in security measures in the immediate aftermath, but the major focus was placed on neutralizing the cult.

Pre-9/11 Land Transportation Security Systems

As terrorism expert Jenkins wrote in 1997, "Because terrorist threats are not easily quantifiable, it is difficult to determine the 'right' level of security. . . . Since the threat of terrorism is murky and security measures are costly, it is hard to justify the expenditures before an attack. Security against terrorism therefore tends to be reactive." This axiom generally holds true across all modes of transportation, both before and after 9/11, but it is certainly the case with respect to pre-9/11 security in the land transportation sector (p. 2).

That reactivity, combined with the localized focus of most attacks on land transportation systems as well as the localized jurisdiction of the authorities deemed responsible for responding to such attacks, produced a highly fragmented, subnational land transportation security "system" that generally lacked the international and national components present in maritime and aviation security.

Jenkins (2001) assessed the state of pre-9/11 security measures for selected land transportation systems.

Security plans: Most systems had written plans for dealing with service interruptions caused by weather, accidents, crimes, terrorist acts, or other causes. Such plans generally emphasized minimizing service disruption and addressed security issues in piecemeal fashion. A formal crisis management plan was included in some, but not all, of the plans.
Threat assessment: Virtually all threat information was provided to transport operators by governmental authorities, with the sources usually being national and local law enforcement agencies.
Station design: In the United Kingdom, and to a lesser extent in France, security considerations were factored into the building or remodeling of stations, and both countries made heavy use of closed-circuit television (CCTV) to monitor public transportation systems. By the latter half of the 1990s, there were 4000 cameras in the Paris subway and commuter rail systems and 3500 in the London counterparts. Such measures were less prevalent in other systems, including those in the United States.
Security force: Most systems had their own dedicated security force, with some of these being private companies and others police department components devoted to transport security, such as the Metropolitan Transportation Authority (MTA) Police Department in New York City. Smaller systems tended to rely on regular police forces, only some of which had training for transportation security incident response.
Security patrols: Some, but not all, systems used a combination of uniformed and plainclothes patrols to deter attacks and reassure passengers.
Training: There was wide variation in the quantity and quality of training provided for security personnel.
Covert testing: The United Kingdom, but not the other jurisdictions, used covert testing of security and response procedures to gauge the performance of security measures and personnel.
Passenger screening: This feature was deemed infeasible by all systems studied because of the delays and costs involved.
Emergency response teams: Several of the systems had some form of mobile response team for rapid response to serious incidents.
Security exercises: Several of the larger systems, including New York's MTA, regularly conducted training exercises to simulate crisis responses (pp. 7–24).

Largely as a consequence of a terrorist campaign waged by the Provisional Wing of the IRA against the British government from 1969 to 1998, which featured attacks against rail transportation targets,[7] the United Kingdom developed the most extensive land transportation security system before 9/11. The Department of the Environment, Transport and Regions (DETR) was the national policymaking agency for railways, mass transit, and the

[7]The IRA campaign included 81 explosive devices placed at transport targets during this time frame. Of those, 79 were hand-placed time bombs, but fully half did not work as intended. In part because of this but also because of the terrorists' efforts to inflict economic and psychological costs rather than casualties and to the rapid response by U.K. authorities to the bomb threats, just three people were killed in all of these attacks (Jenkins and Gersten, 2001, p. 12).

Channel Tunnel. (Buses were not then, nor are they today, subject to security regulation in the United Kingdom.) However, local transportation operators were responsible for most security decision making until 1988, when the Division initiated a "best practices" review to improve security at rail stations. This culminated in the issuance of rail security guidelines (Jenkins and Gersten, 2001, pp. 14–15).

The guidelines, termed the Secure Stations Scheme, encouraged rail operators in England, Scotland, and Wales (with Northern Ireland excluded) to follow the recommended national standards for improving security at both underground and above-ground stations. (The Division had determined that station security was the chief concern of passengers.) The recommendations included good lighting, secure fencing, security staff presence, CCT surveillance, rapid response in emergencies, regular inspection and maintenance, special training for staff, regular passenger surveys, and reporting of results (Jenkins and Gersten, 2001, p. 29).

The DETR took responsibility for setting and enforcing railway security standards for most British rail systems but not the London underground, which continued to develop its own procedures until July 2003 when the renamed Department for Transport assumed regulatory authority for it as well. In all cases, however, the individual transportation system operators were (and still are) "responsible for the day to day delivery of security" (U.K. Department for Transport, n.d.).

The other major national component in the United Kingdom's land transportation security system was the British Transport Police (BTP), which served as the police force for rail systems (including the London Underground) in England, Wales, and Scotland. As of 2000, the BTP had 2106 police officers and 524 civilian support staff. Ongoing IRA attacks led local police to take action to augment the BTF's efforts. In 1998, the Association of Chief Police Officers created the National Terrorist Crime Prevention Unit, which developed a strategy, disseminated "best practices," provided training, and facilitated the deployment of Counter Terrorist Crime Prevention Officers at local police departments (Jenkins and Gersten, 2001, p. 15).

The British rail transportation security system in place by 2000 featured virtually all of the elements contained in the Jenkins assessment of security measures. In addition, the British system included extensive efforts to provide useful information to local and transport police, transportation system operators, passengers, and the general public and to involve the private sector through the sharing of threat information and instructional materials. The large number of incidents aided in the development of improved threat analysis, which in turn allowed the United Kingdom's Security Service to distribute, to police and select others, written assessments of all terrorist threats, including which of four levels of alert and attendant security measures should be deployed in each instance (Jenkins and Gersten, 2001, pp. 16–19).

Jenkins and Gersten (2001) describe the results of these security efforts:

> *Few transport systems experience terrorist events, making it difficult to gauge the effectiveness of security measures. In Britain, however, the persistence of the IRA*

> *campaign allowed such measurement. The evolution of the terrorist campaign indicates that the security measures had a discernible effect. In 1991, IRA terrorist attacks centered on stations in London. By 1992, the attackers were pushed out to suburban stations, and by 1993, they were confined to home counties. The targets of the attackers also shifted from stations to switch boxes and rail lines away from stations. In the later years of the terrorist campaign, there were fewer bombs and more bomb threats. The security measures against terrorism also had the additional effect of reducing ordinary crime. Crime in the Underground, which had been increasing in the late 1980s, reversed direction and declined 54 percent in the 1990s, bucking a national trend.... Another measure of effectiveness was disruption. As the authorities became more familiar with the IRA's* modus operandi, *they were able to develop procedures that reduced response time and the duration of disruptions.*
>
> *Jenkins and Gersten, 2001, pp. 19–20*

In reaction to real and potential terrorist attacks, New York City had also developed a significant mass transit security program before 9/11. However, similar to most other pre-9/11 land transportation security efforts outside the United Kingdom, it lacked national-level institutions to provide support and to help in coordinating the multiple local jurisdictions and agencies that were part of the program (Savage, 1997, pp. 33–48).

Before September 2001, most land transportation systems in North America, Western Europe, and Japan lacked either the experience or perception of security threats to spur the development of extensive security measures. As Jenkins (1997) observed, "The threat faced by the industrialized nations is primarily one of alarm and disruption, the traditional goals of terrorism, and not deaths.... (I)t will be difficult to make persuasive arguments for costly and disruptive security measures unless these are absolutely necessary and promise to be effective in preventing even costlier and more disruptive interruptions of service" (p. 110).

The vast majority of attacks on land transportation systems occurred in places outside of Europe and North America, especially on the Indian subcontinent, in Africa, and in Southeast Asia, and were part of ongoing internal or external political conflicts. Here, the authorities lacked either the resources or the political will (or both) to undertake major security programs (Jenkins and Gersten, 2001, p. 101).

■ ■ Critical Thinking ■

Compare the Aum Shinrikyo and IRA attacks on public transportation systems and the responses of the relevant authorities to those attacks.

Pre-9/11 Aviation Security

The air mode was the last of the major transportation means to be developed, although it has proved to have the highest profile with respect to security measures, both before and after 9/11. Although balloon flights began in the late 18th century, it was not until 1903 and the Wright brothers' demonstration of powered flight that air transportation began to be seen as a practical method of conveyance. The first major application was for mail delivery, which initially proved more practical and profitable than the transport of heavier freight or passengers. Regional and national air transportation services were developed in Europe and North America during the 1920s and 1930s, but these were limited by the capacity and range of the propeller-driven aircraft then in use.

Spurred by technological advances during World War II, the 1940s and 1950s witnessed significant improvements in the speed, range, and carrying capacity of airplanes, with the first jet-powered commercial vehicle, the Boeing 707, entering into service in 1958. From that point, the airplane increasingly became the preferred means of long-distance passenger transportation, at least for those able to afford it. The development of wide-bodied aircraft in the late 1970s and early 1980s marked another milestone in the expansion of passenger service but also ushered in a substantial increase in air freight deliveries beginning in the 1980s (Rodrigue et al., 2006, pp. 23–24).

Pre-9/11 Attacks on Aviation

Documented attacks on airplanes date almost to the beginning of regular air service, with the first hijacking[8] occurring in 1931 and the initial onboard bombing following in 1933. Between then and the end of the 20th century, hijackings were, by far, the main method used in terrorist or other criminal acts against aviation, but bombings produced the most casualties. Before the 1960s, most hijackings were either political (to escape persecution or prosecution) or economic (hostage taking for ransom) but not usually terrorist acts. The limited number of bombings in this period often arose from insurance fraud plots (Price and Forrest, 2009, pp. 43–46).

The takeover of the government of Cuba by Fidel Castro in 1959 and the sharp deterioration in relations between the Castro regime and the United States that set in soon thereafter led to a wave of air hijackings to and from Cuba between 1960 and 1974. In that period, more than 240 such hijackings or attempted hijackings took place, with the hijackers generally seeking political asylum, release of prisoners, or financial gain. Beginning in the early 1970s, the increasing hijacking threat produced the first large-scale response in aviation security measures in the United States, which gradually reduced—but did not eliminate—that threat (with more than 60 Cuba-related hijacking incidents from 1974 to 1989) (Price and Forrest, 2009, p. 47).

Attacks on aviation were not limited to the United States, and Israel was another frequent target. The first hijacking of a commercial Israeli airliner was in July 1968, which inaugurated a series of attacks by the Popular Front for the Liberation of Palestine and related groups upon Israeli and other commercial aircraft, in the air, and on the ground. A key feature of most of these and other pre-9/11 hijackings was that the hijackers intended

[8]*Air hijacking* is synonymous with *air piracy,* which the U.S. Code defines as any seizure or exercise of control of an aircraft, by force or violence, or threat of force or violence, or by any other form of intimidation, with wrongful intent (Price and Forrest, 2009, p. 45n3).

to use the aircraft and passengers as bargaining chips to attain a political or financial objective rather than as targets for destruction (Price and Forrest, 2009, pp. 48–50).

Onboard bombings of aircraft were usually limited in number and impact until the 1980s. The beginning of that decade experienced a temporary resurgence of Cuba-related hijackings, this time primarily Cuban refugees who came to the United States in the 1980 Mariel boat lift seeking to return to Cuba. However, it was the occurrence of two major aircraft bombings—of Air India Flight 182 in June 1985 and Pan Am Flight 103 in December 1988—that produced the most casualties and had the largest effect on the evolution of aviation security measures (Price and Forrest, 2009, pp. 53–54).

While en route from Vancouver, Canada to London, Air India Flight 182 exploded off the southern coast of Ireland, killing all 329 passengers and crew members. Just under an hour earlier, an explosion occurred inside the baggage handling area at Tokyo's Narita Airport, killing two and injuring four others. Evidence linked both explosions to bombs placed in two bags that were checked at Vancouver International Airport—one checked for Air India Flight 182, and the other to be transferred to another Air India aircraft in Tokyo. Canadian authorities suspected Sikh extremists, but a number of difficulties were encountered in the investigation. A Canadian Sikh was convicted in 1991 for his part in obtaining components for the bomb used in the Tokyo bombing, and another plead guilty in 2003 to one count of manslaughter for his role in aiding in the construction of a bomb. Charges against two others, including for the murder of the 329 people on Flight 182, were dismissed by a Canadian judge in December 2004 (Price and Forrest, 2009, p. 59; CBC News Online, 2006).

THE BOMBING OF PAN AM FLIGHT 103

Terrorist Threat to U.S. Aviation Realized

The Pan American flight designated as 103 originated in a Boeing 727 flown from Malta to London with a stopover in Frankfurt, Germany. In London, passengers and luggage were transferred to a larger capacity Boeing 747, which was to convey Flight 103 to its final destination at New York's JFK International Airport. On December 21, 1988, Flight 103 suffered a catastrophic explosion shortly after departing London, over Lockerbie, Scotland. The explosion, which was later determined to be the result of the detonation of plastic explosives in the plane's forward cargo hold, killed all 259 passengers (many of whom were U.S. citizens) and crew on board, as well as 11 individuals who were hit by falling debris on the ground below. Subsequent investigation by Scottish authorities determined that Libyans were the perpetrators, and two Libyan nationals were charged with the crime in 1991. However, the two had fled back to Libya, where the Qaddafi government protected them until, after years of diplomatic negotiations and the imposition of economic sanctions against Libya, Qaddafi handed them over for trial. One was acquitted, and the other was found guilty in 2001 (Price and Forrest, 2009, pp. 59–63).

The families of the American victims on Flight 103 organized themselves and urged "the formation of an independent investigative body to determine the how and why of the final flight of Pan Am 103, and to seek to assure that others could be spared their loss and suffering." In response, President George H.W. Bush created the President's Commission on Aviation Security and Terrorism (often referred to as the Pan Am/Lockerbie Commission), which began its work in mid-November 1989 and issued its report on May 15, 1990 (Report to the President, 1990, p. 1).

The Commission found:

> *The destruction of Flight 103 may well have been preventable. Stricter baggage reconciliation procedures could have stopped any unaccompanied checked bags from boarding the flight at Frankfurt. Requiring that all baggage containers be fully screened would have prevented any tampering that may have occurred with baggage left in a partially filled, unguarded baggage container that was later loaded on the flight at Heathrow [Airport in London]. Stricter application of passenger screening procedures would have increased the likelihood of intercepting any unknowing "dupe" from checking a bomb into the plane at either airport.*
>
> *Report to the President, 1990, pp. ii–iii*

The destruction of Pan Am Flight 103 and the subsequent investigations by the Pan Am/ Lockerbie Commission and by aviation and legal authorities in Scotland and the United Kingdom led to substantial revisions in aviation security procedures, not only in those countries but also throughout the international civil aviation system. One of those changes was to focus the attention and resources of that system away from hijacking toward what was now seen as the greater threat of onboard bombing. Figure 1.5 shows the Lockerbie Memorial Garden of Remembrance in Lockerbie, Scotland.

FIGURE 1.5 Lockerbie Memorial Garden of Remembrance, Lockerbie, Scotland.

Beginning in 1986, the U.S. Federal Aviation Administration's (FAA's) Office of Civil Aviation Security began publishing an annual report called *Criminal Acts Against Civil Aviation*. That publication contained a comprehensive listing and classification of all known terrorist and other criminal actions "against civil aviation aircraft and interests worldwide." It, and subsequent editions, made clear that the vast majority of attacks on civil aviation, including hijackings and bombings, were not the result of terrorism. The 1995 edition provided a review of the initial 10-year period (1986–1995) covered by the reports. In that time period, a total of 601 incidents were recorded:

- A total of 179 hijackings (30% of the total), which were generally committed by individuals or small groups for personal reasons, including seeking transport, asylum, or financial gain. Only 14 were determined to involve political or terrorist motives. There were eight hijackings in the United States. In addition, 23 commandeerings[9] (4%) took place, with personal motivations again the most common factor, although two were determined to be political or terrorist acts.
- A total of 171 off-airport facility attacks (28%), a majority of which involved assaults on airline ticket offices, used bombs and were designed to make a political statement.
- A total of 108 attacks at airports (18%), 67 of which were bombings or attempted bombings
- A total of 58 general aviation or charter aviation incidents (10%).This category includes all forms of attack on general aviation[10] or charter aviation aircraft and facilities, with 27 hijackings, 21 commandeerings, nine destroyed aircraft, and one airport bombing occurring in the 1986 to 1995 period.
- A total of 41 shootings at aircraft (7%). These involved the use of automatic weapons, surface-to-air missiles, mortars, and other projectiles fired at aircraft, which resulted in 13 plane crashes.
- A total of 21 bombings, attempted bombings, or shootings on board aircraft (3%). Although smallest in terms of the number of incidents, the onboard bombings that occurred in this time interval (especially the destruction of Pan Am Flight 103) were unquestionably the most consequential with respect to both the number of casualties (more than 700 deaths) and the impact on aviation security (Federal Aviation Administration, 1995, pp. 38–44).

One of the bombing incidents—the December 1994 bombing of Philippine Airlines Flight 1994, which caused a single casualty when an onboard explosives device detonated—

[9]The FAA reports define *hijacking* as when the aircraft is taken over while in flight ("once the doors are closed"); commandeerings refer to aircraft takeovers that occur on the ground ("when the doors are open"). Commandeerings encompass both aircraft that remain on the ground and those that subsequently take off (Federal Aviation Administration, 2000, pp. 50, 52).

[10]General aviation includes recreational and corporate aircraft, medical services, aerial advertising, aerial mapping and photography, aerial application of seeds or pesticides, other small aircraft, and the supporting facilities. In the United States, it accounts for three-fourths of all takeoffs and landings (Johnstone, 2006, p. 193 n.115).

proved to be of greater significance than was initially recognized. An accidental fire in a Manila apartment in January 1995 led to the discovery of a "bomb factory" as well as a number of related documents. When Filipino authorities called on the United States for assistance, an FAA security official recognized that one of the documents contained flight codes. That along with a laptop computer also found in the apartment led to a determination that a plot had been devised to bomb 12 U.S. airline-owned 747s on flights across the Pacific. The computer was traced to Ramzi Yousef, who had planned the 1993 bombing of the World Trade Center in New York and was associated with Osama bin Laden. Further investigation determined that the 1994 Philippine airline explosion was a trial run of the Yousef plan (called "Bojinka" by its designers) in which he successfully boarded the aircraft carrying components of an explosives device and timer, assembled the device in an onboard lavatory, and then placed it under a passenger seat before departing the plane at an intermediate stop. Yousef was arrested in February 1995 in Islamabad, Pakistan. Among its other effects, the discovery of the Bojinka plot helped alert the FAA to bin Laden's interest in U.S. civil aviation as a potential target (Johnstone, 2006, pp. 17–18).

The 2000 edition of *Criminal Acts Against Civil Aviation* summarized the 1996 to 2000 period, recording a drop in the overall annual rate of events compared with 1986 to 1995, with a total of 146 incidents listed, for an average of 29 per year versus the 60 per year from 1986 to 1995.

However, it noted, "During the past few years, the relatively low number of incidents that were recorded may have been interpreted as an indication that the threat to civil aviation was decreasing. The fact that the number of aviation-related incidents in 2000 increased by 75% [compared to the previous year] proves such an interpretation to be premature." The 146 incidents included:

- Sixty-four hijackings (44% of the total). The Asia–Pacific region experienced the most hijackings (21) followed by Europe (15), the Middle East and North Africa (13), and Latin America and the Caribbean (9). No hijackings occurred in North America in this period. Thirty-eight of the incidents involved personal motives (e.g., asylum), 10 were politically motivated, six were criminal acts, and the motivations for the remainder were unknown. Additionally, 13 commandeerings (9%) took place. Five of these involved either domestic military (Democratic Republic of the Congo) or political (Solomon Islands) conflicts.
- Thirty attacks at airports (21%), including 14 bombings, seven attempted bombings, and nine other incidents. A majority (12) of these occurred in the Asia–Pacific region, and five were considered to be political or terrorist acts.
- Thirteen off-airport facility attacks (9%), all but one of which involved airline ticket offices, with Europe (nine incidents, eight of which were considered to be politically motivated) the primary location.
- Thirteen incidents involving general aviation or charter aviation (9%), eight of which were hijackings and one involved commandeering of the aircraft.

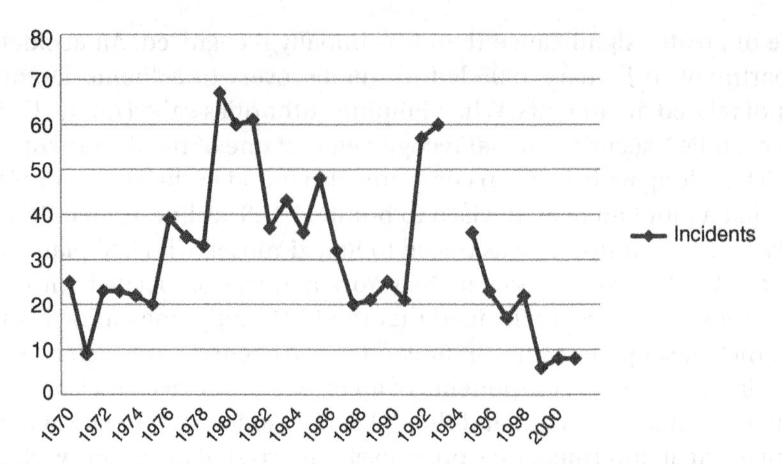

FIGURE 1.6 Aviation terrorism incidents, 1970 to 2000. *(Source: START Global Terrorism Database.)*

- Ten shootings at in-flight aircraft (7%). Nine of the planes crashed, resulting in at least 80 deaths. Only two of the incidents were classified as political or terrorist actions, but the seven that occurred in sub-Saharan Africa were part of ongoing military conflicts between governments and rebel groups.
- One bombing and two attempted bombings on board civil aviation aircraft (2%). None was considered to be politically motivated, and only one fatality occurred in the one actual bombing (pp. 1, 44–53).

Returning to the GTD, a total of 937 incidents of domestic or international terrorism directed against aviation, producing 2416 fatalities, are listed between 1970 and 2000 (Figure 1.6 and Table 1.1). Although these amounts are well above what was recorded for the maritime sector in the same period, they represent approximately one-fourth of the number of incidents and one-third of the casualties from attacks upon the land mode. However, although both the total number of events (55) and fatalities (16) were comparatively low, the United States experienced much greater impact from terrorism directed at the aviation mode than from acts against the other two. The rate of terrorist actions against aviation was fairly constant in the three 10-year intervals (296, or 32%, from 1970 to 1979; 383, or 41%, from 1980 to 1989; 258, or 27%, from 1990 to 2000), but there was a much higher total of fatalities in the middle period (1532, or 63%, from 1980 to 1989)—caused primarily by the large number of casualties from the bombings of Air India Flight 182 and Pan Am Flight 103—but the latest interval had relatively few deaths (107, or just 4%).

■ ■ Critical Thinking ■

Although between 1970 and 2000, far more attacks and casualties occurred in the land transportation mode, much more media and policy attention was focused on the attacks on aviation. Why?

Table 1.1 Comparison of Databases on Terrorist Incidents

	RDWTI	GTD	WITS
Operator	Rand	National Consortium for the Study of Terrorism, University of Maryland	National Counterterrorism Center (NCTC)
Definition of terrorism	Violence calculated to create an atmosphere of fear and alarm to coerce others into actions they would not otherwise undertake or refrain from actions they desired to take. Acts of terrorism are generally directed against civilian targets. The motives of all terrorists are political, and terrorist actions are generally carried out in a way that will achieve maximum publicity.	1970–1997: The threatened or actual use of illegal force and violence by a nonstate actor to attain a political or social goal through fear, coercion, or intimidation 1998–present: An intentional act of violence or threat of violence by a nonstate actor meeting at least two of following criteria: (1) aimed at attaining a political, economic, religious, or social goal; (2) evidence of intention to coerce, intimidate, or convey another message to an audience beyond the immediate victims; and (3) outside context of legitimate wartime activities	Premeditated, politically motivated violence perpetrated against noncombatant targets by subnational groups or clandestine agents
Scope of events	1968–1997: international only 1998–present: international and domestic	1970–present: international and domestic 1993: no data	2004–present: international and domestic
Sources	Primary documents (newspapers, journals, radio, and foreign press)	Primary (news, wire services) and secondary or tertiary (books, journals, existing datasets)	Open sources (news services, local news websites, foreign language sources)
Number of events	~36,000 (through 2009)	~88,000 (through 2008)	~69,000 (through 2010)

GTD, Global Terrorism Database; RDWTI, Rand Database of Worldwide Terrorism Incidents; WITS, Worldwide Incidents Tracking System. (*Source*: Sheehan, I.S., 2012. Assessing and comparing data sources for terrorism research. In: Lum, C., Kennedy, L. (Eds.), Evidence-Based Counterterrorism Policy. Springer, New York, pp. 13–40.)

DIGGING DEEPER
DATA ON TERRORIST AND OTHER ATTACKS ON TRANSPORTATION

Comparability and Reliability of Databases

In describing the level and type of terrorist and other attacks on transportation systems, this chapter has referenced a number of different data sources, with the Global Terrorism Database (GTD) used throughout to allow for trend analyses across the three transportation modes. Making use of such quantitative data is key in attempting to discern the nature and degree of threats to transportation. Yet in using this information, it is important to be aware of its limitations.

In his assessment of various databases used in terrorism research, Sheehan (2012) analyzed five of the most widely cited data sources: the International Terrorism-Attributes of Terrorist Events (ITERATE), Rand Database of Worldwide Terrorism Incidents (RDWTI), Global Terrorism Database (GTD), World Incident Tracking System (WITS), and Terrorism in Western Europe-Events Data (TWEED). Because ITERATE is only partially available online and TWEED applies only to Western Europe, they were not used in this work. A brief comparison of the other three appears in the accompanying table. The GTD was chosen for primary usage because of its greater comprehensiveness in number of incidents included (compared with RDWTI) and time frame (compared with WITS).

In his assessment, Sheehan (2012) found a number of factors that constrained the quality and comparability of the databases:

> *Because of [their] origins,* outside *an academic environment, terrorism data were not always subjected to the kinds of rigorous norms in terms of collecting or coding that are usually expected in academia. This situation led to considerable embarrassment when two Princeton scholars reviewed the data tables at the end of the State Department's annual* Patterns of Global Terrorism *report for 2003 and found that the numbers did not add up and that the conclusion of the report, namely that global terrorism had decreased that year, was in error and that terrorism had actually increased... Terrorism datasets differ from other political and social science data in another important way. Since much of the data is derived from media sources in real time, and since its developers have frequently used different definitions and coding rules, no one dataset is completely comprehensive or exhaustive and there is a great deal of variability across datasets... There is not even a universally accepted definition of terrorism. By one count there are as many as 109 definitions... To complicate matters, terrorism databases have relied almost exclusively on reports in the news media and all too often such reports have been accepted unquestioningly despite their known biases and unreliability.*
>
> *Sheehan, 2012*

To illustrate the consequences of such factors, Sheehan compared the ITERATE and Rand databases for the 1993 to 2004 period. He found "several large discrepancies in quarterly events counts," likely caused by differences in how the databases defined a "transnational" event and furthermore, that "*unique* events, ones covered in only one of the two databases, outnumbered *overlapping* ones at almost every quarterly period," which he attributed to their different sources of data. He goes on to propose six criteria for use in the evaluation of terrorism databases (Sheehan, 2012, pp. 13–24):

- Conceptual clarity in definitions
- Context and immediacy of observations (i.e., data sources)
- Citation transparency
- Coding consistency
- Certainty of the observations
- Conflicts of interest in data gathering and reporting
- Accessibility of the database

Take a closer look at the various databases used in this chapter (GTD; IMO reports on piracy; the Rand chronology of terrorist and other criminal acts against maritime targets; the Mineta Institute chronologies of terrorist and other serious acts against land public transportation targets; the FAA reports on criminal acts against civil aviation). Apply Sheehan's evaluative criteria to each. What are the strengths and weaknesses of each dataset? What could be done to improve them? Sheehan considered only databases on terrorist incidents. From a transportation security standpoint, what are the pros and cons of including nonterrorist criminal acts?

Pre-9/11 Aviation Security Systems

Similar to its maritime counterpart, aviation security had both international and national components as it evolved. From the 1960s onward, however, aviation security systems were more widespread and more detailed and commanded greater resources than those in the maritime or land modes. This increasing attention was—in almost every case—in reaction to specific incidents.

The first major development in aviation security was the 1944 Chicago Convention on International Aviation, which sought to both ensure the growth of international civil aviation and promote flight safety. To further these objectives, the convention established the International Civil Aviation Organization (ICAO), which became a specialized agency of the UN in 1947. The security aspect was incrementally strengthened by the Tokyo Convention of 1963 (which provided minimal guidelines for flight crews and national civil aviation authorities in ensuring the safety of passengers and crew and the return of aircraft in the event of hijackings or other unlawful acts on board), the Hague Convention of 1970 (which specifically defined hijacking, and obligated member nations to apprehend and prosecute or extradite hijackers and to impose severe penalties on those convicted of the crime), and the Montreal Convention of 1971 (which extended the call for apprehension, prosecution, and severe penalties to all unlawful acts against civil aviation, whether in flight or on the ground, including sabotage) (Price and Forrest, 2009, pp. 85–88).

In carrying out its mission, the ICAO creates International Standards and Recommended Practices (SARPs) for member states.[11] By 1973, the organization determined that security threats required more focused attention and promulgated such a guideline in 1974 as Annex 17, Standards and Recommended Practices (Security). Annex 17 outlines a basic security program of minimum requirements and is supplemented by the "Security Manual for Safeguarding Civil Aviation Against Acts of Unlawful Interference," which provides more detailed guidance. As of 2000, Annex 17 obligated member nations to institute measures for, among other things (Price and Forrest, 2009, p. 86):

- Prevention of unauthorized access to airfields
- Training of security personnel
- Isolation of security-processed passengers
- Inspection of aircraft for weapons and other dangerous devices
- Cargo and mail screening
- Background checks for aviation workers
- Reconciliation of passengers and baggage.

After the destruction of Air India Flight 182 in June 1985, the ICAO created the Aviation Security Panel, which was charged with reviewing and improving Annex 17. As described by the former director of security of the International Air Transport Association (IATA),[12] Annex 17 "is a compromise document designed to balance the needs of civil aviation seen

[11]As of 1999, 185 nations had ratified the Chicago Convention and were thus classified as "Contracting States."
[12]The IATA is the trade association of the world's scheduled airlines.

through the eyes of security specialists (the Panel) with political and economic considerations demanded by the wide-ranging membership of ICAO." Furthermore, as an association of sovereign governments, the ICAO has no powers of enforcement over its members, and "whilst ICAO Annexes are 'binding' on governments, provision is made for Contracting States to opt out of regulations which, for one reason or another, they find unacceptable." Thus, primary responsibility for the implementation of aviation security measures has rested at the national level (Wallis, 1999, pp. 84–85).

In the United States, aircraft safety was an intrinsic part of the civil aviation system from its very beginning. The foundational statute for that system, the Air Commerce Act of 1926, was enacted at the urging of the fledgling aviation industry, which believed that governmental action to improve and maintain safety standards was essential if the industry was to reach its full potential. On the other hand, security was viewed by the industry and others as "disruptive" to civil aviation operations (Johnstone, 2006, pp. 31–32).

It was not until 1958 that aviation security received attention via the adoption of the Federal Aviation Act of 1958, which created the basic framework for the U.S. civil aviation system. The statute provided authority to the Federal Aviation Agency (later renamed the Federal Aviation Administration) to conduct investigations and hearings, with its Internal Security component responsible for investigations involving flight crews, other civil aviation workers, and civil aviation facilities, as well as for information security. Over the next 30 years, the security system evolved in piecemeal fashion primarily in response to specific incidents.

- In the wake of the first domestic hijacking in the United States in 1961, President Kennedy directed the FAA to create a force to provide onboard security on a few domestic and international flights. This initiative, which became the Federal Air Marshal Program in 1973, began with 18 officers.
- The Anti-Hijacking Act of 1971 responded to the increasing number of hijackings (primarily to or from Cuba) by establishing substantial penalties for hijackers (including the death penalty and life in prison) and requiring the security screening of passengers (but not carry-on baggage).
- The persistence of the Cuba hijackings and other attacks on civil aviation led to the enactment of the Anti-Hijacking Act of 1974, which authorized the suspension of air service to any country that encouraged hijackings, mandated the screening of all passengers and baggage by weapons-detection technology, required commercial airports to provide a law enforcement presence, and assigned responsibility for passenger screening to the airlines (or their designated contractor). This latter provision resulted in the 1976 Air Carrier Standard Security Program (ACSSP), which was developed jointly by the FAA and the airlines and served as the basis for U.S. air carrier security programs (Price and Forrest, 2009, pp. 91–93).

An analysis of U.K. aviation security in this time period by the chairman of the British Airline Pilots' Association found that although "British aviation security standards were amongst the highest in the world," a relatively low priority was assigned to it because the threat was perceived to be low (with only four "serious" terrorist actions against British aviation in the preceding 20 years). This in turn produced limited resource allocation by the government to both its own security program (with the Department of Transport's aviation security operation usually consisting of just seven officials) and the aviation industry's security costs. "Understandably, industry resented and resisted the continually increasing [security] burden. Thus neither [the British government] nor industry were wholehearted in their pursuit of the highest practicable standards" (Malik, 1999, pp. 112–113, 118).

In the pre-Lockerbie U.K. aviation security program (Malik, 1999, pp. 118–122):

- The Department of Transport's small security staff devoted most of their efforts to drafting security standards, which were generally advisory rather than mandatory and were able to do very little compliance inspection.
- A "negative" passenger-bag reconciliation procedure was used, under which a bag was offloaded from a plane only when a passenger was known to be missing, in contrast to the "positive" reconciliation procedure required (but not always enforced) by the United States, which necessitated a check to ensure that a passenger was on board in order for the bag to remain on the aircraft.
- Passenger and bag screening technology and procedures were "rudimentary."
- Cargo security was very limited.
- As was true of all countries, the security of British aircraft overseas was dependent on the security practices of an airport's host nation or sometimes (especially with respect to passenger and baggage screening) of local police or security agencies. The resulting system was "generally flawed, and in places, non-existent."

The loss of Pan Am Flight 103 produced major reactions in both the United States and United Kingdom. Shortly after the bombing, the Bush Administration undertook a number of initiatives to strengthen anti-explosives security measures for U.S. airlines at high-risk airports, mostly located overseas. These included mandatory x-ray screening of all baggage and a "positive" passenger-bag reconciliation requirement for all passengers and baggage. In addition, plans were commenced to purchase more than $100 million worth of advanced screening equipment specifically designed to detect explosives, unlike the x-ray machines then in use (Johnstone, 2006, pp. 16–17).

The U.S. President's Commission on Aviation Security and Terrorism reported:

> *The U.S. civil aviation security system is seriously flawed and has failed to provide the proper level of protection for the traveling public. This system needs major reform.... The Federal Aviation Administration [is] a reactive agency – preoccupied with responses to events to the exclusion of adequate contingency planning in anticipation of future threats.... Pan Am's apparent security lapses and FAA's failure to enforce its own regulations followed a pattern that existed for months before Flight 103, during the day of the tragedy, and—notably—for nine months thereafter.*
>
> *Report to the President, 1990, p. i*

The commission went on to make a total of 64 recommendations. Many were included in the Aviation Security Act of 1990, which represented the first major revision in the U.S. security system since 1974. The law created new aviation and intelligence positions within the FAA, mandated the FAA to report on aviation system threats and vulnerabilities, authorized the agency to impose security measures at airports, and strengthened procedures for airport access control and aviation personnel identification verification (Johnstone, 2006, p. 17; Price and Forrest, 2009, pp. 93–94).

In the United Kingdom, the most significant responses to the Pan Am Lockerbie tragedy were the creation of the Transport Security (TRANSEC) division within the Department of Transport and the adoption of the Aviation and Maritime Security Act of 1990. TRANSEC was given lead responsibility for aviation, maritime, and Channel tunnel security, and its mandate included inspection, testing, and auditing, in addition to its predecessors' standard setting role. Commensurate with its increased role, the agency was provided with additional resources, reaching a high point of 135 employees in 1993 (although subsequent budget cuts reduced it to a staff of 78 by 1998). The aviation portion of the 1990 Act was primarily designed to expedite the issuance of security directives and to extend their scope to cover all organizations (including air cargo) and persons (including passengers and staff) involved in aviation (Malik, 1999, pp. 122–126).

Initially, the deployment of the new security measures in the United States, the United Kingdom and elsewhere seemed to have been effective in dealing with the threat to civil aviation, especially with respect to bombings and hijackings in North America and Western Europe. The 1993 bombing of the World Trade Center in New York and the discovery of the Bojinka plot in 1995 (both of which were planned and carried out by Islamic extremists and ultimately connected to Osama bin Laden) led to a renewed sense of urgency, especially in the United States, which was the primary target of the attacks. This perception of a heightened threat, and an awareness of certain deficiencies in U.S. security measures, resulted in the establishment by the FAA's Aviation Security Advisory Committee of a Baseline Working Group (BWG), whose goal was "to strengthen the aviation security 'baseline' to a level commensurate with the new threat environment." (Johnstone, 2006, p. 19).

The BWG's most fundamental recommendation in its "Domestic Security Baseline Final Report" of December 1996 was that the principal factor in setting aviation security policy should be "effectiveness" rather than cost or expediency. Among its specific proposals, the report suggested that the airlines use an FAA-approved passenger profiling system to identify individuals who might pose a security threat and whose persons and baggage (both checked and carry-on) would be subjected to security screening. It also called for an expanded role for the FBI in aviation security, expedited deployment of available screening technologies, elevation of the security role within both the FAA and the airlines and airports, and streamlining of the rule-making process for setting security standards. Furthermore, the BWG indicated that the federal government should end the practice of "unfunded mandates" on the aviation industry and pay for "the full cost of implementing and maintaining an improved domestic security baseline" (estimated to be $9.9 billion over a 10-year period) (Johnstone, 2006, pp. 19–20).

With its origin arising from general concerns about security measures in relation to what was seen as an increasing threat and its objective of establishing a sustainable baseline of security, the BWG represented a break from the usual pattern of incident-driven aviation security policymaking. However, on the day of the BWG's very first meeting (July 17, 1996), TWA Flight 800 exploded off the New York coast while en route from New York City to Paris, killing all 230 persons on board. Although the National Transportation Safety Board (NTSB) later determined that the probable cause was the explosion of a fuel tank precipitated by an accidental short circuit, the initial feeling was that the explosion was an act of terrorism. This set in motion a new round of crisis-driven security response, including the August 1996 appointment by President Clinton of the White House Commission on Aviation Safety and Security, chaired by Vice President Al Gore and generally referred to as the Gore Commission. Thus, the work of the BWG was superseded by the presidentially appointed commission almost from the start, with the BWG analyses and recommendations essentially serving as background material for the latter (Johnstone, 2006, p. 20).

The Gore Commission issued its final report and recommendations in 1997. A number of its proposals dealt with the safety side of its mandate. With regard to security, the most significant recommendations addressed the bombing threat, including immediate deployment of explosives detection equipment for baggage screening at airports, partial implementation of positive passenger-bag reconciliation on domestic flights (which was already required on most overseas flights), additional deployment of canine explosives-sniffing teams, establishment of an automated passenger profiling system to identify passengers "who merit additional attention," and requirement of federal certification of screening companies and of minimum training standards for screeners. The Gore Commission stressed that aviation security should be viewed as a national security issue and should thus receive "substantial" federal funding (defined as ~$100 million a year) (Johnstone, 2006, pp. 20-21).

Congress responded favorably to most of the Gore Commission recommendations. The Federal Aviation Reauthorization Act of 1996, which was based on the commission's preliminary proposals (Johnstone, 2006, p. 22):

- Directed the FAA to certify screening companies and improve training and testing of screeners.
- Required background checks of passenger and baggage screeners.
- Directed the FAA and the FBI to conduct periodic joint threat and vulnerability assessments at high-risk airports.
- Required airports and air carriers to conduct periodic vulnerability assessments and the FAA to audit such assessments as well as to conduct periodic and unannounced inspections of airport and air carrier security systems.
- Encouraged the FAA, the Department of Transportation, the intelligence community, and law enforcement agencies to assist air carriers in developing passenger profiling systems.
- Encouraged the FAA to expedite deployment of explosives detection equipment.

Dissatisfied with the implementation of many of the 1996 law's provisions, Congress adopted the Airport Security Improvement Act of 2000, which directed the FAA to work with airports and air carriers to improve airport access controls by January 31, 2001; to issue the final rule for certification of passenger and baggage screening contractors by May 31, 2001; and to maximize the use of explosives detection equipment. Furthermore, the legislation specified that criminal background checks should be done for all airport security applicants and minimum training standards should be established for airport screeners and other aviation security personnel. None of these requirements were fully implemented as of September 11, 2001 (Johnstone, 2006, pp. 22–23; Price and Forrest, 2009, pp. 98–99).

In the United Kingdom, security improvements in the 1990s included more secure airport terminals, audits and inspections by TRANSEC (which improved standards and compliance), nearly 100% screening of checked baggage, implementation of full positive passenger-bag reconciliation, and a beginning of an air cargo security program (Malik, 1999, pp. 128–130).

Despite all of the attention and resources devoted to aviation security, particularly in the United States and United Kingdom after the Pan Am Flight 103 disaster, many assessments of system performance in the late 1990s and 2000 were far from sanguine. Writing about the international level in 1999, the former director of security of the International Air Transport Association indicated, "Unhappily, despite ICAO's efforts, *effective* security exists in only a minority of countries around the world." In the same publication, the chairman of the British Airline Pilots' Association's security committee provided his evaluation of the British aviation security system then in place:

> *In the five years following Lockerbie, there is no doubt that Great Britain showed both commitment and competence. However, [the British government] has not at any time adopted a truly holistic approach to the threat of terrorist and other attacks on aviation. . . . For as long as [it] holds to the convenient fiction that it has no financial responsibility for security countermeasures, industry will resist the argument that its duty to care for its passengers constitutes a blank check for the government. Since Lockerbie, [the British government] has chosen to proceed by a system of edict and enforcement. It has produced a system much inferior to that which would have resulted from a partnership [with industry].*
>
> *Wallis, 1999, p. 87; Malik, 1999, p. 132*

In the United States, where the kind of government–industry partnership in aviation security sought in the United Kingdom was more established, a number of detailed, publicly available evaluations of security performance were issued, principally from the U.S. General Accounting Office (GAO, now called the Government Accountability Office) and the Department of Transportation's Inspector General (DOT IG).

- Beginning in the late 1980s, the GAO documented continuing shortcomings in the performance of security screeners at airport checkpoints, likely caused by rapid

workforce turnover (a product of low wages and difficult working conditions) and inadequate training. In 1987 testing, 20% of the simulated weapons used in the testing of screeners escaped detection, and subsequent testing indicated that performance had worsened throughout the 1990s.
- In the late 1990s, the DOT IG reported serious underutilization of checked baggage explosives detection equipment at airports and inadequate airport access controls (with its covert agents able to gain unauthorized access to secure areas in two-thirds of its tests).
- Summarizing the findings with respect to the overall U.S. aviation security system, Dr. Gerald Dillingham of the GAO testified to a Senate committee in April 2000, "Taken together these problems show the chain of security protecting our aviation system has not one but several weak links. . . . It must be remembered that the responsibility for these problems does not rest with the FAA alone. The aviation industry is responsible for undertaking the security measures at airports and many of the problems identified—such as rapid screener turnover—more appropriately rest with it" (Johnstone, 2006, pp. 32–34).

■ ■ Critical Thinking ■

Why were the aviation security systems in the United States and United Kingdom so reactive in their development of security measures? Consider political, economic, public opinion, and any other factors you deem important.

Conclusion

Violent threats to all means of transportation existed from the very beginnings of each, which is not surprising given the ubiquity and economic value of these systems and the cargo and passengers they conveyed. For the maritime mode, the most frequent danger has come from piracy at sea and in port. However, its generally confined scope (mainly involving economic motives, such as theft), distribution (primarily occurring in southeast Asia and Africa, both far removed from the 20th century centers of economic, political, and media influence), and consequences produced little in the way of enforceable security measures.

Terrorist attacks on maritime transportation were more limited in number, but they produced a greater impact than the acts of piracy because of their higher number of casualties but especially because of the higher visibility, to the news media and to national governments, of certain incidents. The most significant of these was the 1985 seizure of the Italian cruise ship *Achille Lauro*. This led to reevaluations of international and national maritime security measures that were (and continued to be) fragmented and limited. Those reassessments yielded only modest results, including the 1988 Convention for the Suppression of Unlawful Acts against the Safety of Maritime Navigation, which increased the obligation of nations to aid in the prosecution and punishment of those committing crimes involving maritime transportation.

Land (or surface) transportation systems were, by far, the target of the most terrorist and other violent criminal attacks on transportation systems in the 1970 to 2000 period. Although nonterrorist incidents were almost certainly predominant in this mode as well, disentangling transportation-related from other targets is especially challenging in this sector, and other definitional issues (e.g., whether to classify actions arising out of internal armed conflicts as terrorist incidents) also complicate efforts to compile a comprehensive evaluation of attacks.

The localized nature of land transportation systems and of the governing authorities that exercised jurisdiction over them produced a security framework that was fragmented like its maritime counterpart, but with weaker international and national elements. Similar to both the maritime and aviation sectors, security systems for this mode generally evolved in reaction to specific incidents or series of incidents. Thus, the most highly developed land transportation security system was almost certainly the one established by national and local authorities in the United Kingdom to protect the London passenger rail system in the wake of a 30-year bombing campaign by the Provisional Wing of the IRA. But in the period before 2001, most land transportation systems lacked either (in the case of most North American, Western European, and Japanese systems) experience in facing major threats or (in Africa and Southeast Asia and on the Indian subcontinent, where most attacks on land transportation actually occurred) the resources or willingness (or both) to take significant action. The resulting security systems were generally rudimentary or nonexistent.

Although aviation was the last of the major transportation modes to be developed, it attracted the greatest policy and media attention, especially in the United States, which had been a particular target of terrorists and other political actors seeking to stage high-profile incidents. Before the 1960s, aviation security systems had much in common with those involved in maritime and land transportation, being very limited in scope and resources. However, the large number of airplane hijackings between the United States and Cuba during the late 1960s and early 1970s produced a substantial increase in attention to the hijacking threat by the U.S. government via the deployment of weapons-detecting x-ray machines and magnetometers at airport checkpoints and of federal air marshals to provide onboard security on certain flights.

The apparent success of the anti-hijacking measures and the occurrence of a series of high-casualty, onboard explosions (most notably Pan Am Flight 103 in 1988) created a shift in focus to the bombing threat in both the United States and United Kingdom via enlarged government aviation security forces, expanded screening requirements for both passengers and luggage, and expedited deployment of more advanced screening equipment capable of detecting explosives. Concerns about onboard bombings, especially by terrorists, were intensified by the discovery of the 1995 Bojinka plot, which sought to blow up 12 American airliners over the Pacific and the suspected terrorist role in the destruction of TWA Flight 800 in 1996 (which turned out to likely have been a safety-related accident).

In response, the Gore Commission was convened and made a number of security recommendations that were subsequently implemented, including increases in federal funding and the instigation of government threat and vulnerability assessments at high-risk airports, certification of the private companies contracted by the airlines to screen passengers and baggage, and development of a passenger profiling system to identify potentially threatening individuals.

By the end of 2000, maritime security, which had experienced few major incidents, rested on weakly enforced international laws against piracy and other criminal acts against maritime targets and varying national standards and enforcement in ports and territorial waters. Land transportation security systems were very limited, outside a few jurisdictions, including the United Kingdom and New York City, that had experienced or perceived a significant threat. Aviation security was by far the most developed and well-funded of the modal security systems, but its focus on countering previously experienced tactics left it vulnerable to evolving terrorist planning.

Discussion Questions

1. What transportation mode was most frequently attacked before 9/11? Why?
2. What transportation mode received the greatest policy attention before 9/11? Why?
3. How did the terrorist threat evolve in each transportation mode? What were the major similarities and differences?
4. Identify the role played by international organizations in pre-9/11 security for maritime, land, and aviation transportation systems. What was the most serious deficiency in each case?
5. How and why did pre-9/11 transportation security systems in the United States and United Kingdom differ?
6. What was the primary factor in producing major changes in transportation security measures? Why?

References

Bennett, J.C., 2008. Maritime security. In: Bragdon, C.R. (Ed.), Transportation Security. Butterworth-Heinemann/Elsevier, Burlington, MA, pp. 149–181.

CBC News Online. September, 2006. The bombing of Air India Flight 182. Retrieved from <http://www.cbc.ca/news/background/airindia/bombing.html> (accessed 10.17.14.)

Chalk, P., 2008. The maritime dimension of international security: Terrorism, piracy, and challenges for the United States. Rand, Santa Monica, CA.

Federal Aviation Administration, Office of Civil Aviation Security. 1995. Criminal acts against civil aviation. Washington, DC. 1996.

Federal Aviation Administration, Office of Civil Aviation Security. 2000. Criminal acts against civil aviation. Washington, DC. 2001.

Gottschalk, J.A., Flanagan, B.P., 2000. Jolly Roger with an Uzi: The rise and threat of modern piracy. Naval Institute Press, Annapolis, MD.

Government of the United Kingdom. 1990. Aviation and Maritime Security Act 1990. Retrieved from <http://www.legislation.gov.uk/ukpga/1990/31/contents> (accessed 10.17.14.)

International Maritime Organization. 2000. Piracy and armed robbery at sea. London. 2000.

Jenkins, B.M., 1983. A chronology of terrorist and other criminal actions against maritime targets. Rand Corporation, Santa Monica, CA.

Jenkins, B.M., 1997. Protecting surface transportation systems and patrons from terrorist activities, case studies of best security practices and a chronology of attacks. Norman Y. Mineta International Institute for Surface Transportation Policy Studies, San Jose State University, San Jose, CA.

Jenkins, B.M., 2001. Protecting public surface transportation systems against terrorism and serious crime: An executive overview. Mineta Transportation Institute, San Jose, CA.

Jenkins, B.M., Gersten, L.N., 2001. Protecting public surface transportation systems against terrorism and serious crime: Continuing research on best security practices. Mineta Transportation Institute, San Jose, CA.

Johnstone, R.W., 2006. 9/11 and the Future of Transportation Security. Praeger, Westport, CT.

Malik, O., 1999. Aviation security before and after Lockerbie. In: Wilkinson, P., Jenkins, B. (Eds.), Aviation Terrorism and Security. Frank Cass, London, pp. 112–133.

National Counterterrorism Center. 2011. Report on terrorism. Washington, DC. 2012.

Price, J.C., Forrest, J.S., 2009. Practical aviation security: Predicting and preventing future threats. Butterworth-Heinemann/Elsevier, Burlington, MA.

Report of the Interagency Commission on Crime and Security in U.S. Seaports. 2000. Washington, DC.

Report to the President. 1990. The Commission on Aviation Security and Terrorism, Washington, DC.

Rodrigue, J., Comtois, C., Slack, B., 2006. The geography of transport systems. Routledge, New York.

Savage, T., 1997. The New York City Transit Authority: Contingency planning for emergency response. In: Jenkins, B.M. (Ed.), Protecting Surface Transportation Systems and Patrons from Terrorist Activities, Case Studies of Best Security Practices and a Chronology of Attacks. Norman Y. Mineta International Institute for Surface Transportation Policy Studies, San Jose State University, San Jose, CA, pp. 33–48.

Sheehan, I.S., 2012. Assessing and comparing data sources for terrorism research. In: Lum, C., Kennedy, L. (Eds.), Evidence-Based Counterterrorism Policy. Springer, New York, pp. 13–40.

Simon, J.D., 1986. The implications of the Achille Lauro hijacking for the maritime community. Rand corporation, Santa Monica, CA.

Sweet, K.M., 2006. Transportation and cargo security: Threats and solutions. Pearson/Prentice Hall, Upper Saddle River, NJ.

The 9/11 Commission report: final report of the National Commission on Terrorist Attacks upon the United States. 2004. Norton, New York.

U.K. Department for Transport. n.d. Responsibilities of Transport Security's Land Transport Division. <http://www.dft.gov.uk/publication/responsibilities-of-dft-transport-security-land-transport-division> (accessed 10.17.14.)

U.S. Department of Defense. January, 2001. Navy announces results of its investigation on USS Cole. <http://www.defense.gov/releases/release.aspx?released=2814> (accessed 10.17.14.)

U.S. Department of Homeland Security, U.S. Coast Guard. n.d. Missions: Maritime security. <http://www.uscg.mil/top/missions/MaritimeSecurity.asp> (accessed 10.17.14.)

Wallis, R., 1999. The role of the international aviation organizations in enhancing security. In: Wilkinson, P., Jenkins, B. (Eds.), Aviation Terrorism and Security. Frank Cass, London, pp. 83–100.

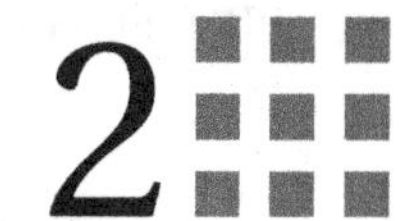

The 9/11 Watershed

CHAPTER OBJECTIVES:

In this chapter, you will learn about the September 11, 2001, aircraft hijackings in the United States, including:

- Key attributes of U.S. aviation security as of 9/11
- Aviation security layers and how the hijackers evaded them
- The evolution of transit security in New York City and the impact of 9/11 on New York's public transportation systems
- The security response in the United States and elsewhere to the terrorist attacks

Introduction

Global transportation systems continued to face attacks as the 21st century began, but those threats seemed to have abated, with annual incidents involving the maritime and aviation sectors numbering in the single digits in the last years of the 20th century and most of the 50 or so yearly assaults on land transportation being a part of ongoing armed internal conflicts in the Philippines, Russia, Sri Lanka, and elsewhere. Although the United States had experienced attacks against its transportation systems for many years before 2001, especially in civil aviation, the terrorist hijackings of September 11, 2001, had a profound effect on how the American government perceived the terrorist threat and prioritized and organized what it termed *homeland security*. Furthermore, the magnitude of that event and the central position of the United States in the field of aviation produced major reactions worldwide, as other countries reviewed and revised their own security systems in view of their newly perceived vulnerability.

Transportation Security in 2001

In 2001, maritime terrorism remained at a relatively low level, but the threats to both land transportation and aviation accelerated; however, with a very notable exception, most of these were consistent with previous patterns. According to the Global Terrorism Database, in 2001, there were five terrorist incidents involving maritime transportation, producing a total of 14 fatalities, and 100 attacks on land transportation, resulting in 465 deaths. More than half of these (259) stemmed from an August 11 assault in Angola by UNITA (National Union for the Total Independence of Angola) rebels. Of the 27 aviation-related terrorist attacks, 16 had no casualties; in two instances, the casualties were unknown; and of the remaining nine, five yielded a total of 59 deaths. The other four were the September 11 hijackings of commercial aircraft within the United States (Figure 2.1).

FIGURE 2.1 World Trade Center, New York, March 2001. *(From Wikipedia, http://en.wikipedia.org/wiki/File:Wtc_arial_march2001.jpg.)*

U.S. Aviation Security in 2001

The aviation security system in place in the United States in September 2001 was essentially the same one in existence back in 1974. It consisted of shared and "complementary" responsibilities in which the Federal Aviation Administration (FAA) set and enforced minimum security standards for airports and air carriers; the air carriers (or their contractors) screened passengers, baggage, and cargo, protected aircraft from unauthorized access, and trained their personnel in emergency procedures; the airports provided law enforcement and facility security; and Congress made laws, including providing such funding as it deemed necessary, and performed oversight (Johnstone, 2006, p. 24).

Implementation of the system was strongly influenced, and in some ways inhibited, by a number of long-standing attributes.

- The shared roles created, in the words of the 1990 President's Commission on Aviation Security and Terrorism, "a division of security responsibilities that leaves no entity accountable."
- The reactive nature of the system—in which substantial changes generally occurred only in the aftermath of a serious incident—was reinforced by the means through which operational modifications had to be made: rule making, a cumbersome and time-consuming process for making permanent revisions (and which experience indicated could only be speeded up after a major disaster), and Security Directives, used to make "temporary" changes (but usually issued in response to a specific event or detailed intelligence).
- The fact that, unlike in many other countries where passenger airlines were directly or indirectly controlled by national governments, civil aviation in the Unites States was (and is) a privately owned, for-profit enterprise had significant security implications. The Department of Transportation's (DOT's) Inspector General told the 9/11 Commission that there were great pressures within the aviation security system to control costs and to "limit the impact of security requirements on aviation operations, so that the industry could concentrate on its primary mission of moving passengers and aircraft. . . . Those counter-pressures in turn manifested themselves as significant weaknesses in security."
- The approach of the FAA was significantly shaped by its so-called "dual mandate" under which the agency was tasked to both regulate and promote civil aviation. This "dual mandate" was formally abolished by the Congress in 1996. However, under laws still in effect in 2001, the FAA was directed to consider "encouraging and developing civil aeronautics" in carrying out its safety and security responsibilities.
- Although safety was made intrinsic to the operation of civil aviation from the very start (with the foundational Air Commerce Act of 1926 "passed at the urging of the aviation industry, whose leaders believed the airplane could not reach its full potential without federal action to improve and maintain safety standards"), security was generally regarded by the industry as "disruptive" to its regular operations. A National Research Council report highlighted the differences between aviation safety in which "one agency (the FAA) has a dominant role in ensuring safety through multiple, coordinated means" and aviation security in which "tactics and techniques emerged piecemeal, in reaction to a series of individual security failures" (Johnstone, 2006, pp. 24–32).

Similar to other human endeavors, U.S. aviation security did not operate in a vacuum, and any understanding of how that system performed on September 11, 2001, must consider the context. The 9/11 Commission staff described the pre-9/11 mindset:

Before September 11, 2001, the aviation security system had been enjoying a period of relative peace. No U.S. flagged aircraft had been bombed or hijacked in over a decade. Domestic hijacking in particular seemed like a thing of the past, something that could only happen to foreign airlines that were less well protected. The public's own "threat assessment" before September 11 was sanguine about commercial aviation safety and security. In a Fox News/Opinion Dynamics survey conducted at the end of the 1990's, 78 percent cited poor maintenance as "a greater threat to airline safety" than terrorism. Demand for air service was strong and was beginning to exceed the capacity of the system. Heeding constituent calls for improved air service and increased capacity, Congress focused its legislative and oversight attention on measures to address these problems… The leadership of the Federal Aviation Administration also focused on safety, customer service, capacity and economic issues. The agency's security agenda was focused on efforts to implement a three-year-old Congressional mandate to deploy explosives detection equipment at all major airports and complete a nearly five-year-old rulemaking effort to improve checkpoint screening.

9/11 Commission, Staff Statement No. 3, 2004

Jane Garvey, the head of the FAA in 2001, told the Commission:

On September 10, based on intelligence reporting, we saw explosive devices on aircraft as the most dangerous threat. We were also concerned about what we now think of as traditional hijacking, in which the hijacker seizes control of the aircraft for transportation, or in which passengers are held as hostages to further some political agenda.

9/11 Commission, Jane Garvey Testimony, 2003

Aviation Security Layers

The pre-9/11 U.S. aviation security system was premised on the notion that its multiple layers of defense would provide overall protection against hijackings and sabotage even if one or more of these were successfully breached. There were six such layers in operation in 2001: intelligence, passenger prescreening, airport access control, passenger checkpoint screening, checked baggage and cargo screening, and onboard security. Two of these were not relevant to the tactics used by the 9/11 hijackers. There is no evidence that the terrorists used baggage or cargo checked into the planes' cargo holds, nor is there any indication that airport access controls were implicated in carrying out the plot (9/11 Commission, Staff Statement No. 3, 2004).

1. *Intelligence.* The FAA relied on the intelligence community (primarily the CIA for overseas and the FBI for domestic intelligence) to discover specific plots as well as general threats to civil aviation. It then used that information, as processed by its own intelligence unit, to devise and justify the security measures it imposed on the airlines and airports. Whereas a large amount of threat information was received in 2001 (nearly 200 pieces per day), little of it pertained to the presence and activities

of terrorists within the United States. With all parts of the U.S. aviation system (including the industry, Congress, the travelling public, and the FAA itself) more concerned about checkpoint and flight delays than a hijacking threat that seemed to have all but vanished, especially within the United States, attempts to impose tightened security absent specific threat information were almost certain to be rejected. The FAA was aware of security warnings in the summer of 2001, including a potential threat to civil aviation posed by Osama bin Laden and al Qaeda, and issued alerts to the U.S. aviation industry about possible near-term terrorist attacks against aviation, particularly in the Middle East. However, in the words of the 9/11 Commission staff, "No major increases in anti-hijacking security measures were implemented in response to the heightened threat levels in the spring and summer of 2001, other than general warnings to the industry to be more vigilant and cautious. . . . Without actionable intelligence information to uncover and interdict a terrorist plot in the planning stages or prior to the perpetrator gaining access to the aircraft in the lead-up to September 11, 2001, it was up to the other layers of aviation security to counter the threat."

2. *Passenger prescreening.* The second key antihijacking defense in 2001 were the measures used to identify potentially threatening passengers before their reaching the security checkpoint. On 9/11, these included two major components: a "no-fly" list of individuals identified as threats to civil aviation by the FAA (based on information supplied by the intelligence community), which directed the airlines to prevent such individuals from boarding flights, and the FAA-approved and airline administered Computer-Assisted Passenger Prescreening System (CAPPS), which was a computerized formula for selecting for additional security scrutiny passengers whose profiles (largely derived from ticketing information) suggested a potential threat to the aircraft. As of 9/11, neither of these measures was of much use in defending against hijackings because the "no-fly" list contained only 12 names (including 9/11 mastermind Khalid Sheikh Mohammed), and the consequences for selection by CAPPS were limited to having the individual's checked baggage screened for explosives or held off the aircraft until the individual boarded the plane.
3. *Passenger checkpoint screening.* According to the 9/11 Commission Final Report:

Checkpoint screening was considered the most important and obvious layer of security. Walk-through metal detectors and X-ray machines operated by trained screeners were employed to stop prohibited items.[1] *Numerous government reports indicated that checkpoints performed poorly. . . . Many deadly and dangerous items did not set off metal detectors, or were hard to distinguish in an X-ray machine from*

[1]In September 2001, the checkpoint metal detectors were calibrated to detect guns and large knives. Most of the screeners were employed by security companies under contract with the individual airlines (9/11 Commission, Staff Statement No. 3, 2004).

innocent everyday items. While FAA rules did not expressly prohibit knives with blades under 4 inches long, the airlines' checkpoint operations guide (which was developed in cooperation with the FAA), explicitly permitted them. . . . A proposal to ban knives altogether in 1993 had been rejected because small cutting implements were hard to detect and the number of innocent 'alarms' would have increased significantly, exacerbating congestion problems at checkpoints.

4. *Aircraft and onboard security.* Once onboard the plane, there were two final potential obstacles to hijackers. First, the Federal Air Marshal Program of armed, onboard security personnel initiated by President Kennedy back in 1961 was still in existence, but as of 9/11, there were only 33 Air Marshals, and none of them were assigned to domestic flights within the United States. Second, onboard security procedures for the aircraft's crew were governed by the FAA-approved "Common Strategy," which, according to the 9/11 Commission, "taught flight crews that the best way to deal with hijackers was to accommodate their demands, get the plane to land safely, and then let law enforcement or the military handle the situation. The strategy operated on the fundamental assumption that hijackers issue negotiable demands . . . and that, as one FAA official put it, 'suicide wasn't in the game plan' of hijackers. FAA training material provided no guidance for flight crews should violence occur"[2] (*9/11 Commission Report*, 2004, pp. 83–85; 9/11 Commission, Staff Statement No. 3, 2004; 9/11 Commission, Third Monograph, 2006, pp. 638–641).

■ ■ Critical Thinking ■

In its Final Report, the 9/11 Commission cited a lack of "imagination" as one of the chief causes of the security failures leading up to and on the day of September 11, 2001. Name three specific instances of a failure in imagination with respect to pre-9/11 U.S. aviation security and indicate why they may have occurred.

Pre-9/11 Aviation Security Performance

Although there were some individual examinations of the performance of specific elements of aviation security measures in different countries (most notably the studies undertaken in the United States by the Government Accountability Office [GAO] and the DOT's inspector general), there was very limited information about how these measures functioned overall or at the international level. One attempt to do so was by Israeli terrorism expert Ariel Merari (1999). He performed an analysis of the effectiveness of the

[2]Another potential onboard security measure, used for many years by Israel, was "hardened" cockpit doors that could not be easily breached by intruders. This was not done in the United States before 9/11, but under the prevailing "Common Strategy" of accommodation and an FAA regulation mandating "ready access" into and out of cockpits in emergencies, it is doubtful that such a measure would have presented a significant obstacle to hijackers (9/11 Commission Report, 2004, p. 85).

pre-9/11 international aviation security system in which he calculated the system's success rate in thwarting[3] attempted hijackings and bombings. He found that in the 1947 to 1996 period, 31% of all hijackings, 19% of terrorist hijackings, and 17% of all bombing attempts were foiled. Furthermore, "despite the long-accumulated experience with attacks on commercial aviation, and notwithstanding the immense investment in security measures and procedures, the effectiveness of aviation security measures has not improved" in recent years. Between 1987 and 1996, the security system's success rate was just 19% for all hijackings, 15% for terrorist hijackings, and 24% for bombings. Merari attributed the security shortcomings to authorities' "lack of foresight" in anticipating evolving terrorist methods of attack and "inherent limitations of an airline companies-based security system," themes that would reemerge in the analyses of the 9/11 hijackings (Merari, 1999, pp. 23–26).

Transit Security in New York City in 2001

Aviation was not the only transportation mode heavily impacted by the 9/11 attacks. The collapse of the WTC towers caused massive damage and disruption to New York's transit system. In the aftermath, that system played a central role in evacuating civilians from the affected area and transporting emergency personnel into it.

Whether in New York or elsewhere, transit operators faced a number of constraints with respect to their security systems. The GAO (September 2002) reported:

> *Further complicating transit security is the need for transit agencies to balance security concerns with accessibility, convenience, and affordability. Because transit riders often could choose another means of transportation, such as a personal automobile, transit agencies must offer convenient, inexpensive and quality service. Therefore, security measures that limit accessibility, cause delays, increase fares, or otherwise cause inconvenience could push people away from transit and back into their cars (p. 9).*
>
> *GAO, September 2002, p. 9*

And as was true in aviation, transit operators "fully accept responsibility for safety but see security as a domain of shared responsibility involving agencies and resources beyond their control—especially in dealing with a threat such as terrorism. As these officials see it, transportation operators keep the system running; others must deal with terrorists" (Jenkins and Edwards-Winslow, 2003, p. 13).

The federal role in transit security continued to be limited in 2001. The DOT's Federal Transit Administration (FTA), the chief federal agency responsible for mass transit issues, was precluded by statute from regulating transit safety or security, but it was permitted to undertake nonregulatory safety and security initiatives, including training, research, and demonstration projects. In addition, the agency was allowed to include safety or security

[3]In the Merrari study, a hijacking attempt was considered thwarted only if the hijackers failed to gain control of the aircraft for any length of time, and bombings were considered thwarted only if the explosive device was discovered and rendered harmless before it was designed to explode.

requirements as conditions for the receipt of federal grants. Before 9/11, the FTA had offered voluntary security assessments, sponsored security training, and issued emergency response planning guidelines. However, similar to the local transit operators, it saw its primary focus as safety rather than security (GAO, September 2002, p. 15).

In its capacity as global financial center, host to the United Nations, and home to a very large number of immigrants and descendants of immigrants, New York City had the most experience in the United States in dealing with major terrorist incidents. The most significant of these was the 1993 bombing of the WTC.

1993 WORLD TRADE CENTER BOMBING

"A New Terrorist Challenge. . . Whose Rage and Malice Had No Limit"

On February 26, 1993, a large bomb contained in a truck parked in the underground garage between the Twin Towers was detonated via a timing device. The resulting explosion created a seven-story-high hole, killed six, and wounded more than 1000 people. In the ensuing investigation, the Justice Department and FBI quickly identified, apprehended, and successfully prosecuted a number of individuals, each of whom had ties to the "Blind Sheikh," Sheikh Omar Abdel Rahman, an extremist Muslim cleric then residing in Brooklyn. During this investigation, a plot to bomb a number of major New York landmarks, including the Holland and Lincoln tunnels, was uncovered. The "landmarks plot" was thwarted, and Rahman was arrested in June 1993.

One of the chief perpetrators of the World Trade Center bombing—in fact, the one who planted the bomb—was Ramzi Yousef. Yosuef was arrested in Pakistan in 1995 after the discovery of the Bojinka plot to blow up American airliners over the Pacific. Yousef indicated that his intent had been to bring down the Twin Towers and kill as many as 250,000 people.

The 9/11 Commission Final Report concluded, "(T)he bombing signaled a new terrorist challenge, one whose rage and malice had no limit. . . . (A)lthough the bombing heightened awareness of a new terrorist danger, successful prosecutions contributed to widespread underestimation of the threat" (*The 9/11 Commission Report*, 2004, pp. 71–73).

New York City's land transportation networks were the scene of other major violent incidents during the 1990s. For example, 53 individuals were injured in a December 1994 subway firebombing that was the result of an extortion attempt, and New York City police were able to foil a 1997 plot for suicide bombings in the subway system. As Jenkins and Edwards-Winslow (2003) observed, such incidents "underscored the reality of the terrorist threat and provided transportation operators with additional experience in evacuating injured passengers, rerouting trains, working with police and fire crews, and providing information to the public." They also served as the impetus for a number of institutional and policy changes.

- Creation by the City of New York of the Office of Emergency Management (OEM) with personnel drawn from the city's police, fire, emergency medical services, and other departments. The OEM was charged with developing emergency plans and coordinating the city's response to major emergencies.

- Merger of the various transit organization police departments into a new Metropolitan Transportation Authority (MTA) Police Department[4] to facilitate the coordination of security planning and emergency preparedness.
- Establishment within the MTA Police Department of an Emergency Management Office to provide liaison with all local emergency planning groups, coordinate emergency planning through an Emergency Operations Center, conduct training exercises, establish an intelligence network, and conduct risk and vulnerability assessments (pp. 11–12).

Additional post-1993, pre-9/11 preparations included development of multiagency task forces involving the various transportation authorities as well as the local police and fire departments and coordinated through the OEM, increased investment in emergency response and communications infrastructure, and the documentation and practicing of emergency response procedures (with the latter termed the "backbone" of New York's pre-9/11 preparedness) (*Effects of catastrophic events*, 2002, pp. 31, 34, 38).

A city official stated in a 2004 interview, "Prior to 9/11, security on the New York subway system was very limited. Mostly, the measures we took were through police, and were crime-related. New York had had terrorist attacks such as the first World Trade Center. . . . So we were aware of it, but didn't really have much going on" (Taylor, 2005, p. 122).

■ ■ Critical Thinking ■

Compare and contrast the pre-9/11 security systems for U.S. civil aviation and New York City mass transit. Consider organization, outlook, and any other factors you deem important.

9/11

On the morning of Tuesday, September 11, 2001, 19 operatives of the terrorist organization al Qaeda successfully hijacked four airliners departing on transcontinental flights from airports in the northeastern United States. Three of the hijackings succeeded in their mission of using the aircraft as, in effect, missiles to be detonated upon impact into major symbols of American economic and military power. The fourth failed as a result of a passenger-led effort to take control of the aircraft that ended in the plane's crash into an empty field near Shanksville, Pennsylvania. In addition to the 19 hijackers, 2977 victims lost their lives,

[4]New York City's transportation system was (and is) composed of a complex collection of local, state, and regional public authorities, with some privately owned components as well. In 2001, the MTA included the Long Island Rail Road, Metro-North Railroad, New York City Transit (subways and buses), Long Island buses, and seven bridges and two tunnels. The other two major transportation authorities were (1) the Port Authority of New York and New Jersey, which operated the three major area airports as well as two tunnels, four bridges, the PATH (Port Authority Trans-Hudson) interstate rail system, two interstate bus terminals and seven marine cargo terminals, and (2) the NYC Department of Transportation, which managed the city's streets, highways, parking facilities, four bridges, six tunnels, and a ferry service while overseeing five private ferry companies and seven private bus companies. The MTA Police Department was not responsible for security on New York City subways, which was under the jurisdiction of the New York Police Department's Transit Bureau (*Effects of catastrophic events*, 2002, pp. 3–4).

including all 246 passengers and crew on the four aircraft, 2606 in and around the WTC towers in New York City, and 125 at the Pentagon (New York reduces 9/11 death toll by 40, 2003; First video of Pentagon 9/11 attack released, 2006; Green, 2011).

The Hijackings

The 9/11 plot had its genesis after the 1993 WTC bombing, which convinced Khalid Sheikh Mohammed of the need to develop a novel and more effective way of damaging the U.S. economy, especially its economic capital of New York City. Mohammed then worked with Ramzi Yousef in 1994 in planning the unsuccessful Bojinka plot, and it was at that point that he conceived of using "airplanes as weapons" to strike at the American economy. At a 1996 meeting in Afghanistan with Osama bin Laden, Mohammed presented a potential operation that would train pilots to crash aircraft into buildings in the United States, but the plan was not approved at this point. However, after the successful 1998 al Qaeda bombings of the U.S. embassies in Kenya and Tanzania, bin Laden gave his consent, and Mohammed began supervising the planning and preparations for what was called the "planes operation" (*The 9/11 Commission Report*, 2004, pp. 148–155).

Late in 1999, the al Qaeda leadership recruited Mohamed Atta, Ziad Jarrah, and Marwan al Shehhi for the operation, with Atta designated as the leader. They were presented with a preliminary list of targets (including the WTC, the Pentagon, and the U.S. Capitol) and told to enroll in flight training courses. A fourth individual was identified in 2000; this was Hani Hanjour, who already possessed a U.S. commercial aviation certificate, which he obtained in March 1999. These were the suicide pilots on 9/11. Both Atta and Shehhi completed their flight training at a Florida flight school and obtained FAA commercial pilot licenses in December 2000. Jarrah trained at a different Florida flight school and received his FAA license in November 2000. During the summer of 2000, they each undertook surveillance flights as passengers on transcontinental flights across the U.S., generally flying on the same type of aircraft they would pilot on 9/11 (Boeing 767s for Atta and Shehhi; Boeing 757s for Jarrah and Hanjour) (*The 9/11 Commission Report*, 2004, pp. 163–168, 225–227, 241–243).

The remaining 15 hijackers, the "muscle" who were to storm the cockpits and control the passengers on the four flights (four on Flights 11, 175, and 77; three on Flight 93) arrived in the United States between late April and late June 2001 (*The 9/11 Commission Report*, 2004, pp. 231–237).

To carry out their plan, the terrorists booked four early-morning flights with near-simultaneous departures from northeastern airports close to their intended targets.[5] To achieve maximum impact upon crash, they sought wide-bodied, heavily fueled aircraft (namely the Boeing 757s and 767s they had been training for) that typically flew transcontinental routes. Tickets were purchased for the morning of September 11, 2001, on American Airlines Flights 11 (Boston to Los Angeles) and 77 (Washington-Dulles to Los Angeles) and United Airlines Flights 175 (Boston to Los Angeles) and 93 (Newark to San Francisco) (*The 9/11 Commission Report*, 2004, p. 451n1) (Table 2.1).

[5]For unknown reasons, hijackers Mohamed Atta and Abdul Aziz al Omari began September 11 with a flight (Colgan Air Flight 5930) from Portland, ME to Boston, where they joined three other hijackers in boarding American Airlines Flight 11 (*The 9/11 Commission Report*, 2004, p. 451n1).

Table 2.1 Chronology of the 9/11 hijackings (including connecting Colgan Air Flight 5930 for Flight 11)

AA Flight 11 (Boston to Los Angeles)	UA Flight 175 (Boston to Los Angeles)	AA Flight 77 (Washington, DC to Los Angeles)	UA Flight 93 (Newark to San Francisco)
5:43: Hijackers Atta and Omari check in at Portland, ME, Jetport for Colgan Air Flight 5930 bound for Boston.			
5:45: Atta and Omari pass through checkpoint without incident.			
6:00 Flight 5930 takes off.			
	6:25–7:00: Hijackers pass through security checkpoint without reported incident.		
6:45: Flight 5930 arrives in Boston.			
6:50–7:30: Hijackers pass through Boston security checkpoint without reported incident.			
			7:04–7:40: Hijackers pass through security checkpoint without reported incident.
		7:18: Hijackers Moqed and al Mihdhar set off alarm of first walk-through metal detector at checkpoint, and Moqed also sets off alarm in second detector, after which he is screened with a handheld metal detector. Both are permitted through the checkpoint.	
		7:29: Hijackers Nawaf and Salem al Hazmi check in at the AA ticket counter. They arouse suspicion of the ticket agent, who makes them both CAPPS selectees.	
		7:35: Hijacker Hanjour passes through the checkpoint without incident.	

(Continued)

Table 2.1 Chronology of the 9/11 hijackings (including connecting Colgan Air Flight 5930 for Flight 11) *(cont.)*

AA Flight 11 (Boston to Los Angeles)	UA Flight 175 (Boston to Los Angeles)	AA Flight 77 (Washington, DC to Los Angeles)	UA Flight 93 (Newark to San Francisco)
		7:36: Salem al Hazmi passes through the checkpoint without incident; Nawaf al Hazmi sets off alarms on first and second walk-through detectors. After being hand wanded and having his shoulder bag swiped for explosives, he is permitted to pass through the checkpoint.	
7:59: AA 11 takes off.			
8:14: Likely takeover by hijackers.	8:14: UA 175 takes off.		
8:19: Flight attendant Betty Ong notifies AA of hijacking.			
		8:20: AA 77 takes off.	
8:25: Boston ATC hears two transmissions from flight. The first ("We have some planes. Just stay quiet, and you'll be okay. We're returning to the airport") is not immediately understood. The second ("Nobody move. Everything will be okay. If you try to make any moves, you'll endanger yourself and the airplane. Just stay quiet") leads to awareness of the hijacking.			
8:34: Boston ATC receives a third transmission ("Nobody move, please. We are going back to the airport. Don't try to make any stupid moves.")			
	8:37: Boston ATC asks flight crew to look for AA 11.		
	8:38: Flight crew reports to Boston ATC it has spotted AA 11 and is told to turn aircraft to avoid it.		

	8:41: Flight crew reports to New York ATC, "We heard a suspicious transmission [from another aircraft] on our departure out of Boston—like someone keyed the mike and said, 'Everyone stay in your seats.'"		
			8:42: UA 93 takes off after a 42-minute delay caused by traffic congestion.
8:46:40: AA 11 crashes into WTC North Tower.	8:42–8:46: Likely takeover by hijackers.		
		8:51-8:54: Likely takeover by hijackers.	
	8:52: Flight attendant notifies UA of hijacking. 9:03:11: UA 175 crashes into WTC South Tower.		
			9:24: UA 93 pilot receives warning from UA: "Beware any cockpit intrusion—Two a/c hit World Trade Center."
		9:25: FAA Herndon Command Center orders a nationwide halt to takeoffs.	
			9:28: Likely takeover by hijackers. Cleveland ATC receives two transmissions from aircraft: "Mayday" followed by "Hey get out of here—get out of here—get out of here." 9:32: Cleveland ATC receives third transmission: "Ladies and gentlemen, here the captain. Please sit down; keep remaining sitting. We have a bomb on board. So sit."

(Continued)

Table 2.1 Chronology of the 9/11 hijackings (including connecting Colgan Air Flight 5930 for Flight 11) *(cont.)*

AA Flight 11 (Boston to Los Angeles)	UA Flight 175 (Boston to Los Angeles)	AA Flight 77 (Washington, DC to Los Angeles)	UA Flight 93 (Newark to San Francisco)
			9:36: Flight attendant notifies UA of hijacking
		9:37:46 AA 77 crashes into Pentagon.	
			9:39: Cleveland ATC receives fourth transmission: "Uh, is the captain. Would like you all to remain seated. There is a bomb onboard and are going back to the airport and to have our demands. Please remain quiet."
			9:42: Herndon Command Center orders all aircraft to land at the nearest airport.
			9:57: Attempt by passengers to take control of the aircraft begins.
			10:03:11: UA 93 crashes near Shanksville, PA.

(all times are AM Eastern Standard Time)

AA, American Airlines; ATC, air traffic control; CAPPS, Computer-Assisted Passenger Prescreening System; FAA, Federal Aviation Administration; UA, United Airlines; WTC, World Trade Center.

As mentioned earlier, the 9/11 hijackers had successfully avoided being listed on the "no fly" list,[6] and although between April 1, 2001 and September 10, 2001, the FAA's intelligence unit issued 52 daily summaries mentioning bin Laden, al Qaeda, or both, none contained specific information indicating that al Qaeda or any other group was plotting to hijack airplanes in the United States, and none identified any of the 9/11 hijackers. The "planes operation" had successfully evaded the key intelligence layer of aviation security (9/11 Commission, Third Monograph, 2006, pp. 632, 634).

The CAPPS system did indeed work as intended in identifying a majority of the hijackers as potential threats warranting additional security attention. Seven of the 19 were selected by the CAPPS algorithm (including three on Flight 11, three on Flight 77, and one on Flight 93); two more (on Flight 77) were added at the discretion of an airline counter customer representative who found their behavior suspicious; and Atta (on Flight 11) was flagged by the system's random selection feature when he checked in for his connecting flight in Portland, Maine. However, pursuant to the aviation security system's focus on bombs placed in cargo holds (the tactic used on Pan Am Flight 103), their selection by CAPPS led only to the screening of their checked luggage for explosives or at airports such as Dulles where such equipment was not available, to a delay in loading it until they boarded the aircraft. Because access to their checked bags had no relevance to their planned tactics, the prescreening process was rendered moot in foiling the 9/11 plot. Thus, even before their arrival at the airports, the 9/11 hijackers were only faced with defeating the final two layers of aviation security: checkpoint screening and onboard security (9/11 Commission, Third Monograph, 2006, pp. 581, 584, 597, 606–607, 614).

The 9/11 Commission staff testified:

Of the checkpoints used to screen the passengers of Flights 11, 77, 93 and 175 on 9/11, only Washington Dulles International Airport had videotaping equipment in place. Therefore the most specific information that exists about the processing of the 9/11 hijackers is information about American Airlines Flight 77.... The staff has also reviewed testing results for all the checkpoints in question, scores of interviews with checkpoint screeners and supervisors who might have processed the hijackers, and FAA and FBI evaluations of the available information. There is no reason to believe that the screening on 9/11 was fundamentally different at any of the relevant airports.

9/11 Commission, Staff Statement No. 3, 2004

At the Dulles checkpoint, three of the five hijackers set off an alarm as they walked through the checkpoint's metal detector. These three were then directed to a second metal detector, which two of the three again alarmed. The two who triggered the second alarm were then searched via a metal-detecting wand held by a security screener, and one of the two had his shoulder bag swiped by an explosives trace detector. All five of the hijackers

[6]Two of the hijackers, Nawaf al Hazmi and Khalid al Mihdar, had been placed on the State Department's TIPOFF terrorist watch list in August 2001, but the FAA was unaware of this information (9/11 Commission, Third Monograph, 2006, p. 648).

were ultimately able to successfully pass through the security checkpoint, as were the other 14 terrorists bound for the other three flights (9/11 Commission, Third Monograph, 2006, p. 606).

Purchase records and other evidence (including reports from passengers and flight crew on all four flights and wreckage found at the site of the Flight 93 crash) indicate that knives with blades of less than 4 inches were the primary weapons used by the hijackers.[7] Because such items were specifically permitted by the airlines' checkpoint operations guide, they would likely have been allowed to go through the checkpoint even if detected. On the other hand, Mace or pepper spray (reported by passengers or flight crew on at least Flights 11 and 175) were prohibited and should have been confiscated if found by screeners. In any event, there is no record of any item being taken from any of the hijackers at the checkpoints on 9/11. Thus, through inadequate regulations, poor performance by the screeners, or some combination of both, checkpoint screening also failed to stop the 9/11 plot, and all 19 hijackers were able to board their flights carrying the items they needed to gain control of the aircraft (9/11 Commission, Staff Statement No. 4, 2004).

The final layer of antihijacking security, onboard security, was similarly unsuccessful. None of the limited number of Federal Air Marshals was assigned to any of the four hijacked flights, and the "Common Strategy" of accommodating hijackers who were assumed to be nonsuicidal actually aided the hijackers' cause.

> *The hijackers likely gained control of the forward section of the aircraft after the aircraft's seatbelt sign was turned off, the flight attendants had begun cabin service, and passengers were allowed to move around the cabin. The hijackers took over the aircraft by force or threat of force, as reported on all four flights.... The hijackers gained access to the cockpit and sealed off the front of the aircraft from passengers and cabin crew, moving them to the back of the aircraft.... The hijackers also used announcements on Flight 11 and Flight 93 that the aircraft was returning to the airport to make passengers believe they were in no immediate danger if they cooperated. Initially, these tactics, techniques and communications resembled those of a traditional hijacking for the purpose of taking hostages or transportation. This was a scenario that the "Common Strategy" was designed to address.*
>
> 9/11 Commission, Third Monograph, 2006, pp. 627–628

The existing onboard security measures had also failed to prevent the hijackings, and the hijackers were able to successfully complete their missions in the cases of Flights 11 and 175, which crashed into the Twin Towers of the WTC, and of Flight 77, which crashed into the Pentagon. However, growing awareness on board all of the flights that something

[7]The presence of a bomb was reported on three of the flights (Flights 11, 175, and 93, with the latter case including announcements to that effect by one of the hijackers), but the 9/11 Commission staff concluded this was done as a threat "to frighten and control passengers." In the absence of evidence of bomb-related purchases by the hijackers and given their objective of flying intact aircraft into ground targets, it is not likely that bombs were actually brought on board (9/11 Commission, Third Monograph, 2006, p. 627).

more than a traditional hijacking was taking place culminated in a "passenger revolt" on Flight 93 that resulted in the crash of that plane into an open field in Pennsylvania, less than 200 miles from its intended target in Washington, DC (either the U.S. Capitol or White House). As former FAA Administrator Jane Garvey told the 9/11 Commission, "No one had to order [the Common Strategy] changed. The men and women on the fourth airplane that crashed in Pennsylvania changed that policy. It will never be our country's policy again" (9/11 Commission, Third Monograph, 2006, p. 628; *The 9/11 Commission Report*, 2004, p. 14; 9/11 Commission, Jane Garvey Testimony, 2003).

DIGGING DEEPER

9/11 U.S. AVIATION SECURITY DEFENSES

Could They Have Prevented the Hijackings?

Serious flaws existed in each of the security layers relevant to the 9/11 plot. The consequences of selection by CAPPS were confined to preventing explosives from being placed in checked baggage. Passenger checkpoint screening was of limited use because its detection equipment was geared to finding guns and large knives, its human screeners performed poorly, and its procedures explicitly permitted the kind of small knives likely used by the hijackers. Onboard security was guided by a "Common Strategy" that instructed flight crew to accommodate what were assumed to be nonsuicidal hijackers. Thus, it is possible to view the 9/11 hijackings as a "systems failure" in which a properly designed plan adequately carried out by the 19 terrorists was "likely to be successful" (9/11 Commission, Staff Statement No. 3, 2004).

But consider the following:

1. As originally conceived by the FAA's Baseline Working Group in 1996, CAPPS was supposed to be more robust, with airlines required to apply an "FAA-approved passenger profile to all passengers enplaning at U.S. airports to identify selectees, whose persons and property (checked baggage and carry-on bags/items) will receive additional security scrutiny." In attempting to explain the subsequent watering down of these requirements, the 9/11 Commission staff cited "the desire to limit the purchase of expensive explosives detection technology [that would be necessary to screen carry-on items], concerns about customer dissatisfaction with delays and 'hassle,' the need to avoid operational delays, and the fear of potential discrimination or the appearance of it" in the selection process. Even so, in its January 2004 testimony to the 9/11 Commission, the staff asked, "Was it wise to ease the consequences of being a prescreening selectee at a time when the U.S. government perceived a rising terrorist threat, including domestically, and when the limits of detection technology and shortcomings of checkpoint screening were well known?" (9/11 Commission, Third Monograph, 2006, pp. 650–651; 9/11 Commission, Staff Statement No. 4, 2004).
2. Although all firearms and knives with blades 4 inches long or longer were prohibited from being carried past the security checkpoint, per the checkpoint operations guide developed by the airlines in cooperation with the FAA, "Knives with blades under 4 inches, such as Swiss Army knives, scout knives, pocket utility knives, etc. may be allowed to enter the sterile areas [beyond the checkpoint]. However, some knives with blades under 4 inches could be considered by a reasonable person to be a 'menacing knife' and/or illegal under local law and should not be allowed to enter the sterile area." In implementing these

guidelines, screeners were advised to use "common sense" in determining what would be allowed past the checkpoint. This conditional permit for small-bladed knives was based on the FAA's belief that such weapons were not "menacing," the fact that such items were not prohibited under most local laws in the United States, and the inability of metal detectors to detect such items unless their sensitivity levels were greatly increased (which would have significantly raised the number of false alarms and slowed down checkpoint processing times) (9/11 Commission, Third Monograph, 2006, pp. 652–653).

3. In considering both CAPPS and checkpoint operations, the 9/11 Commission staff conjectured:

Had CAPPS required selectees to be subject to a secondary search of their person, carry-on bags, or both, perhaps screeners could have found and confiscated the prohibited items (e.g., Mace or pepper spray), perhaps an alert screener would have identified the component parts of a fake bomb, perhaps the additional screening would have exposed a rattled hijacker, or perhaps any knives found by the screeners would have been confiscated as they used the "common sense" urged of them by FAA rules and the discretion provided them by the airline's checkpoint operations guide to prohibit menacing items.

9/11 Commission, Third Monograph, 2006, pp. 651–652

What do you think? Could the existing security system have prevented the 9/11 hijackings? If so, what would have had to be changed? Considering all of the factors in play (including mindset and system performance indicators), *should* that existing system have been able to thwart the plot?

New York Land Transportation Systems

The two aircraft struck the North and South Towers of the WTC "during the morning rush hour when the city's roads, bridges, and transit system were operating at peak capacity." A 2002 study by the Volpe National Transportation Center summarized the initial challenges facing New York's land transportation system:

Transportation officials were immediately faced with the need to make critical decisions on how to respond in order to protect the safety of the traveling public. The decisions were made more difficult because of the circumstances that were unfolding at a rapid pace. Adding to the difficulty were the lack of accurate, immediate information about the implications and extent of the event, the inability to quickly communicate agency actions internally and externally, and the need to ensure the safety of their own transportation facilities in the event of possible follow-up attacks. . . . As fires raged in the two buildings, vital utility and communications systems began to fail. Communications failures included the loss of numerous radio and communications towers located on top of the towers. . . . As electrical power was lost to the area, traffic signals no longer worked, hindering traffic movement. . . . The loss of electricity made it more difficult to fight the fires resulting from the attack and to begin pumping operations to prevent flooding of underground transit and utility facilities.

Effects of catastrophic events, 2002, pp. 7–8

Despite the challenges, within the first hour of the initial crash into the WTC (at 8:46 AM), New York's transportation systems took a number of steps.

- At 8:47 AM, an MTA subway operator alerted the MTA Subway Control Center of an explosion at the WTC, which led to the implementation of emergency procedures that rerouted trains and buses around the damaged area and dispatched MTA personnel to the affected stations, tracks, and tunnels to ensure that they had been completely evacuated. As a result of the prompt action, none of the 300 NYC transit employees or 60,000 passengers who were in the immediate vicinity of the WTC when the first plane hit lost their lives that day.
- At 8:52 AM, the Port Authority-Trans Hudson (PATH) activated emergency procedures that included evacuating the PATH WTC station, having all PATH trains proceed to the end of the line in New Jersey and remain there and stopping in-bound trains from New Jersey.
- At 9:00 AM, city officials activated the OEM Emergency Operations Center, located within the WTC complex. After the second plane crash into the WTC at 9:03 AM, OEM staff immediately relocated to an office with functioning phone lines about a block away. (The OEM Emergency Center was moved twice more that morning, finally setting up at the New York City Police Academy 2 miles north of the WTC.)
- At 9:10 AM, the Port Authority of New York and New Jersey closed all its bridges and tunnels into Manhattan.

At 9:59 AM, the South Tower of the WTC collapsed followed by the North Tower at 10:29 AM, "spreading thousands of tons of debris and ash over Lower Manhattan. Visibility was diminished and breathing became difficult. . . . Electrical and communications failures spread throughout Lower Manhattan as the collapsing . . . towers took down surrounding infrastructure."

- Between the tower collapses, at 10:20 AM, all subway service in New York City was suspended. The subway system in Lower Manhattan had sustained particularly severe damage, with 1400 feet of tunnel and track destroyed, one station completely destroyed, and one tunnel flooded. However, the decision to shut down the entire system was based on security considerations rather than as a consequence of damage, with officials mindful of a 1993 terrorist plan to attack tunnels. Thus, it was decided that service would only resume after all subway tunnels had been searched and secured.
- At 10:45 AM, PATH operations were suspended. A PATH station beneath the WTC had been destroyed.
- At 11:02 AM, New York City Mayor Rudy Giuliani called for the total evacuation of Lower Manhattan. The city's Department of Transportation and Police Department began closing all highways in that area.
- At 12:48 PM, subway service was resumed on a very limited basis, reaching 6% of its normal level. By 1:40 PM, 33% of normal service was restored, and it reached 65% by the end of the day.

- Other transportation systems resumed operations over the course of the afternoon, including Long Island Rail Road eastbound service at 1:15 PM and PATH service from Newark to Manhattan at 4:12 PM. Some area bridges opened to outbound traffic at 7:02 PM. Bus service north of Lower Manhattan was maintained throughout the day. (*Effects of catastrophic events*, 2002, pp. 8–13; Jenkins and Edwards-Winslow, 2003, pp. 23, 27, 51).

Jenkins and Edwards-Winslow (2003) described the land transportation system's response in the evacuation process:

> *Emergency operators at the MTA ordered all available equipment into Grand Central and Penn Stations. Three lines on Metro-North and three lines on the Long Island Rail Road were used for evacuation. Trains operated on a load-and-go basis, departing as soon as they were filled. This procedure had been used before, and interviewed sources said that commuters were familiar with it. Buses loaded passengers and headed north. No one paid attention to fares or routes. NYPD's harbor unit ferried 5,000 people to New Jersey and Staten Island, and the commercial ferry transports and tugboats moved victims and fleeing people to New Jersey.*
>
> *Jenkins and Edwards-Winslow, 2003, p. 24*

In addition to its role in evacuation, MTA and other public transportation agencies played a significant part in aiding other emergency operations, including transporting thousands of police officers and other emergency personnel to the stricken area, supplying personnel and equipment for rescue and debris removal, providing generators to supply power for traffic signals, assisting in moving patients to medical care facilities, using internal communications systems to augment police and fire emergency communications networks, and assisting in providing information to the public (pp. 31–34).

In its analysis of how area transportation agencies responded to the events of 9/11, the Volpe report concluded: "Key players were partially prepared because of actions taken in response to the terrorist attack of the World Trade Center in 1993 and other subsequent major and minor events. . . . Although the key players were prepared for the standard emergency operations, agencies were not prepared for a disaster of the magnitude of the attack." The report also observed that, "The main focus of transportation operators on September 11 was safety at the expense of mobility." Among specific responses that were found to have worked well were:

- The activation of emergency procedures by the transit agencies to ensure the safety of its customers
- The mobilization of transportation resources, including heavy machinery, mobile generators, and skilled personnel, that helped reestablish vital communications and utility links

- The ability of field staff to quickly make good decisions on their own in the absence of headquarters personnel
- The provision of alternative transportation options

On the other hand, shortcomings were noted in the transportation systems' reliance on emergency management centers that were concentrated in a high-threat location (namely the WTC complex); inadequate redundancy in the communications, utility, and emergency response systems; and a lack of sufficient real-time information for the public on transportation status and options (*Effects of catastrophic events*, 2002, pp. 31–33).

■ ■ Critical Thinking ■

Describe the pros and cons of the decision by New York transportation agencies to prioritize safety over mobility on 9/11. Do you think they made the right choice? Why?

Security Response to 9/11

The impact of the 9/11 hijackings was enormous. In addition to the nearly 3000 fatalities, thousands were injured, and 10 years after the event, nearly 60,000 first responders, office workers, and local residents who had been exposed to the heat, dust, and other environmental consequences of the WTC crashes and tower collapses had been or were being treated for respiratory and other health problems stemming from that exposure. The economic harm—which had been one of the terrorists' chief objectives from the beginning—was massive, with a 10th anniversary calculation by *The New York Times* estimating physical damages in New York City at $55 billion (including $6 billion to the city's transportation and utilities infrastructure) and an economic impact of $123 billion (including $39 billion in reduced air travel) (*2011 Annual Report on 9/11 Health*, 2011, p. 13; Carter and Cox, 2011).

Beyond the tangible effects, the 9/11 trauma exacted a high psychological cost. In commenting on a nationwide study she helped conduct shortly after the 9/11 attacks, Jennifer Lerner of Carnegie Mellon University reported the random survey found that individuals estimated there was a 20% chance they themselves would be directly affected by a terrorist attack in the next year, and they put the risk for "average Americans" at 48%. She concluded, "There was an overwhelming overestimation of risk. For even the 20 percent estimate to be accurate, we would have had to have September 11 every day and then some" (Vedantam, 2003).

Under these circumstances and given the historical tendency for transportation security systems to be highly reactive to major incidents, it is not surprising that the response to 9/11 produced far-reaching changes not only in the United States (within which a wholly new discipline, termed *homeland security*, came into existence) but also around the world.

Post-9/11 U.S. Aviation Security

The most immediate reaction of the U.S. aviation system to the hijackings was the FAA's unprecedented order instructing all planes in U.S. airspace to land at the nearest airport. This instruction, which came shortly after the crash of Flight 77 into the Pentagon and just under 30 minutes before Flight 93's crash in Pennsylvania, led to the safe landing of the other 4500 commercial and general aviation aircraft then in flight over the United States. By September 13, 2001, commercial aircraft were able to resume service at airports meeting newly established interim security standards, and general aviation aircraft were able to resume some operations on the following day (9/11 Commission, Third Monograph, 2006, p. 613; *The 9/11 Commission Report*, 2004, p. 327).

At the policy level, a number of initiatives were undertaken in response to the hijackings both in the immediate aftermath and for several years thereafter. Even as this process got underway, another incident added to the high level of concern about aviation security. On December 22, 2001, Richard Reid—who was flying on American Airlines Flight 63 bound to Miami from Paris—attempted to detonate explosives concealed in his shoes. Illustrating the change in the "Common Strategy" that had already taken hold, the attempt was foiled by a flight attendant who noticed Reid seeking to light a match near his shoe and was assisted in subduing him by another flight attendant and several passengers. Although the failure of the plot and the subsequent conviction and sentencing of the al Qaeda linked Reid represented a security success of sorts, the fact that he was able to board the aircraft despite a number of preflight suspicious behaviors and to pass through the security checkpoint with concealed explosives demonstrated ongoing vulnerabilities in aviation security. An immediate consequence and continuing legacy of this incident was the requirement for passengers to remove their shoes for screening at airport checkpoints (Price and Forest, 2009, pp. 75-76).

Aviation and Transportation Security Act of 2001

Moving at unaccustomed speed, the U.S. Congress adopted the Aviation and Transportation Security Act (ATSA) little more than a month after 9/11, with President Bush signing the measure into law on November 19, 2001 (PL 107-71). ATSA made a number of significant changes in the U.S. aviation security system, including:

- Establishing within the DOT a new Transportation Security Administration (TSA), which was given overall responsibility for transportation security
- Transferring responsibility for checkpoint and checked bag screening at airports from the airlines to the TSA, with the agency directed to hire and deploy federal checkpoint screeners by November 19, 2002, and checked bag screeners by December 31, 2002
- Mandating that, within 60 days of enactment, all airports must implement a system to screen all checked bags and cargo carried on passenger aircraft for explosives using explosive detection systems (EDS), manual searches, explosive-sniffing canines, or positive passenger-bag reconciliation, with the further requirement that all such screening must be performed by EDS by no later than December 31, 2002

- Significantly expanding the Federal Air Marshal program, requiring that a marshal must be deployed on every "high-risk" flight (both domestic and foreign), and transferring the program to the TSA
- Requiring the strengthening of the cockpit doors on passenger aircraft and directing that the doors remain locked except when needed for the entrance or exit of authorized personnel
- Directing the FAA to develop a mandatory training program for flight and cabin crews in dealing with all hijack situations, including suicide hijackings
- Authorizing a passenger fee of $2.50 per enplanement (capped at $5 per one-way ticket) to help pay for the expanded aviation security program
- Authorizing passenger aircraft pilots to carry a firearm into the cockpit subject to the approval of the TSA and the airline
- Requiring that within 60 days of enactment, airlines operating flights into the United States must provide a passenger and crew manifest to the Customs Service before landing (Kirk, 2001, pp. 6–9).

By the end of 2002, the newly established TSA had met the federalized screening workforce mandates by hiring and deploying more than 40,000 checkpoint screeners and more than 20,000 checked bag screeners, made "substantial progress" in expanding the Federal Air Marshal program, and completed the strengthening of the cockpit door on 80% of commercial aircraft.[8] However, although the agency reported that it was screening approximately 90% of all checked baggage by EDS or explosives trace detection systems, it had only completed installation of one-fifth of the EDS machines and one-third of the explosives trace detectors needed to meet the 100% requirement. Also, the GAO found vulnerabilities in cargo security on both passenger and all-cargo planes. (The ATSA stipulated that the TSA should, as soon as practicable, implement a system to screen, inspect, or otherwise secure cargo on all-cargo aircraft) (9/11 Commission, Gerald Dillingham Testimony, 2003).

Vision 100

Vision 100—Century of Aviation Reauthorization Act was enacted on December 12, 2003 (PL 108-176). This legislation addressed a variety of aviation issues, including several that were security related. First, the law modified the ATSA requirements for security training of flight and cabin crews by explicitly making the airlines responsible for providing mandatory basic security training and directing the TSA to develop and provide voluntary advanced self-defense training for these crew members. Second, it created the Aviation Security Capital Fund to help finance security-related capital improvements at airports, such as the integration of baggage explosives detection systems with baggage conveyor systems (in-line screening). The fund was authorized at a level of up to $500 million a year, with the first $250 million collected each year in aviation security fees from passengers and the aviation

[8]The cockpit door hardening effort was completed in April 2003, by which time more than 10,000 aircraft (both foreign and domestic) serving the United States had met the requirement (Federal Aviation Administration, 2003).

industry to be deposited into the fund. Third, the TSA was directed to implement security programs for larger charter aircraft operators. Fourth, the GAO was instructed to review the proposed CAPPS II system, which the TSA was developing as a replacement for the CAPPS program, and the TSA was enjoined from implementing the program until a number of issues concerning civil liberties, data protection, performance, and oversight had been adequately addressed. Finally, the act strengthened requirements for background checks for foreign pilots seeking flight training in the United States (Elias, 2005, pp. 11–13) (Figure 2.2).

9/11 Commission

Largely because of the efforts of families of the victims of 9/11, in November 2002, Congress passed legislation creating the National Commission on Terrorist Attacks Upon the United States (generally referred to as the 9/11 Commission). The 9/11 Commission was composed of 10 members, equally divided between Democratic and Republican

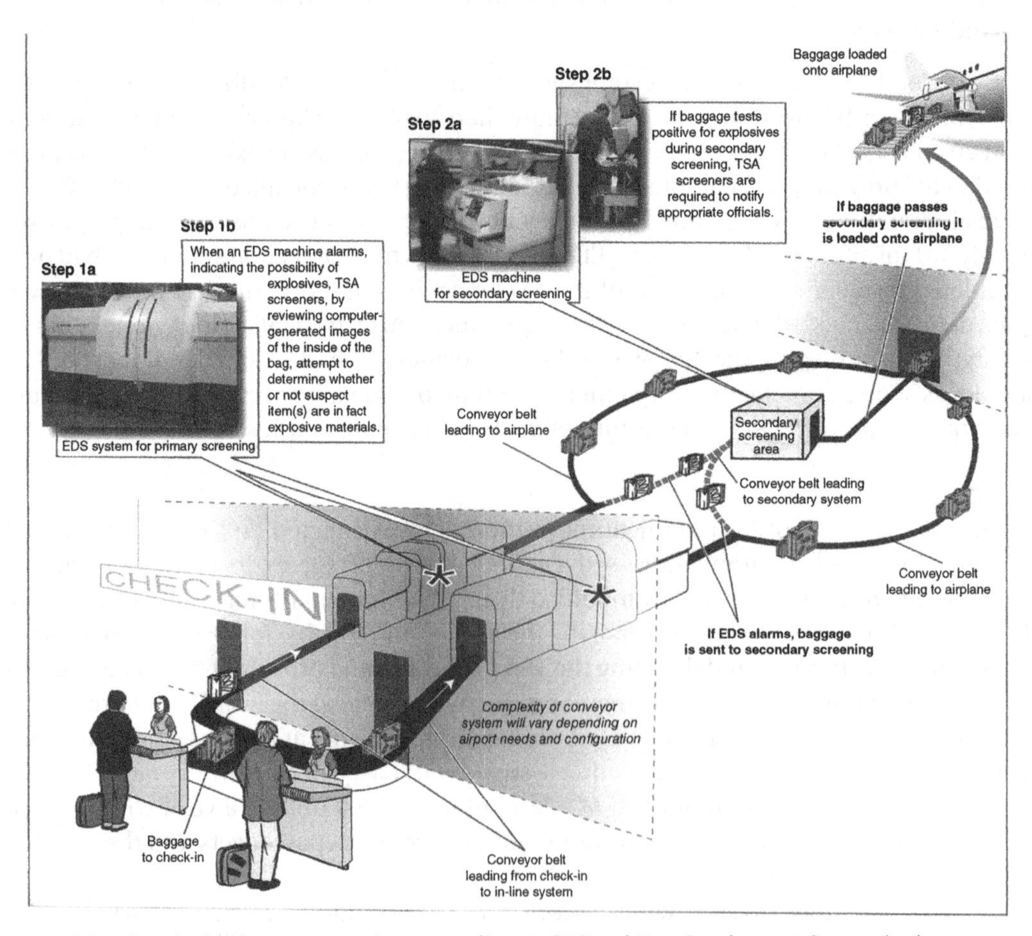

FIGURE 2.2 In-line checked baggage screening system. *(Source: GAO and Nova Development Corporation.)*

appointees, and chaired by former Republican Governor Thomas Kean of New Jersey, with former Democratic Congressman Lee Hamilton of Indiana as the vice chair. Its mandate was to "make a full and complete accounting of the circumstances surrounding the [September 11] attacks, and the extent of the United States' preparedness for, and immediate response to, the attacks; and investigate and report to the President and Congress on its findings, conclusions and recommendations for corrective measures that can be taken to prevent acts of terrorism."

The Commission's Final Report was issued in July 2004, and dealt with a range of issues, including intelligence, border control, emergency response, national air defense, and others. With respect to transportation security, the report focused primarily on aviation, and the Commission made the following recommendations:

- A transportation security plan should be developed to clearly define security roles of the governmental and private entities involved in operating the country's transportation systems and identify the means for funding and implementing the plan.
- The TSA should immediately take over from the airlines the administration of the "no fly" list while questions about the CAPPS II program were being resolved.
- The TSA and Congress should give priority attention to improving the detection of explosives at airport checkpoints, and the TSA should require that all passengers selected for additional security scrutiny be screened for explosives.
- The TSA should conduct a detailed study to identify "human factors" that might be impairing the performance of security screeners.
- Explosives detection equipment should be removed from airport lobbies and placed "in line" with airport baggage conveyor systems to enhance security and improve efficiency. To finance this, there should be appropriate cost sharing between the federal government and industry.
- Each passenger aviation plane that also carries cargo should be equipped with a hardened container to carry the cargo.
- The TSA should intensify its efforts to identify, track, and screen potentially dangerous maritime and air cargo (Johnstone, 2006, pp. 49–50).

Intelligence Reform and Terrorism Prevention Act of 2004

The Intelligence Reform and Terrorism Prevention Act (PL 108-458) was designed to address the 9/11 Commission's recommendations, as well as other perceived deficiencies in security programs. It was signed into law by President Bush in December 2004. For the most part, the law adopted the Commission's transportation-related recommendations, including those concerning the "no fly" list, prioritization of explosives detection, the human factors study, and in-line screening. It went beyond the 9/11 Commission's call for improved cargo security by establishing a detailed program for air cargo security. On the other hand, the act weakened the transportation security plan by deleting the funding language and the requirement for specific assignment of security roles, and it changed the hardened cargo container requirement into a pilot program.

In addition, PL 108-458 incorporated proposals developed by the 9/11 Commission staff that were transmitted to Congress but were not part of the Commission's formal recommendations:

- Providing for use of biometrics in confirming the identity of individuals seeking access to secure areas of airports
- Requiring airport employees seeking unescorted access to secure areas of airports to be prescreened against terrorist watch lists
- Prioritizing development of more advanced airport checkpoint screening devices
- Directing the Federal Air Marshal program to take steps to ensure the operational anonymity of its officers
- Requiring cruise ship passengers to be prescreened against terrorist watch lists (Johnstone, 2006, pp. 51–52)

Finally, the law directed TSA to begin testing of a more advanced passenger prescreening system by January 1, 2005, while also developing redress procedures for passengers falsely selected by CAPPS and its successor (Elias, 2005, pp. 13–14).

Other Post-9/11 U.S. Transportation Security Measures

Although most post-9/11 policy attention was directed toward the aviation sector, other transportation-related initiatives were also undertaken by the U.S. government in the wake of the hijackings.

Homeland Security Act of 2002

The Homeland Security Act (HSA) was enacted in November of 2002 (PL 107-296) and represented a far-ranging change in the American approach to securing its territory and people. As Morag (2011) explains:

> *Homeland security is a uniquely American concept. It is a product of American geographic isolation and the strong tendency throughout American history to believe that there was a clear divide between events, issues and problems outside U.S. borders and those inside U.S. borders... In the aftermath of the terrorist attacks of September 11, 2001, American leaders realized that they would need new tools to deal with large-scale terrorist threats and yet they were constrained by the Constitution, legislation, and federalism. Consequently, they largely could not apply tried and tested national security tools and methodologies to the domestic arena. Homeland security policies, institutions and methodologies thus developed to fill this void between what the U.S. could do overseas and what it was unable to do domestically.*

The HSA represented a key component of this evolving approach by creating a new Department of Homeland Security (DHS) composed of 22 existing federal departments and agencies, including—in the field of transportation security—TSA, the Coast Guard and the Customs Service. Whereas the TSA and the Customs Service were placed within the DHS Directorate of Border and Transportation Security, the Coast Guard was transferred as a stand-alone agency, outside of the directorate. Furthermore, although there was no change in the ATSA directive assigning TSA responsibility for all modes of transportation, the HSA specifically designated the Coast Guard as the responsible entity for ports, waterways, and coastal security (Krouse, 2002, pp. 1, 14–16).

Maritime Transportation Security Act of 2002

The Maritime Transportation Security Act (MTSA), which also became law in November 2002 (PL 107-295), established a national maritime security system and contained a number of requirements for federal agencies, local port authorities, and maritime vessel owners:

- The DOT was mandated to conduct vessel and port facility vulnerability assessments, administer a grant program to finance security upgrades, develop standardized training for crew and port workers, and control access to security-sensitive areas through the development of a transportation worker security identification card.[9]
- The Coast Guard was directed to develop national and regional area maritime security plans, and its Sea Marshal program for placing armed personnel on at-risk vessels was explicitly authorized.
- Customs and Border Protection (formerly the Customs Service) was authorized to require cargo manifest information be provided to it before the cargo's arrival or departure.
- Ports, waterfront terminals, and maritime vessels were required to develop security and incident response plans, subject to Coast Guard approval, and local port security committees were established to coordinate the activities of all governmental and nongovernmental stakeholders (Frittelli, 2003, p. 14; McNicholas, 2008, pp. 117–118)

Customs and Border Protection Initiatives

The Customs and Border Protection agency started three new programs in 2002 designed to improve the security of cargo containers. The Cargo Security Initiative (CSI), begun in February 2002, used an automated system to identify and prescreen high-risk containers bound for the United States from major foreign ports. The Customs-Trade Partnership Against Terrorism (C-TPAT) was established in April 2002 and created government–shipping industry partnerships that offered expedited customs processing for companies that reduce their security vulnerabilities. Last, in November 2002, Operation Safe Commerce was created with the objectives of verifying the contents of seaborne containers at their point of loading, preventing tampering in transit, and tracking their movement through to their final destination. By the end of 2004, the CSI was operating in 34 overseas ports, and enrollment in C-TPAT exceeded 4000 certified partners (Johnstone, 2006, p. 87).

[9]The security functions assigned by MTSA to the DOT were subsequently transferred to the new DHS.

U.S. Land Transportation Initiatives

In contrast to aviation and maritime security, no overarching policy was developed for the land transportation mode in the aftermath of 9/11. Even though the TSA was given responsibility for land transportation security, it focused little of its attention and resources on the land sector (in large part because of congressional mandates and funding that required it to undertake a host of aviation security measures), and the various DOT modal agencies, as well as state, local, and private entities, retained a sizeable role. In the period after 9/11:

- The FTA completed threat and vulnerability assessments of the 37 largest and most at-risk local transit agencies, deployed technical assistance teams to assist local agencies in implementing security programs, awarded grants for emergency response drills, accelerated the deployment and testing of chemical detection systems in subways, and developed a training course on security awareness.
- The Federal Railway Administration commissioned a comprehensive security review of Amtrak's security posture, and Amtrak received $100 million from the federal government for safety and security enhancements of New York City's rail tunnels.
- The Association of American Railroads, representing the freight rail industry, developed a classified analysis of freight rail risks and potential countermeasures.
- The TSA conducted rail security training exercises, issued general pipeline security guidelines, and began to inspect the largest pipeline operators.
- Customs inspectors screened high-risk rail cargo entering the United States.

Overall, though, as noted in a February 2003 report by the Dartmouth Institute for Security Technology Studies, "A non-prohibitive, cost efficient security strategy for surface transportation has yet to be developed. . . . Efforts to secure surface transportation systems have only minimally reduced the threat of terrorist attacks." In explanation, another analysis stated, "The least emphasis has been placed [on land transportation security] because it was perceived as least pressing, and also because it is the hardest to protect" (Johnstone, 2006, pp. 92–93, 98–99).

U.S. Funding for Transportation Security

The 9/11 attacks also produced a massive upsurge in governmental expenditures:

- Just 3 days after the hijackings, Congress passed the 2001 Emergency Supplemental Appropriations Act for Recovery and Response to Terrorist Attacks on the United States (PL 107-38). This legislation authorized $40 billion (called the Emergency Response Fund), of which $20 billion was available for allocation before the end of 2001, and the other $20 billion was made available with the enactment on January 10, 2002, of the fiscal year (FY) 2002 Department of Defense and Emergency Supplemental Appropriations for Recovery from and Response to Terrorist Attacks on the United States (PL 107-117) (Riehl, 2003, pp. 1–2).

- An additional $23.9 billion was authorized in the FY2002 Supplemental Appropriations Act for Further Recovery from and Response to Terrorist Attacks on the United States (PL 107-206), which was signed into law by the president on August 2, 2002. This measure included $5.5 billion in recovery aid for New York City, bringing the total assistance for the city to $21.5 billion under the three emergency appropriations bills (Belasco and Nowels, 2002, pp. 1–2).
- Congress and the Bush Administration also provided substantial assistance to the U.S. airline industry, whose financial situation was severely impacted by 9/11, through the damage and flight cancellations precipitated by the attacks and the subsequent drop in airline travel that persisted until the summer of 2004. Under the Air Transportation Safety and System Stabilization Act of 2001 (PL 107-42), $5 billion in direct aid was granted, and another $10 billion in government-backed loans was authorized (although less than $2 billion of the latter was actually committed). The Emergency Wartime Supplemental Appropriations Act of 2003 (PL 108-11), which was aimed primarily at funding the wars in Iraq and Afghanistan but also provided $5.1 billion for homeland security programs, included $2.4 billion to reimburse the airlines for increased security costs and created a 4-month "tax holiday" during which passenger and airline security fees would not be collected (Peterman, 2004, pp. 15–16).
- State and local governments also incurred significant security costs. Whereas the National Governors Association estimated that the states spent a minimum of $650 million for protection of critical infrastructure and other security-related missions in the first year after 9/11, the U.S. Conference of Mayors put local security spending at approximately $525 million during the same period (*Securing the Homeland*, 2002, p. 5).

Funding for transportation security claimed a substantial share of the federal resources made available after the events of 9/11, although ascertaining precise amounts is difficult. Before the establishment of the DHS and the development of its first regular budget for FY2004, most agencies did not maintain separate funding lines for homeland security activities, including transportation security, making determination of spending for these programs problematic. For example, the FTA was the major federal agency involved in assisting security activities by local public transit authorities before the creation of DHS, and it retained a significant role thereafter. Yet until its budget request for FY2005,[10] the agency did not disaggregate security funding from within the much larger safety account. Furthermore, congressional appropriations categories did not always match up with newly defined homeland security missions. The Coast Guard's transportation security activities were reported in the president's budget under the mission heading of "Ports, Waterways and Coastal Security" starting with FY2002, but no such line item is to be found in the DHS appropriations bills enacted by Congress. Finally, within DHS, transportation and border

[10]The FY2005 FTA budget reported that $37.8 million was spent on transit security in FY2004 and was being requested for FY2005 (Johnstone, 2005, p. 8).

security programs are housed in the same division, complicating efforts to distinguish the two for agencies such as Customs and Border Protection that have major, and often overlapping, responsibilities for each.

The main exception to such accounting complications was in aviation, in which both the FAA and the TSA that took over its security functions maintained specific aviation security accounts, which were reflected in both presidential budget submissions and congressional appropriations bills. Thus, it is likely that efforts to determine the modal distribution of federal funding for transportation security programs understate the portion expended for maritime and land transportation. Nonetheless, based on the available evidence, it is clear that (1) before 9/11, federal spending for transportation security was limited (including in the appropriations measures adopted for the fiscal year that ended on September 30, 2001), with the lion's share going to aviation; (2) shortly after the hijackings, aviation security expenditures increased by orders of magnitude, as to a lesser extent, did appropriations for Coast Guard security programs; and (3) funding for land transportation security remained relatively small in both periods (Table 2.2).

■ ■ Critical Thinking ■

The ATSA gave the TSA responsibility for securing all modes of transportation, yet from its inception, the agency focused almost all of its attention on passenger aviation. Why?

Congressional Oversight

The 9/11 Commission recommended that the House and Senate streamline and improve their oversight of homeland security programs by each creating an appropriations subcommittee and a single authorizing committee for homeland security. In 2003, both houses established the appropriations subcommittees, and by 2005, both had formed homeland security–authorizing committees. However, the full effect of consolidated oversight was limited in all cases because (1) the appropriations subcommittees on homeland security provided funds for only programs within the DHS and thus shared that department's limitations in not having jurisdiction over homeland security functions that were performed outside of DHS as well as being responsible for non–homeland security programs within it, and (2) the authorizing committees shared jurisdiction over various homeland security programs with a number of other committees. For example, many transportation security activities were primarily overseen by the House and Senate transportation committees (Grimmett, 2006, pp. 55–56).

The International Response

Previous aviation disasters, including the bombing of Pan Am Flight 103, had also produced a global reaction but one that was largely limited to the field of civil aviation and did not persist for an extended period in increasing security measures or the level of resources

Table 2.2 Transportation Security Appropriations for Fiscal Years 2000 to 2003 (Amounts in Millions of Dollars)*

	FY2000	FY2001	ETR	FY2002	FY02 Supp	FY2003	FY03 Supp	Post-9/11 Total
Aviation Security								
FAA	286.2	281.7	922.0	294.2	0	144.0	0	1360.2
Airport reimbursement	0	0	175.0	0	0	0	0	175.0
TSA	0	0	1031.5	1250.0	3370.0	4486.9	645.0	10,783.4
Subtotal	*286.2*	*281.7*	*2128.5*	*1544.2*	*3370.0*	*4630.9*	*645.0*	*12,318.6*
As % of total	*82.2%*	*82.1%*	*88.1%*	*76.4%*	*90.3%*	*75.5%*	*70.3%*	*80.9%*
Maritime Security								
Coast Guard security	60.0	60.0	59.3	473.0	209.0	1254.0	218.0	2333.3
Port security grants	0	0	93.3	0	125.0	150.0	20.0	388.3
Safe Commerce	0	0	0	0	28.0	30.0	0	58.0
Container security	0	0	0	0	0	12.0	35.0	47.0
Port security research and development	0	0	0	0	0	10.0	0	10.0
Subtotal	*60.0*	*60.0*	*152.6*	*473.0*	*362.0*	*1456.0*	*273.0*	*2716.6*
As % of total	*17.2%*	*17.5%*	*6.3%*	*23.4%*	*9.7%*	*23.8%*	*29.7%*	*17.9%*
Land Transportation Security								
FTA	2.0	1.4	28.7	5.1	0	8.5	0	45.7
Rail security	0	0	6.0	0	0	0	0	6.0
Amtrak security	0	0	100.0	0	0	0	0	100.0
Trucking security	0	0	0	0	0	25.0	0	25.0
Intercity bus security	0	0	0	0	0	10.0	0	10.0
Subtotal	*2.0*	*1.4*	*134.7*	*5.1*	*0*	*43.5*	*0*	*183.3*
As % of total	*0.6%*	*0.4%*	*5.6%*	*0.2%*	*0*	*0.7%*	*0*	*1.2%*
Grand total	**348.2**	**343.1**	**2415.8**	**2022.3**	**3,732.0**	**6130.4**	**918.0**	**15,218.5**

*Figures do not include aid to airlines.
FY2000 = Department of Transportation and Related Agencies Appropriations Act, 2000 (PL 106-69).
FY2001 = Department of Transportation and Related Agencies Appropriations Act, 2001 (PL 106-346).
ETR = Emergency Terrorism Response fund, composed of 2001 Emergency Supplemental Appropriations Act for Recovery from and Response to Terrorist Attacks on the United States (PL 107-38) and Department of Defense and Emergency Supplemental Appropriations for Recovery from and Response to Terrorist Attacks on the United States (PL 107-117).
FY2002 = Department of Transportation and Related Agencies Appropriations Act, 2002 (PL 107-87).
FY02 Supp = 2002 Supplemental Appropriations Act for Further Recovery from and Response to Terrorist Attacks on the United States (PL 107-206).
FY2003 = Consolidated Appropriations Resolution, 2003 (PL 108-7).
FY03 Supp = Emergency Wartime Supplemental Appropriations Act, 2003.
Post-9/11 Total = Combined total from ETR, FY2002, FY02 Supp, FY2003, and FY03 Supp.
FAA (Federal Aviation Administration) aviation security includes civil aviation security operations, explosive detection equipment, system security technology, and Air Marshals.
Airport reimbursement = FAA reimbursement to airports for security investments.
TSA = aviation security
Coast Guard security = ports, waterways, and coastal security as included in the president's budget beginning for FY2002 (figures for FY2000 and FY2001 are estimates for security share of marine safety and security account).
Safe Commerce = Operation Safe Commerce, administered by Customs and Border Protection.
Container security = Container Security Initiative, administered by Customs and Border Protection.
FTA = Federal Transit Administration combating terrorism, as calculated by OMB 2002 Report to Congress on Combating Terrorism.
Rail security = funding provided to Federal Railway Administration.
Trucking security = security grants made to trucking industry.

devoted to them. The 9/11 attacks provoked a more widespread and long-lasting response in international and national transportation security policies.

International Civil Aviation Organization Security Standards

In November 2001, the International Civil Aviation Organization (ICAO) adopted a Declaration on Misuse of Civil Aircraft as Weapons of Destruction directing the ICAO council to convene a high-level international conference on aviation security to strengthen the organization's role in securing the adoption by member states of security standards and recommended practices. At the February 2002 conference, the ICAO adopted an Aviation Security Plan of Action, which included a requirement for regular security audits to evaluate aviation security in all member states while reiterating the principle that each nation has exclusive responsibility for its aviation security system. In addition, in December 2001, Annex 17 on security was amended to make the annex applicable to domestic flights and to add guidance to national aviation authorities on airport access control standards, the screening of passengers and carry-on and checked baggage, in-flight security personnel (e.g., Air Marshals), and protection of the cockpit (Price and Forrest, 2009, pp. 86–90).

International Ship and Port Facility Security Code

The International Maritime Organization (IMO) also moved to enhance its security efforts after the 9/11 hijackings. In December 2002, the International Convention for the Safety of Life at Sea (SOLAS) was amended to create the International Ship and Port Facility Security Code (ISPS Code). McNicholas (2008) writes: "The purpose of the ISPS Code was to establish an international framework of 'standards' to be achieved involving governments, government agencies, local administrations, and the shipping and port industries to detect and assess security threats and standardize the requirements of the maritime industry in taking preventive measures against potential security incidents that could affect ships or port facilities used in international trade." The ISPS Code included provisions requiring ships and port facilities to:

- Properly train personnel to carry out their security duties.
- Gather and assess information.
- Restrict access.
- Prevent the introduction of unauthorized weapons.
- Provide for relevant training and the conduct of training exercises.
- Establish threat-based security levels and associated countermeasures.
- Address a range of threats, including (in addition to terrorism) stowaways, piracy, smuggling, sabotage, hijacking, cargo tampering, hostage taking, and vandalism, among others (p. 90).

National-Level Responses

At the request of the government of the Netherlands, the European division of the Rand Corporation undertook a *Quick Scan of Post 9/11 National Counter-terrorism Policymaking and Implementation in Selected European Countries.* The report, which covered Finland, France, Germany, the Netherlands, Spain, and the United Kingdom, was issued in May 2002. Among its major conclusions pertinent to transportation security were the following:

1. There were no significant differences in the initial responses of the six countries. "Top ministerial committees and task forces were quickly put together in all countries to provide leadership and a focal point in the confusion that followed the attacks. In particular the security and surveillance of commercial aviation, designated objects and components of critical infrastructures, dignitaries and, to a lesser extent, country borders were immediately strengthened."
2. "Differences among countries largely stem from previous national experiences with domestic terrorism and the (often associated) national institutional structures. Spain, the UK, France and Germany have more significant and more recent experience with domestic terrorism than [the other countries]."
3. "None of the analyzed countries specifically track government spending on counter-terrorism. As a result it is difficult to compare or analyze expenditures…Increased aviation security is partly being paid for by a levy on airline tickets in all countries. From the limited data that we have collected in this quick scan, we conclude that counter-terrorism expenditures in France, Germany, the UK and Spain are higher than in the Netherlands, Belgium and Finland."
4. "None of the countries have a centralized national body to organize and orchestrate counter-terrorism, although France is moving towards it."
5. "One particular issue that seems not to have received balanced and worldwide security attention yet is container transport – certainly not in the countries we surveyed."
6. All six countries were judged to have been "well underway" in implementing 100% luggage and passenger checks on all flights, although none had "accomplished" those objectives.
7. "Another gap is the involvement of the private sector…Those issues are being addressed necessarily in the aviation industry; in other areas, such as container transport, the responsibility of the private sector has not been adequately addressed in any of the researched countries" (van de Linde et al., 2002, pp. 4–8, 24–29).

The United Kingdom took a number of steps in the aftermath of the 9/11 hijackings. The Anti-terrorism, Crime and Security Act of 2001 provided for the forcible removal of unauthorized persons from aircraft or restricted areas at airports. The Nationality, Immigration, and Asylum Act of 2002 included a "right to carry" measure that requires air and sea carriers to check passenger names against a database to confirm that passengers do not pose a known security or immigration risk before they are permitted to board. Other aviation security–related efforts initiated in late 2001 or 2002 (and similar to parallel undertakings in the United States) included increasing the screening of passengers and baggage (especially on flights bound for the United States or Canada), expanding the list of prohibited items not allowed to be carried on board, installing hardened cockpit doors, and training and deploying sky marshals for United Kingdom–registered aircraft. In addition, the United Kingdom enhanced the stop-and-search powers of customs officials, strengthened the ability of law enforcement authorities to request information about passengers and goods from air and sea carriers, and installed radioactivity detectors at all U.K. ports and airports. Finally, several British ports participated in the U.S. Container Security Initiative, and most U.K. ports complied with the strengthened IMO maritime security requirements (Archick, 2006, pp. 44–45).

Conclusion

Before 9/11, transportation security was a relatively inexpensive and unobtrusive undertaking, becoming apparent in the immediate aftermath of a major incident and for a short time thereafter. In the case of the United States, such incidents were almost always far removed from American soil, further diminishing their impact. 9/11 changed all that, with its economic and psychological damage provoking a massive and sustained reaction that almost instantly produced a series of congressional mandates on security measures and enormous increases in funding for aviation security that made the agents of federal transportation security efforts—especially the TSA's checkpoint screeners—one of the most recognized representatives of the federal government. This was followed shortly thereafter by one of the largest governmental reorganizations in U.S. history through the creation of the DHS and by a significant expansion in the maritime security role of the Coast Guard. And these changes proved durable, persisting not for days or months but years.

The international protocols for aviation and maritime security were strengthened in the aftermath of the 9/11 disaster, and—although not to the same degree as in the United States—many other nations undertook a major reworking of their own approaches to counterterrorism and transportation security.

Almost certainly, the revisions in security measures after September 11, 2001—especially with respect to aviation onboard security—made a repeat of the 9/11 hijackings very difficult, if not impossible. However, the speed with which these security changes occurred raised questions for the future about their appropriateness in view of ever-evolving threats; their long-term sustainability in an era of constrained governmental resources; and their relationship to other societal priorities, including civil liberties and economic efficiency, that would inevitably reassert themselves. But from the end of 2001 to the present, the world remade by 9/11 remains the operative environment for transportation security.

Discussion Questions

1. What were some of the institutional and contextual factors that contributed to the unpreparedness of the U.S. civil aviation security system for the 9/11 attacks?
2. Name the layers of aviation security and discuss why they failed to prevent the 9/11 hijackings.

3. What were the terrorists' objectives? Which one was not achieved, and why?
4. What was the most important reason for the New York transit system's relatively high level of preparedness in coping with the events of 9/11?
5. What were the key legislative, institutional, and budgetary responses to the 9/11 attacks?

References

Archick, K., July, 2006. European approaches to homeland security and counterterrorism. Congressional Research Service, Washington, DC.

Belasco, A., Nowels, L., August, 2002. Supplemental appropriations for FY2002: Combating terrorism and other issues. Congressional Research Service, Washington, DC.

Carter, S., Cox, A. One 9/11 Tally: $3.3 Trillion. September 8, 2011. *The New York Times, 9/11 The Reckoning.* Retrieved from http://nytimes.com/interactive/2011/09/08/us/sept-11-reckoning/cost-graphic.html.

Effects of catastrophic events on transportation system management and operations, New York City—September 11, 2001. 2002. U.S. Department of Transportation, ITS Joint Program Office, Washington, DC.

Elias, B., March, 2005. Aviation security-related findings and recommendations of the 9/11 Commission. Congressional Research Service, Washington, DC.

Federal Aviation Administration. April, 2003. Airlines meet FAA's hardened cockpit door deadline. Retrieved from http://www.faa.gov/news/press_releases/news_story.cfm?newsId=5623.

First video of Pentagon 9/11 attack released. March, 2006. CNN.com. Retrieved from http://edition.cnn.com/2006/US/05/16/pentagon.video/index.html.

Frittelli, J.F., December, 2003. Port and maritime security: Background and issues for Congress. Congressional Research Service, Washington, DC.

Government Accountability Office (GAO). September, 2002. Mass transit: Challenges in securing transit systems. Washington, DC.

Green, J., August, 2011. Accounting of the dead. *New York Magazine, The Encyclopedia of 9/11.* Retrieved from http://nymag.com/news/9-11/10th-anniversary/number-of-deaths/.

Grimmett, R.F., December, 2006. 9/11 Commission recommendations: implementation status. Congressional Research Service, Washington, DC.

Jenkins, B.M., Edwards-Winslow, F., 2003. Saving city lifelines: Lessons learned in the 9-11 terrorist attacks. Mineta Transportation Institute, San Jose, CA.

Johnstone, R.W., 2005. New strategies to protect America: Terrorism and mass transit after London and Madrid. Center for American Progress, Washington, DC.

Johnstone, R.W., 2006. 9/11 and the Future of Transportation Security. Praeger, Westport, CT.

Kirk, R.S., December, 2001. Selected aviation security legislation in the aftermath of the September 11 attack. Congressional Research Service, Washington, DC.

Krouse, W.J., December, 2002. Department of Homeland Security: Consolidation of border and transportation security agencies. Congressional Research Service, Washington, DC.

McNicholas, M., 2008. Maritime Security. Butterworth-Heinemann/Elsevier, Burlington, MA.

Merari, A., 1999. Attacks on civil aviation: Trends and lessons. In: Wilkinson, P., Jenkins, B. (Eds.), Aviation Terrorism and Security. Frank Cass, London, pp. 9–26.

Morag, N., 2011. Does homeland security exist outside the United States? Homeland Security Affairs Journal 7, The 9/11 Essays, September 2011.

National Commission on Terrorist Attacks Upon the United States (9/11 Commission). 2006.

New York reduces 9/11 death toll by 40. October, 2003. CNN.com. Retrieved from http://articles.cnn.com/2003-10-29/us/wtc.deaths.

National Commission on Terrorist Attacks Upon the United States (9/11 Commission). 2003. Gerald Dillingham Testimony. New York.

National Commission on Terrorist Attacks Upon the United States (9/11 Commission). 2003. Jane Garvey Testimony. Washington, DC.

National Commission on Terrorist Attacks Upon the United States (9/11 Commission). 2004. Staff Statement No. 3. Washington, DC.

National Commission on Terrorist Attacks Upon the United States (9/11 Commission). 2004. Staff Statement No. 4. Washington, DC.

Peterman, D.R., October, 2004. Transportation issues in the 108th Congress. Congressional Research Service, Washington, DC.

Price, J.C., Forrest, J.S., 2009. Practical aviation security: Predicting and preventing future threats. Butterworth-Heinemann/Elsevier, Burlington, MA.

Riehl, J.R., January, 2003. Combating terrorism: First Emergency Supplemental Appropriations – distribution of funds to departments and agencies. Congressional Research Service, Washington, DC.

Securing the homeland, strengthening the nation. 2002. The White House, Washington, DC.

Taylor, B.D., 2005. Designing and operating safe and secure transit systems: Assessing current practices in the United States and abroad. Mineta Transportation Institute, San Jose, CA.

The 9/11 Commission Report: final report of the National Commission on Terrorist Attacks Upon the United States. 2004. Norton, New York.

Third Monograph: The four flights and civil aviation security. In 9/11 Commission Report (pp. 579-699). Barnes & Noble, New York.

2011 Annual Report on 9/11 Health. 2011. World Trade Center Medical Working Group of New York City, New York.

van de Linde, E., O'Brien, K., Lindstrom, G., de Spiegeleire, S., Vayrynen, M., de Vries, H., 2002. Quick scan of post 9/11 national counter-terrorism policymaking and implementation in selected European countries. Rand Europe, Leiden, Netherlands.

Vedantam, S., 2003. Science Notebook: More afraid than we should be. The Washington Post, p. A. 06. March 31, 2003.

3

Transportation Systems and Security Risks

CHAPTER OBJECTIVES:

In this chapter, you will learn about transportation systems and the security risks they face, including:

- Characteristics of transportation systems
- Major security incidents since 2001
- Risk management and its application to transportation security
- Vulnerability, threat, and consequence assessments

Introduction

Transportation is a central component of the increasingly global economy of the 21st century, moving raw materials, finished goods, and people within and between countries and providing individuals with access to a growing array of goods and services, as well as job opportunities. All of these factors have been magnified in the past few years as "new opportunities arose with the convergence of telecommunications and information technologies, supporting a higher level of management of production, consumption and distribution" (Rodrigue et al., 2006, pp. 76–77).

In a study of 34 countries, Kauppila (2011) found that transportation ranked as one of the top three categories of household spending in almost all of them. Housing ranked first in every country, and food came second in most nations followed by transportation, but in 10 of the countries (including the United States, where such spending accounted for 18% of total household expenditures, and the United Kingdom, where it represented 14%) transportation expenditures placed second (p. 2).

In the United States, transportation-related goods and services accounted for $1.6 trillion (10.2%) of the $15.6 trillion U.S. gross domestic product in 2011, and the transportation sector produced more than 12 million jobs in 2010 (9.3% of the labor force). In 2011, transportation systems carried $3.7 trillion worth of U.S. international merchandise trade, of which 47% was conveyed by the maritime mode, 25% by aviation, 17% by truck, 4% by rail, 2% by pipeline, and 5% by other or unknown modes (U.S. Department of Transportation, 2013, pp. 29, 32, 34).

Transportation Systems

Global transportation systems convey people, food, fuel, medicines, and many other commodities through an expansive and interconnected network of waterways, roads, tracks, pipelines, and airways. The United States—by a considerable margin—has the most extensive network of roadways, railways, and pipelines and the greatest number of airports; China and Russia maintain the largest systems of waterways (Table 3.1).

The U.S. Transportation Research Board (2002) identified five key common characteristics of transportation systems that have security implications:

1. *Open and accessible: Designed and organized for the efficient, convenient, and expeditious movement of large volumes of people and goods, transportation systems must have a high degree of user access. In some cases—highways, for instance—access is almost entirely open. Many transportation facilities, such as train stations, are public places, open by necessity. In other cases, such as commercial aviation, access is more limited but still not fully closed; access to most airport lobbies, ticket lines, and baggage check-in areas remains unrestricted.*
2. *Extensive and ubiquitous: Transportation systems require vast amounts of physical infrastructure and assets... Most of this infrastructure is unguarded and sometimes unattended. Distributed over the networks are millions of vehicles and containers.*

Table 3.1 Extent of transportation systems in select countries, 2008 (Ranked by Roadways)

Country	Roadways (km)	Railways (km)	Waterways (km)	Pipelines (km)	Airports (number)
United States	6,465,799	226,427	41,009	793,285	5146
India	3,316,452	63,327	14,500	22,773	251
China	1,930,544	77,834	110,000	58,082	413
Brazil	1,751,868	28,857	50,000	19,289	734
Japan	1,196,999	23,506	1770	4082	144
Canada	1,042,300	46,688	636	98,544	514
France	951,500	29,213	8501	22,804	295
Russia	933,000	87,157	102,000	246,855	596
Australia	812,972	37,855	2000	30,604	462
Spain	681,224	15,288	1000	11,743	154
Germany	644,480	41,896	7467	31,586	331
Italy	487,700	19,729	2400	18,785	101
Turkey	426,951	8697	1200	11,191	103
Sweden	425,300	11,633	2052	786	249
Poland	423,997	22,314	3997	15,792	126
United Kingdom	398,366	16,454	3200	12,759	312
Indonesia	391,009	8529	21,579	13,752	669
Mexico	356,945	17,516	2900	40,016	243
Saudi Arabia	221,372	1392	Unavailable	8662	215
Belgium	152,256	3233	2043	2023	42

(*Source*: U.S. Department of Transportation, Research and Innovative Technology Division. 2013. http://www.rita.dot.gov/bts/sites/rita.dot.gov.bts/files/publications/freight_transportation/table_03.html (accessed 10.21.14.))

These vehicles and containers are repeatedly moved from one location to another, complicating the task of monitoring, safeguarding, and controlling them.

3. *Emphasis on efficiency and competitiveness: Although much of the transportation infrastructure in the United States is owned by the public sector, the development of this infrastructure has been driven largely by the demands of private users. Widespread use of private cars and motor carriers, for instance, has spurred greater investment in the highway system relative to public transit and railroads... The economic deregulation that swept through the transportation sector during the last quarter of the 20th century led to even greater emphasis on efficiency as a criterion for transportation investments and, to a certain degree, to a loss of redundancy and excess capacity. The dynamism of the U.S. transportation sector is unmatched in the world and is a major reason for the nation's high productivity and mobility. Another consequence of the increased emphasis on efficiency, however, is that costly security measures that promise unclear benefits or impede operations are likely to be resisted or eschewed, but those that confer economic benefits are apt to be deployed and sustained.*

4. *Diverse owners, operators, users, and overseers: Much of the physical infrastructure of transportation—from highways and airports to urban rail networks—is owned and administered by the public sector. Although the federal government helps fund construction, it owns and operates very little of this infrastructure. Most of it is controlled by thousands of state and local governments. Private companies and individuals own some fixed infrastructure (as with freight railroads), but they function mainly as service providers and users, controlling most of the vehicles and containers that use the networks. These public and private owners and operators are largely responsible for policing and securing the system, with the help of state and local law enforcement authorities and, for movements outside the country, foreign governments and international organizations.*

5. *Intertwined with society and the global economy: Trucks of all sizes distribute to retail outlets nearly all the products purchased by consumers and many of the goods and supplies used by industry and government. The rail, pipeline, and waterborne modes, along with large trucks, move products and commodities long distances among utilities, refineries, suppliers, producers, and wholesalers, as well as to and from ports and border crossings... At the same time, airlines have become indispensable in connecting our increasingly diffuse nation, and passenger airline service is essential to many areas of the country that depend on tourism and business travel. At the more local level, a quarter or more of the workers in some large cities commute by public transit... The highway system pervades the lives of Americans, who use motor vehicles for most daily activities and for much of their longer-distance vacation travel. Highways are also used by emergency responders, and both the highway and public transit systems are vital security assets for evacuating people in crises and moving critical supplies and services. Consequently, disruptions to transportation networks can have far-reaching effects not only on transportation operations but also on many other unrelated functions and activities.*

U.S. Transportation Research Board, 2002, pp. 12–15

The 2010 Department of Homeland Security (DHS) *Transportation Systems Sector-Specific Plan,* which is a component of the National Infrastructure Protection Plan and "describes collaboratively developed strategies to reduce risks to critical transportation infrastructure from the broad range of known and unknown terrorism threats," defines six major transportation modes: (1) maritime; (2) aviation; and within the land transportation sector, (3) highways and motor carriers, (4) mass transit and passenger rail, (5) freight rail, and (6) pipelines (2010, pp. 1, 15–16).

Maritime Mode

The U.S. maritime transportation sector "is a network of maritime operations that interface with shoreside operations at intermodal connections and as part of global supply chains or domestic commercial operations" and includes:

- Approximately 95,000 miles of coastline
- More than 10,000 miles of navigable waterways
- 361 ports
- More than 1400 intermodal connections (pp. 15, 173, 176).

Seaports and Marine Terminals

The United States has approximately 70 deep-water seaports, 40 of which handle 10 million tons or more of cargo each year. These ports include approximately 2000 major marine terminals, most of which are owned by local port authorities[1] and operated by the private sector. Many of the terminals and their associated berths handle specific types of cargo or passengers. Those handling cargo containers are usually found within larger port complexes, with ports in six geographic areas (Long Beach/Los Angeles, New York/Newark/Elizabeth, San Francisco/Oakland, Hampton Roads, Charleston/Savannah and Seattle/Tacoma) accounting for just under two-thirds of all container ship calls in the United States (Figure 3.1). Petroleum tanker calls are concentrated in ports on the Gulf Coast, Delaware Bay, New York Harbor, San Francisco Bay, and San Pedro Harbor (pp. 176–177).

Vessels

The major categories of oceangoing vessels—accounting for two-thirds of U.S. seaport traffic each year—are tankers (which transport liquid cargo, primarily in the form of oil, liquefied natural gas and chemicals), container ships (which carry standardized containers typically holding higher-value finished goods and component parts), and dry bulk carriers (which convey such materials as iron ore, grain, and coal). Much of the transport within inland waterways is by barges (most of which have to be towed by another vessel or vehicle) and a variety of boat types. Passenger carriers include ferries, which can transport cars and

[1]In many other nations, governance of seaport facilities is handled by the national government. However, in the United States, port authorities "are instrumentalities of state or local government established by enactment or grants of authority by the state legislature" (Sherman, n.d., p. 2).

FIGURE 3.1 Container terminal at Port of Los Angeles. *(Source: Port of Los Angeles, http://www.portoflosangeles.org/newsroom/photo_gallery.asp.)*

trucks as well but usually for relatively short distances, and cruise ships, which carried an estimated 13.2 million passengers from U.S. ports in 2008 (p. 178).

Inland Waterways

The three components of the U.S. inland waterway network are *inland river systems* (of which the Mississippi River system is the largest), *coastal and intracoastal waterways* (primarily the Gulf Intracoastal Waterway running for 1300 miles from Texas to the Florida Gulf Coast and the intracoastal waterway along a portion of the U.S. Atlantic seaboard), and the *Great Lakes System* (composed of six ports and approximately 350 terminals situated on the U.S. shoreline of the Great Lakes) (pp. 178–179).

Intermodal Systems

Intermodal transportation systems link the various transportation modes and allow cargo and passengers to complete trips by using more than one mode. Because such transport most often involves containers moved by oceangoing vessels to seaports where they are transferred to another mode (typically land mode conveyances such as trucks or railways), intermodal systems are generally considered along with maritime transportation systems, which is the approach adopted by the DHS *Transportation Systems Sector-Specific Plan* (pp. 180–181).

Land Mode

The U.S. land transportation network is composed of several major parts:

- More than 260,000 miles of mass transit and passenger rail systems
- More than 140,000 miles of active freight railways

- More than 4 million miles of roadways and 600,000 bridges and tunnels
- More than 1.7 million miles of pipeline (U.S. Department of Homeland Security, 2010, pp. 15–16, 216; U.S. Department of Transportation, 2013, pp. 10–11)

Mass Transit and Passenger Rail

The U.S. mass transit and passenger rail sector encompasses a wide variety of means designed to transport passengers on local and regional routes, including municipal transit buses, subways, commuter rail, long-distance rail (mainly Amtrak), trolleys, and demand-response systems primarily for senior citizens and persons with disabilities. (In the DHS's division of security responsibilities, interstate buses, school bus systems, and over-the-road private shuttle services are not included here but are considered as part of the highway sector.) (U.S. Department of Homeland Security, 2010, p. 216.)

In 2010, more than 10 billion passenger trips were taken on transit and commuter rail systems, accounting for more than 54 billion passenger miles.

- Buses and trolleybuses: 52% of the trips and 39% of the miles
- Heavy rail subway systems: 35% of the trips and 30% of the miles
- Commuter rail systems: 4.5% of the trips and 20% of the miles
- Light rail trolleys and streetcars: 4.5% of the trips and 4% of the miles
- Demand-response systems: 2% of the trips and 3% of the miles
- Other: 2% of the trips and 4% of the miles

The New York City area by itself accounted for 40% of the total trips and 39% of the total miles travelled (American Public Transportation Association, 2012, pp. 9, 13–14).

In fiscal year 2011 (October 2010 to September 2011), the Amtrak passenger rail system carried more than 30 million passengers on its 21,200-mile network. Approximately two-thirds of the ridership was within the Washington, DC–Boston "Northeast Corridor" (Amtrak, 2012, p. 1).

Freight Rail

Freight railroads deliver goods and commodities to virtually all industrial, wholesale, and retail segments of the American economy. They are composed of a diverse array of 558 privately owned carriers of various sizes. In the absence of any nationwide freight rail operator, the companies have developed a series of arrangements that allow for the transfer of rail cars between carriers and for one carrier's trains to operate on the tracks of another railroad. The freight rail companies are divided into categories based either on their revenues (class I, class II, and class III) or the size of their rail networks (class I, regional and local or short line), with the resulting divisions being very similar in each system.

- There are currently seven class I operators (same designation in both classification systems), with a minimum operating revenue of $401 million. Although they represent less than 1% of all freight operators, they operate on 69% of the track, use 90% of the industry's workforce, and generate 94% of its revenue.

- Class II freight railroads are those with revenues of between $40 million and $400 million, and the related category of regional railroads must operate on at least 350 miles of track.
- Class III freight railroads have revenues of under $39 million, and local or short line railroads operate on less than 350 miles of track. A subcategory of the latter is switching or terminal railroads, which primarily provide connecting services between freight carriers in major cities (U.S. Department of Homeland Security, 2010, p. 283; Association of American Railroads, 2012, p. 1).

Highways, Bridges, and Tunnels

The highway network is fundamental to the entire U.S. transportation system, with all modes relying on its infrastructure to one degree or another. As of 2008, the highway system consisted of 164,095 miles of the National Highway System,[2] of which 47,011 miles is part of the Eisenhower Interstate Highway System, and 3,895,244 miles of other roads. Although the federal government has played a major role in funding highway construction and regulating interstate commerce using the highway network, local governments own and operate most of the nation's roads (77% local, 20% state, and 3% federal) and a majority of U.S. bridges (51% local, 48% state, and 1% federal) (U.S. Department of Transportation, 2013, p. 10; U.S. Department of Homeland Security, 2010, pp. 253–254).

Most of the vehicles operating on the highway system are owned and operated by private individuals and companies:

- Approximately 212 million noncommercial, light-duty vehicles (including automobiles and light trucks) and 8 million motorcycles, the vast majority of which are privately owned
- 29 million privately owned commercial trucks
- 460,000 school buses, 70% of which are owned by local school districts, with the remaining 30% privately owned by for-profit companies
- 29,325 privately owned motorcoaches (interstate buses) (U.S. Department of Transportation, 2013, p. 12; U.S. Department of Homeland Security, 2010, pp. 255–256).

Pipelines

Pipelines transport almost all of the natural gas and nearly two-thirds of all hazardous liquids (including crude and refined petroleum) in the United States. Most of these pipelines are underground and are privately owned and operated. There are three major types:

1. *Natural gas distribution*: 1.23 million miles. The largest pipeline network in the United States is that used to transport natural gas from transmission pipelines to residential and commercial customers.

[2]The National Highway System was designated by the U.S. Department of Transportation in consultation with the Department of Defense and state, local, and regional authorities and is composed of roadways deemed important to the U.S. economy, defense, and mobility (U.S. Department of Homeland Security, 2010, p. 254).

2. *Natural gas transmission and storage*: 324,600 miles. These move natural gas from its sources to the local companies operating the distribution network. Included are more than 400 storage facilities.
3. *Hazardous liquid pipelines and tanks*: 177,600 miles. Most of these carry crude oil to refineries or refined petroleum products (e.g., gasoline or diesel fuel) to product terminals and airports.

In addition, there are more than 100 *liquefied natural gas (LNG) processing and storage facilities* that store LNG either processed on site or received from elsewhere (U.S. Department of Transportation, 2013, p. 11; U.S. Department of Homeland Security, 2010, p. 317).

Aviation Mode

The U.S. aviation system is designed to safely and efficiently transport passengers and cargo within and beyond U.S. borders. Its main divisions are commercial aviation, air cargo, general aviation, and flight schools. It includes just under 20,000 airports (of which approximately 450 are commercial airports, another 4500 are other public use facilities, more than 14,000 are private airfields, and 271 are military) and 231,000 aircraft (7500 passenger and freight carriers and 223,500 general aviation aircraft) (U.S. Department of Transportation, 2013, pp. 10, 12).

Key to the entire sector is the National Airspace System (NAS).

> *The NAS is the dynamic network of facilities, systems, services, airspace, and routes that support flights within U.S. airspace, including the international airspace delegated to the United States for air navigation services. The Federal Aviation Administration (FAA) regulates and operates this system. Specifically, the NAS includes more than 690 air traffic control facilities with associated systems and equipment to provide radar and communications services; more than 19,800 general aviation and commercial aviation airports capable of accommodating an array of aircraft operations; and volumes of procedural and safety information necessary for users to operate in the system. In addition, the NAS includes more than 11,000 air navigation facilities and approximately 13,000 flight procedures.*
>
> *U.S. Department of Homeland Security, 2010, p. 129*

Commercial Aviation

Commercial airports in the Unites States have regularly scheduled commercial passenger service or public charter flights. Commercial airlines are defined as those that engage in regularly scheduled or public charter operations and include domestic air carriers and foreign carriers operating within, from, to, or over the United States (Figure 3.2).

Air Cargo

The air cargo sector includes all freight transported by air (including on passenger aircraft). The U.S. air cargo network is composed of the 450 domestic commercial airports,

FIGURE 3.2 Runway at Hartsfield-Jackson Atlanta International Airport. *(Source: Mmann, 1988, http://en.wikipedia.org/wiki/File:Delta_plane_and_Atlanta_skyline.jpg.)*

airports in the 98 countries where cargo is transported to and from the U.S., more than 300 domestic and foreign air carriers, more than 4000 indirect air carriers (also known as freight forwarders), and more than 1 million shippers from all over the world.

General Aviation

General aviation uses virtually all of the approximately 19,000 nonmilitary airfields in the United States, and general aviation aircraft include all that are not either military or regularly scheduled commercial aircraft. This sector accounts for more than three-quarters of all flights in the United States and includes such diverse components as private-use recreational aircraft (by far the largest segment), corporate and business jets, and emergency medical helicopters.

Flight Schools

All pilot schools, flight training centers and air carrier flight training facilities, and all those providing instruction in aircraft operation or aircraft simulation are considered part of the flight school sector (U.S. Department of Homeland Security, 2010, p. 130).

■ ■ Critical Thinking ■

What are the advantages and disadvantages of the U.S. system of mixed funding, ownership, and operation of transportation assets? Consider economic, governance, and security factors.

Attacks on Transportation Systems Since 2001

The vast and economically vital systems of transportation have continued to be frequent targets of terrorists and others despite the significant increase in security measures put in place after the September 2001 hijackings. Although far less numerous than other forms of attack, terrorist incidents have remained the driving force in further security adjustments.

Terrorism

Certain attributes of modern transportation systems have made them particularly attractive targets for terrorists:

- Passenger vehicles and facilities can contain significant numbers of individuals (and thus potential victims) in confined spaces.
- Vehicles travelling at high speeds are susceptible to great damage—and can cause considerable collateral damage—when subjected to relatively limited force.
- National flag-carriers and landmark sites, such as certain bridges and tunnels, provide important symbolic targets.
- Some facilities are key nodes for handing a large proportion of the people or goods transported by a given system. For example, a limited number of seaports process a major share of international trade, and commuter rail and mass transit systems are critical to people movement in most of the world's largest cities.
- With their onboard fuel, mobility, range, and ubiquity, transportation vehicles and containers themselves offer a ready means for delivering terrorist weapons (Transportation Research Board, 2002, pp. 9–10).

■ ■ Critical Thinking ■

Which characteristics of transportation systems do you think have been the most important in their targeting by terrorists in the 21st century? Why? You may include factors not mentioned in the 2002 report.

The multitude of new transportation security measures put in place in the aftermath of 9/11 did not put a stop to terrorist attacks on transportation systems, though they appear to have had a significant impact in reducing the number of incidents in the aviation sector, where most of the new measures were concentrated. According to figures from the Global Terrorism Database (GTD), between 2002 and 2011, there were 60 incidents involving maritime transportation, producing 205 fatalities. These results are similar to those in the 1980s and 1990s. Land transportation systems continued to attract the large majority of attacks (1607 incidents) and to sustain the most casualties (3828 fatalities). Again, these figures are very similar to the ones for that mode in the two decades before 2001. However, in aviation, the 10 years after 9/11 experienced just 137 incidents, the lowest of any 10-year period dating back to the beginning of the GTD compilation. (The 276 fatalities were well above the 1990s total but far below the deaths recorded in the 1970s and 1980s.) The most recent data

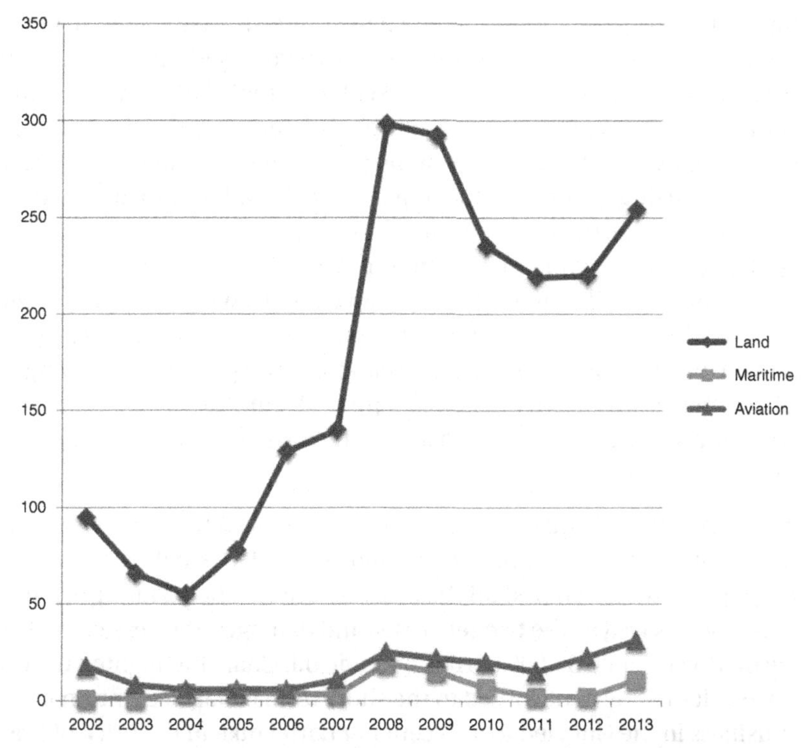

FIGURE 3.3 Transportation terrorism incidents, 2002 to 2013. *(Source: START Global Terrorism Database.)*

(for 2012 and 2013) largely conform to these same patterns, although the incident rate in aviation (but not fatalities) has trended up slightly (Figure 3.3 and Table 3.2).

Post-2001 Terrorist Attacks on Maritime Transportation

Although it continued to be the mode least targeted by terrorists, the maritime sector experienced a "modest yet highly discernible spike in high-profile terrorist incidents at sea" in the years after 9/11 (although none of the 73 incidents recorded in the GTD from 2002 to 2013 took place within the United States or its territorial waters). Writing in 2008, Chalk

Table 3.2 Incidents and Fatalities from Terrorist Attacks on Transportation Systems

	Maritime		Land		Aviation	
Period	**Incidents**	**Fatalities**	**Incidents**	**Fatalities**	**Incidents**	**Fatalities**
1970–1979	20	13	251	265	296	777
1980–1989	60	407	1,753	3,372	383	1,532
1990–2000*	65	223	1,742	3,738	258	107
2002–2011**	60	205	1,607	3,828	137	276
2012–2013	13	5	474	718	54	59
TOTAL	218	853	5,827	11,921	1,128	2,751

*No data for 1993
**2001 excluded
(*Source*: START Global Terrorism Database)

attributed the rise to the greater attention given to countermeasures in the land and aviation sectors (leaving maritime transportation as a less hardened target), the growth of the maritime sports and recreation industry (e.g., scuba diving) that provided would-be terrorists with more access to the training and equipment necessary for maritime operations, and the increased realization by terrorists of the potential for significant economic destabilization (through closing a port or blocking a critical sea lane) or mass casualties (via attacks on cruise ships or ferries) (pp. 20–26).

Other important recent trends in maritime terrorism include the use of small boats as the primary platform for launching attacks and the growing linkage between terrorist organizations and drug smuggling and other transnational criminal enterprises, with the terrorists using this connection to obtain weapons, transportation, training, resources, and funding from the latter (McNicholas, 2008, pp. 248, 258–261).

The following are some of the major post-9/11 terrorist attacks involving the maritime mode:

- On October 6, 2002, the oil tanker MV *Limburg* was attacked by suicide terrorists associated with al Qaeda, who rammed an explosives-laden, small fiberglass boat into the *Limburg* while it was anchored off the coast of Yemen. The resulting explosion killed one crew member, as well as the two terrorists, and damaged the vessel, resulting in the spillage of an estimated 50,000 barrels of crude oil. Chalk highlighted the economic consequences: "It directly contributed to the short-term collapse of international shipping business in the Gulf, led to a 48 cent per barrel hike in the price of Brent crude oil, and due to the tripling of war risks premiums levied on ships calling at Aden, resulted in a 93 percent drop in container terminal throughput that cost the Yemeni economy an estimated $3.8 million a month in port revenues" (Chalk, 2008, pp. 23–24, 49).
- On February 27, 2004, the Filipino ferry vessel *SuperFerry 14* was victimized by the detonation of 20 sticks of dynamite that had been brought on board in a hollowed-out television set. The resulting fire killed at least 116 passengers—which represents more than half of the total number of fatalities caused by maritime terrorism in the 2002 to 2013 period—and has been called the most destructive act in the history of maritime terrorism. The Filipino terrorist organization Abu Sayyaf Group claimed responsibility (Chalk, 2008, pp. 26, 51).
- On October 18, 2006, the Sri Lankan rebel group Members of the Liberation Tigers of Tamil Eelam launched a suicide attack on the Sri Lankan naval base in the port city of Galle. The attackers were disguised as fishermen on five vessels that attempted to enter the base. Three of the boats were destroyed immediately upon coming into range because the Sri Lankan authorities had advance word of the plot. However, the other two boats exploded, killing 15 rebels and one other individual. Fourteen civilians were injured from grenade attacks launched from the boats before their destruction (START, Global Terrorism Database).

There were no recorded incidents of maritime terrorism in the United States in the 2002 to 2013 period (START, Global Terrorism Database).

2008 TERRORIST ATTACKS ON MUMBAI

"INDIA'S 9/11"

India in general and Mumbai in particular had been the scene of many terrorist incidents before November 2008. However, the attack of November 26, 2008, on a series of targets in Mumbai stood out because of "its audacious and ambitious scope, the complexity of the operation, and the diversity of its targets" (Rabasa et al., 2009, p. 1).

According to evidence collected by the Indian government, the attackers started from Karachi, Pakistan on November 22, 2008, in a small boat and then were transferred to a larger Pakistani cargo vessel to continue the voyage toward Mumbai. On November 23, they took over an Indian fishing trawler. The trawler's crew members were killed, except for the captain, who was compelled to navigate the fishing boat the rest of the way to Mumbai. Use of the Indian boat helped them avoid close scrutiny by the Indian coast guard. As they neared shore in late afternoon on November 26, the terrorists killed the fishing boat captain and boarded an inflatable dinghy, which landed in the southern part of Mumbai later that night. There were a total of five two-man assault teams. Each was armed with automatic assault rifles, grenades, and other weapons. They took taxis to their target destinations, leaving behind on two of the vehicles explosive devices that were later detonated, killing the drivers (pp. 4–5).

Five locations were attacked between 9:20 and 10:25 PM local time on November 26: the CST Railway Station (Mumbai's main passenger train station), Leopold Café and Bar, Taj Mahal Hotel, Oberoi-Hilton Hotel, and Nariman House (a five-story building owned by the orthodox Jewish organization Chabad Liberation Movement of Hasidic Jews). Many were killed and injured at each of the sites, and hostages were taken at the latter three. In addition, a portion of the Taj Mahal Hotel was set on fire by the attackers, and explosive devices were detonated at the Oberoi-Trident and Nariman House.

Local police assumed initial security responsibility, exchanging fire with the terrorists (which resulted in the killing of one terrorist and the capture of the other who had been involved in the attack on the train station), cordoning off the hostage sites, and rescuing some of the hostages from Nariman House. The Indian National Security Guards (NSG) assumed control of rescue operations on the next morning (November 27), and these operations were not concluded until the afternoon of November 28 at the Oberoi-Trident and the morning of November 29 at the Taj Mahal. Many of the hostages were rescued (just under 450 at the Taj Mahal), and the remaining eight terrorists were all killed (pp. 5–9).

In all, 165 civilians and security personnel were killed, and 304 were injured in the attacks (p. 1).

The one captured terrorist, Mohammed Ajmal Kassab, supplied many of the details about the planning and execution of the attacks. His testimony and other evidence strongly suggested that the Pakistani-based terrorist group Lashkar-e-Taiba (LeT) was responsible. In 2010, Kassab was convicted of murder, conspiracy and waging war on India and sentenced to death, and he was executed in November 2012 (Rabasa et al., 2009, p. 1; CNN, 2012b).

A 2009 Rand analysis of the Mumbai attacks identified a number of shortcomings in India's security system:

1. *Intelligence failures*, including a lack of specific threat information (even though both Indian and U.S. sources indicated a major attack was likely) and inadequate coordination and information-sharing between the national security agencies and the local police

2. *Gaps in coastal surveillance,* reflecting insufficient resources provided for monitoring the Indian coastline (under 100 boats to cover over 5000 miles of coast)
3. *Inadequate protective measures at the train station,* including unreliable metal detectors and poorly armed and trained Railway Protection Force officers
4. *Inadequate preparation for security response to the attacks,* including lack of training and poor planning for the rapid deployment of the NSG, which was designed to be the country's main rapid-reaction force in major security incidents
5. *Inadequate training and equipment for the local police*
6. *Poor communication and information management,* as evidenced by "Throughout the crisis, the central government and security forces failed to project an image of control, with the words 'chaos' and 'paralysis' used repeatedly to describe events as they unfolded. . . . More seriously, breaches of basic information security protocols provided the terrorists with operational intelligence" (pp. 9–11).

In the aftermath of the attacks, the Indian government took steps to address many of the security weaknesses. NSG offices were opened in all state capitals to facilitate their more rapid deployment. A Coastal Command was formed with the specific mandate to secure India's coastline. Twenty counterterrorism schools were established. A National Investigation Agency was created as the central counterterrorism investigation and law enforcement agency in India (although the existing intelligence and law enforcement agencies retained "concurrent" jurisdiction in many areas). The Unlawful Activities (Prevention) Amendment Act of 2008 was passed, providing for the detention of terror suspects for up to 180 days (Rabasa et al., 2009, pp. 11–12; Matthew, 2009).

In addition to the successful attacks, many planned terrorist operations against maritime transportation were thwarted before their execution, including "bombings of U.S. naval ships sailing in Singaporean, Malaysian, and Indonesian waters, suicide strikes against Western shipping interests in the Mediterranean, small boat ramming of supertankers transiting the Straits of Gibraltar, and attacks on cruise liners carrying Israeli tourists to Turkey" (Chalk, 2008, pp. 20–21).

Post-2001 Attacks on Land Transportation

According to the Mineta Transportation Institute (MTI) Database,[3] terrorist attacks on land transportation systems have continued at the higher rate that began in the mid-1990s, with fatalities following a similar trend. Other key characteristics of such attacks also reflected patterns that emerged in the 1990s, with buses the most frequent target (more than 40% of all attacks) followed by subway and passenger trains (~20%) and highways and roads the least targeted (less than 2%). Bombings continued to account for the majority of weapons used (over 60%), and the Indian subcontinent was the scene of the most attacks

[3]The MTI Database figures differ somewhat from those in the GTD. Whereas the former focuses solely on land transportation systems and provides more detailed information on those attacks, the GTD allows for comparison across all transportation modes. The differences between the two, which are significant for some years, highlight the difficulty of arriving at consistent and reliable data on terrorist attacks.

(~30% of the total in India, Pakistan, and Sri Lanka). Only one of the attacks occurred within the United States (a 2009 bombing of a rail signal device in Illinois) (Jenkins and Butterworth, 2010, pp. 12, 19–20, 24–26, 30–31).

Although the overall trends were similar to those from the preceding decade, land transportation systems experienced several major terrorist attacks after 2001. Two of those—the March 2004 attacks on commuter rail trains in Madrid, Spain and the July 2005 bombings on London's public transit system—were among the highest profile terrorist events of the post-9/11 era:

- On December 5, 2003, Chechen rebels detonated explosives on a commuter train in southern Russia. Forty-seven people were killed, and 170 people were injured.
- On February 6, 2004, a bomb exploded in a subway car on the Moscow Metro, killing 40 and injuring 122 people. Russian authorities blamed Chechen rebels, but the latter denied involvement.

2004 TERRORIST ATTACKS ON MADRID COMMUTER TRAINS

ISLAMIC EXTREMISTS TARGET SPAIN

Spain had a long history of dealing with terrorism before 2004, with most of that experience resulting from the national authorities' confrontation with the Basque separatist group ETA (Euskadi Ta Askatasuna). However, as one Spanish transportation official explained, "There wasn't an assumption that something so major could occur. Always there was a possibility that somebody crazy could do something like start a fire in a station, but we never thought there was a capacity to do something like this" (Taylor, 2005, p. 133).

On March 11, 2004, a total of 10 bombs exploded during morning rush hour on four commuter trains in Madrid, resulting in 191 fatalities and more than 1800 injuries. The trains had all originated or passed through a station 12 km east of the city within a 15-minute interval beginning at 7:00 AM. That station is where the bombs were loaded, as evidenced by a stolen van containing seven detonators later found near the station. At 7:39 AM, three bombs exploded as the first of the trains entered Madrid's Atocha Station followed almost simultaneously by the explosion of four bombs on the second train as it neared the same station. Three other unexploded bombs were discovered at the station by Spanish authorities and were subsequently detonated in controlled explosions. Investigators believed that the terrorists' intention was to set off all 10 bombs inside the station to maximize casualties and damage. At 7:41 AM, two bombs exploded in the third train as it passed through the El Pozo Station, and at 7:42 AM, one bomb was set off on the fourth train as it passed through the Santa Eugenia Station. The discovery of the unexploded bombs packed in backpacks and using a mobile phone as a detonator provided police with vital information about the means of attack. By 8:00 AM, emergency workers began to arrive at the three stations to treat the injured and move the most serious cases to area hospitals. At 10:27 AM, all trains incoming to Madrid were stopped because of concerns about further explosions (BBC News, 2004).

With the national elections only 3 days away, the "11-M" attacks, as they were called in Spain, immediately became a major political issue, with the incumbent Popular Party maintaining that the ETA was responsible despite mounting evidence pointing to Islamic extremists linked with al Qaeda.

The police investigation and subsequent trial uncovered no evidence of a link to ETA. The bombings were carried out by a group of young men, mostly from North Africa, who were, according to prosecutors, inspired by a tract on an al Qaeda–affiliated website that called for attacks in Spain. The tract called for "two or more attacks . . . to exploit the coming general elections in Spain in March 2004," saying that they would ensure the "victory of the Socialist Party and the withdrawal of Spanish forces [from Iraq]"[4] (Hamilos, 2007).

On April 3, 2004, seven suspects in the case blew themselves up in an apartment outside of Madrid rather than submit to police who were closing in. Two years later, on April 11, 2006, 29 individuals (including 15 Moroccans and nine Spaniards) were indicted for their role in 11-M attacks. On October 31, 2007, a Spanish judge found 21 of the defendants guilty: three for the murders and 18 others on lesser charges, including membership in a terrorist group and trafficking in weapons (Associated Press, 2007; Hamilos and Tran, 2007).

A parliamentary commission was formed to investigate the March 11 attacks, and it focused on the need for Spain to increase the resources it devoted to dealing with the specific threat from Islamist terrorism, as well as to improve coordination within and among the country's security and intelligence agencies (Archick, 2006, p. 32).

Despite Spain's long-term experience with ETA and other terrorist attacks and the existence of "a substantial body of law and institutional capacity to fight terrorism," no specific transit security measures were in place in March 2004 because of the expense and difficulty of implementation. After the events of 11-M, the Spanish government took a number of actions to bolster its security system:

- Funding for counterterrorism programs was significantly boosted, to 350 million Euros ($417 million) in 2005.
- Police forces (including intelligence units) were expanded and redeployed to provide better protection for rail passengers, and the army was given authority to help police rail lines.
- The national rail system began to procure canine explosives detection teams, x-ray machines, and fixed and mobile scanners. It also expanded its existing network of closed-circuit television cameras.
- However, to avoid raising the public's level of fear and anxiety, Spanish transportation authorities decided against posting messages or making announcements urging passengers to be vigilant for signs of a possible attack.
- A National Antiterrorism Coordination Centre was created to conduct "regular assessments of terrorist risks and threats to Spain, assessments that provide high-quality strategic intelligence that includes possible scenarios for intervention and operational recommendations for dealing with such risks and threats" (Archick, 2006, pp. 32–35; Taylor, 2005, pp. 133–135; Reinares, 2008, p. 7).

[4] The Socialists did win the March 14 elections in large part because of the sitting government's inaccurate assignment of blame for the attacks to ETA and the unpopularity of the Iraq war and subsequently carried out the withdrawal of Spanish troops from Iraq (Hamilos, 2007).

- On July 7, 2005, three near-simultaneous explosions occurred within the London Underground at around 8:50 AM, and a fourth went off on a transit bus at 9:47 AM. Fifty-six individuals were killed (including the four suicide bombers, three of whom were British citizens of Pakistani descent), and more than 700 people were injured. The official U.K. investigation of the attacks concluded, "[The bombers'] motivation appears to be typical of similar cases: fierce antagonism to perceived injustices by the West against Muslims and a desire for martyrdom. The extent of al Qaeda involvement is unclear. [Two of the bombers] may have met al Qaeda figures during visits to Pakistan or Afghanistan. There was contact with someone in Pakistan in the run up to the bombings. Al Qaeda's deputy leader has also claimed responsibility."
- On June 15, 2006, a passenger bus set off a land mine in northern Sri Lanka, resulting in at least 63 fatalities and 71 injuries. The Sri Lankan government blamed the Liberation Tigers of Tamil Eelam, but no claim of responsibility was made.
- On July 12, 2006, a series of bombs was detonated at rush hour on seven passenger trains within the Mumbai train system. The attacks resulted in more deaths (at least 187) and injuries (817) than did the maritime-launched terrorist assaults on Mumbai that occurred 2 years later. As in that case, the terrorist group Lashkar-e Taiyiba was likely responsible (START, Global Terrorism Database; Government of the United Kingdom, 2006, pp. 2, 26).

Jenkins and Trella (2012) examined 15 failed terrorist plots against land transportation targets, all but one of which took place after 2001. To concentrate on the plots that had the greatest potential consequences, 14 of the 15 selected were directed against major city mass transit and commuter rail systems, 10 involved bombs (with at least five of those being suicide bombings), and four contemplated the use of chemical or biological agents. The authors described how the various plots had been thwarted.

> *Intelligence was a key factor in foiling most of the plots. Eleven plots were uncovered by intelligence operations. One was uncovered when a frightened roommate of the bombers told police about the impending attack, and one was reportedly called off by the terrorists' leader abroad. Two plots had progressed to an actual attempt (the attempted Tube and bus bombings in London and the attempted German train bombings). Both attempts failed because of faulty bomb construction. The fact that so many of the plots hardly got beyond the discussion phase limits what is known about how the plotters viewed security. The overall evidence, however, suggest that they were undeterred by the security measures in place. Where awareness of security does appear in the plots, it is a cause for caution, perhaps a reason to modify a date or location, not a reason to call off the attack.*
>
> *Jenkins and Trella, 2012, pp. 3–5*

Post-2001 Attacks on Aviation

As previously noted, there has been a substantial decline in successful terrorist attacks on aviation since 9/11. Indeed, since 2001, the major terrorist incidents in North America and

Western Europe have been foiled plots against the aviation mode. However, the long-term pattern of reactive security continued, with even failed attempts eliciting significant changes in security policies.

Just five major incidents accounted for 179 (or 53%) of the GTD's total count of 335 post-2001 fatalities from aviation-related terrorism. The most recent three were all in Russia and involved Islamic separatists from southern Russia.

- On February 17, 2002, Maoist rebels attacked an airport in Nepal, killing 27 policemen, with an unknown number injured.
- On March 4, 2003, a suicide bomber killed himself along with 20 others and injured more than 150 when he detonated an explosive device near the main terminal of the Davao International Airport in the Philippines. The al Qaeda-linked Abu Sayyaf Group initially claimed responsibility, but the Filipino government placed the blame on the Moro Islamic Liberation Front.
- On August 24, 2004, two Russian airliners crashed, killing all 90 on board the two aircraft. Russian authorities suspected that the crashes were the result of the actions of a female suicide bomber on each flight associated with Chechen rebels. Both suspects had boarded at Moscow's Domodedovo International Airport, which was using the type of metal detectors and x-ray machines for passenger screening that were in use in the United States before 9/11. As a result of the two crashes, Russian authorities required passengers to submit to a prescreening security interview and to remove bulky clothing, shoes, and belts before passing through the screening devices (Price and Forrest, 2009, pp. 77–78).
- On January 24, 2011, a suicide bomber set off an explosion inside Moscow's Domodedovo Airport, killing himself and 37 others and wounding 168. The Dagestan Front of the Caucasus Empire, the largest Islamist terrorist organization within the Russian republic of Dagestan, claimed responsibility (START, Global Terrorism Database).

Only four of the 2002 to 2013 incidents recorded by the GTD took place in the United States. The first was a July 4, 2002, shooting by a native Egyptian at the Israeli airliner El Al's ticket counter at Los Angeles International Airport, which resulted in three fatalities (including the shooter) and four wounded. The victims were all Israelis, and the perpetrator reportedly espoused anti-Israeli views. Two other security events took place in 2013. On April 18 an explosive device was discovered at the McCook, NE airport, but was safely defused and no group or individual claimed responsibility. More seriously, on November 1 a lone gunman opened fire at Los Angeles International Airport, killing a TSA agent and wounding two TSA agents and five civilians. The perpetrator was also wounded during the attack and subsequently apprehended. From evidence uncovered in his apartment, he was apparently motivated by intense hostility toward TSA (START, Global Terrorism Database).

The other U.S. incident took place on December 25, 2009, on board Northwest Airlines Flight 253 from Amsterdam as it was approaching its destination in Detroit. Nigerian Umar Farouk Abdulmutallab attempted to detonate an explosives device sewn into his underwear. The device ignited but did not detonate, and the perpetrator was subsequently subdued by other passengers. Abdulmutallab and one other passenger were injured, and although the plane sustained damage, no one was killed. He was linked to Al Qaeda in the Arabian Peninsula (AQAP) and was convicted and sentenced to life in prison in February 2012. A White House review of the incident faulted U.S. aviation security, which allowed the suspect to board a U.S.-bound aircraft in spite of the government having "sufficient information . . . to have potentially disrupted the AQAP plot . . . by identifying Mr. Abdulmutallab as a likely operative of AQAP and potentially preventing him from boarding Flight 253." In addition, although U.S. officials expressed concerns about the inability of the airport checkpoint screening in Amsterdam to detect the explosives, the U.S. Government Accountability Office (GAO) subsequently testified to a congressional committee that "it remains unclear whether [the newly deployed technology at U.S. airports] would have detected the weapons used in the December 2009 incident." In response to these and other vulnerabilities revealed by the AQAP plot, the president proposed and Congress ultimately appropriated approximately $800 million in additional funding for advanced checkpoint screening technology, portable explosives trace detection systems, explosives detection canine teams, and Air Marshals (Carafano et al., 2012, p. 14; The White House, 2009, p. 2; GAO, 2010, Highlights; U.S. Senate Committee on Appropriations, 2011).

Two other significant plots involving U.S. commercial aviation were foiled:

- On August 9, 2006, law enforcement authorities in the United Kingdom announced that they had arrested 24 individuals suspected of plotting to detonate liquid explosives on 10 commercial airliners bound for the United States. The similarities to the 1994 to 1995 Bojinka plot led to a suspicion of al Qaeda involvement, but the ensuing investigation and court proceedings—which eventually resulted in a September 2008 jury decision that found just three of the defendants guilty of conspiracy to commit murder—failed to produce any concrete link to the perpetrators. In response to the plot, the Transportation Security Administration (TSA) banned passengers from carrying any liquids or gels on board an aircraft, with the action widely reported to be necessary because existing screening equipment was unable to detect the kind of explosives involved in the plot (Carafano et al., 2012, p. 9; Johnstone, 2007, p. 51).
- In October 2010, two packages originating in Yemen and bound for Chicago were discovered to be containing explosive materials hidden within printer cartridges. The plot was first uncovered by Saudi intelligence officials, who alerted their U.S. counterparts. The packages were intercepted while en route at stopovers, one in the United Kingdom and the other in the United Arab Emirates. Apparently, they were designed to detonate in mid-air, presumably over Chicago or another U.S. city. AQAP claimed responsibility, but further investigation has, as yet, not led to any arrests. In reaction, the United States temporarily banned all cargo shipments originating in Yemen or Somalia and modified prescreening procedures for cargo in-bound to the United States (Carafano et al., p. 15; U.S. Department of Homeland Security, 2011, p. 27).

■ ■ Critical Thinking ■

What factors have most distinguished successful from unsuccessful major terrorist attacks since 9/11?

Other Criminal Attacks

Although post-9/11 terrorist incidents have generally been more consequential—at least in terms of casualties, media attention, and security response—a large majority of attacks on and against transportation systems have been nonterrorist in motivation. Furthermore, the sheer number and nature of some of these have produced a greater economic impact. For example, the GTD reports 2345 worldwide terrorist incidents involving transportation between 2002 and 2013, but in the United States alone, there were an estimated 737,142 motor vehicle thefts, resulting in more than $4.5 billion in economic losses to owners, in the single year of 2010 (FBI, 2011).

Nonterrorist crimes against each transportation mode span a wide range, from petty theft to violent, life-threatening assaults. The following are a sampling of some of these criminal acts.

In the maritime sector, piracy continues to be a particularly serious threat, with the number of incidents far exceeding those of maritime terrorist actions and generally exhibiting an upward trend in recent years. The International Maritime Organization reported 4040 acts of piracy and armed robbery against ships in the period from 2002 through 2012. Particularly since 2007, most of the increase can be attributed to Somali-based pirates, whose use of "motherships" and increased range resulted in the spread of their operations from the East African coast to the Indian Ocean and Arabian Sea (Figure 3.4) (International Maritime Organization, 2013, pp. 1–2).

Another particularly consequential form of nonterrorist attack on transportation is cargo theft, which the FBI defines as the stealing of any commercial shipment, moving by any

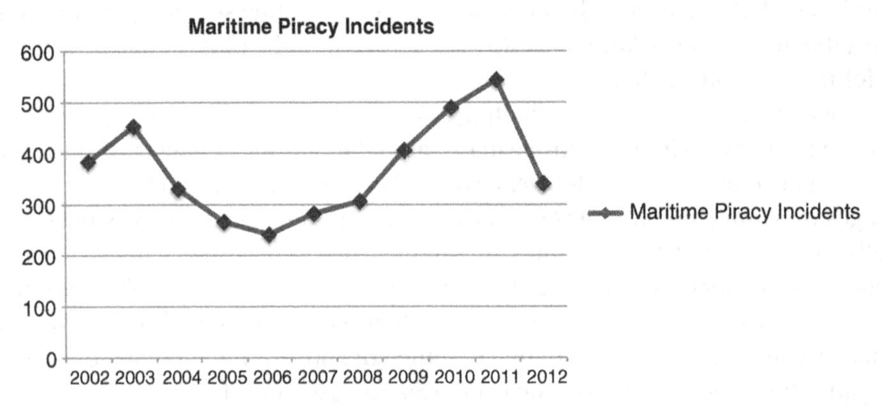

FIGURE 3.4 Maritime piracy incidents, 2002 to 2012. *(Source: International Maritime Organization. 2003–2013. Reports on acts of piracy and armed robbery against ships: Annual report—2002–2012. London.)*

transport means (including trucks, ships, rail cars, and airplanes), at any place between point of origin and final destination.

> *Because cargo theft statistics have never, until recently, been a separate reportable category in the [FBI's] Uniform Crime Report (UCR) and because many companies do not report cargo crimes (to avoid bad publicity, higher insurance rates, damage to reputation, embarrassment, and so on), the exact dollar losses are not known. Industry experts estimate that the total of cargo thefts rings up as much as $30 billion in losses each year. Cargo theft has many victims, from employees (i.e., drivers, warehouse workers) who can be hurt during an armed hijacking or robbery, to retailers who lose merchandise, to consumers who pay as much as 20% more to make up for cargo theft, to state and local governments who lose sales tax revenue, and even to insurance companies, manufacturers, and shipping companies.*
>
> *FBI, 2010*

As with terrorist incidents, the land mode experiences the most nonterrorist attacks, but the impact is usually highly localized, and the security responsibilities generally are held by local law enforcement authorities. In the U.S. highway sector, in addition to motor vehicle thefts, the FBI's UCR for 2010 identified 158,765 robberies on streets and highways and more than 2.1 million incidents of larceny from motor vehicles (FBI, 2011).

Mass transit systems also experience significant numbers of nonterrorist crimes. According to U.S. data collected by the Federal Transit Administration for 2003, there were:

- Two homicides, 23 rapes, 1027 robberies, 648 aggravated assaults, 348 burglaries, 6242 acts of larceny or petty theft, 1640 instances of vehicle thefts, and 14 cases of larceny involving mass transit rail systems
- Two homicides, two rapes, 369 robberies, 957 aggravated assaults, 79 burglaries, 1846 acts of larceny or petty theft, 149 instances of vehicle theft, and 9 cases of arson involving transit buses (Federal Transit Administration, 2005, pp. 120, 122).

Property theft at airports is the most common form of crime in commercial aviation, and includes baggage theft, burglaries, and vehicle theft from parking lots. Drug smuggling via commercial and cargo aircraft and airports is another major type of aviation crime (Price and Forrest, 2009, pp. 344–347).

Although increased security measures have produced a noticeable decline in terrorist attacks on aviation, the number of "air rage" incidents, which involve aggressive and disruptive behavior by unruly passengers during flight, has been on the increase in recent years. The International Air Transport Association (representing approximately 240 airlines worldwide) reported a 29% increase in such incidents in 2010 compared with the previous year, following a 27% rise between 2008 and 2009. The organization estimates there are currently an average of approximately 10,000 air rage cases per year internationally. Verbal or sometimes physical confrontations between the disruptive passenger and another passenger or member of the flight crew are the most typical examples (CNN, 2012a).

Risk Management

The essential characteristics of transportation systems—including their openness, scale, diversity, and importance in the global economy—create a vast array of vulnerabilities and targets, which terrorists and other criminal actors have continued to exploit in the post-9/11 era. Given such circumstances, it is impossible for those charged with securing transportation systems to eliminate all vulnerabilities or protect all targets. To address this challenge, the U.S. DHS and its component agencies have adopted a risk management approach, which is defined as "a planning methodology that outlines the process for setting goals and objectives; identifying assets, systems, and networks; assessing risks; prioritizing and implementing protection programs and resiliency strategies; measuring performance; and taking corrective action" (U.S. Department of Homeland Security, 2010, p. 83).

At the core of this risk management framework is risk assessment, for which DHS has adopted the formulation developed by GAO in 2001.

> *An effective risk management approach includes a threat assessment, a vulnerability assessment, and a criticality[5] assessment. A threat assessment identifies and evaluates threats based on various factors, including capability and intentions as well as the potential impact of an event. Nonetheless, we will never know whether we have identified every threat or event and may not have complete information about the threats that we have identified. Consequently, two other elements of the approach, vulnerability assessments and criticality assessments, are essential to prepare better against threats. A vulnerability assessment is a process that identifies weaknesses that may be exploited and suggests options to eliminate or mitigate those weaknesses. A criticality assessment is a process to systematically identify and evaluate an organization's assets based on a variety of factors, including the importance of its mission or function, whether people are at risk, and the significance of a structure or system. Criticality assessments are important because they provide a basis for prioritizing which assets require higher or special protection.*
>
> *GAO, 2001, p. 3*

As the DHS Transportation Sector Specific Plan (2010) notes, a number of factors complicate the risk assessment process:

- Uncertainty as to the types of threats to transportation systems
- Difficulty in predicting the likelihood and consequences of known threats
- The inestimable nature of unknown risks
- A wide spectrum of risks, which may require different assessment methodologies
- The greater complexity of risk assessments involving intentional, human threats (including terrorism)
- The creative and adaptive capabilities of terrorists and other human threats to transportation

[5]DHS and others currently use the term "consequence" in place of "criticality."

- Widely varying preparedness and response capabilities within the different transportation systems.

Such considerations "preclude any single assessment methodology" and necessitate the use of general principles rather than a standard formula. The resulting transportation risk assessments "examine the probability and consequences of an undesirable event affecting or resulting from, sector assets, systems, or networks" and include both risks to transportation systems and risks from those systems. The governmental and private sector components involved in transportation security, in principle, use the assessments to establish strategic priorities, inform the selection of appropriate countermeasures, develop risk reduction measures, and determine budget and resource allocation priorities (pp. 31–32).

Vulnerability Assessment

DHS defines transportation system vulnerabilities as "the physical, cyber, human, or operational attributes that render it open to exploitation or susceptible to hazards. Vulnerabilities are weaknesses that diminish preparedness to deter, prevent, mitigate, respond to, or recover from any hazard that could incapacitate or disable the infrastructure. The physical, cyber, and human elements of the sector are often co-dependent and additional vulnerabilities may result from their interaction." Vulnerability assessments also include a description of the protective measures currently in place and their effectiveness. By identifying the security weaknesses of potential targets and gauging the probability of a successful attack on them, these assessments seek to provide options for reducing the vulnerabilities (U.S. Department of Homeland Security, 2010, pp. 37–38).

Transportation systems in all modes share vulnerabilities to natural disasters (including hurricanes, tornadoes, and flooding), aging infrastructure, cyber risks (from transportation systems' growing interconnectivity with and reliance on computerized information and communications networks), and human factors (deliberate or inadvertent actions by transportation workers).

CONDITION OF U.S. INFRASTRUCTURE

ASCE 2013 Report Card

The American Society of Civil Engineers (ASCE) periodically issues an evaluation of U.S. infrastructure. The latest assessment, released in March 2013, indicated that some slight improvement had occurred since the last analysis in 2009, but that overall, American infrastructure warranted a grade of only D+ (compared with D in 2009). The following are the grades given to various transportation modes, along with additional comments from the report (with A = exceptional, B = good, C = mediocre, D = poor and F = failing).

- *Aviation: D. The FAA estimates that the national cost of airport congestion and delays was almost $22 billion in 2012. If current federal funding levels are maintained, the FAA anticipates that the cost of congestion and delays to the economy will rise from $34 billion in 2020 to $83 billion by 2040.*

- *Bridges: C+. Over 200 million trips are taken daily across deficient bridges in the nation's 102 largest metropolitan regions. In total, one in nine of the nation's bridges are rated as structurally deficient, while the average age of the nation's 607,380 bridges is currently 42 years. The Federal Highway Administration estimates that to eliminate the nation's bridge deficient backlog by 2028, we would need to spend $20.5 billion annually, while only $12.8 billion is being spent currently.*
- *Inland waterways: D-. In many cases, the inland waterways system has not been updated since the 1950s, and more than half of the locks are over 50 years old. Barges are stopped for hours each day with unscheduled delays, preventing goods from getting to market and driving up costs. There is an average of 52 service interruptions per day throughout the system.*
- *Ports: C. While port authorities and their private sector partners have planned over $46 billion in capital improvements from now until 2016, federal funding has declined for navigable waterways and landside freight connections needed to move goods to and from ports.*
- *Rail: C+. Both freight and passenger rail have been investing heavily in their tracks, bridges, and tunnels as well as adding new capacity for freight and passengers. In 2010 alone, freight railroads renewed the rails on more than 3100 miles of railroad track, equivalent to going from coast to coast. Since 2009, capital investment from both freight and passenger railroads has exceeded $75 billion, actually increasing investment during the recession when materials prices were lower and trains ran less frequently.*
- *Roads: D. Forty-two percent of America's major urban highways remain congested, costing the economy an estimated $101 billion in wasted time and fuel annually. While the conditions have improved in the near term and federal, state, and local capital investments increased to $91 billion annually, that level of investment is insufficient and still projected to result in a decline in conditions and performance in the long term.*
- *Transit: D. Unlike many U.S. infrastructure systems, the transit system is not comprehensive, as 45% of American households lack any access to transit, and millions more have inadequate service levels... Although investment in transit has increased, deficient and deteriorating transit systems cost the U.S. economy $90 billion in 2010, as many transit agencies are struggling to maintain aging and obsolete fleets and facilities amid an economic downturn that has reduced their funding, forcing service cuts and fare increases (American Society of Civil Engineers, 2013).*

The poor condition of many of these systems creates security vulnerabilities by increasing their susceptibility to damage and reducing their resiliency and ability to aid in recovery from attacks.

The wide geographic dispersal, international character, and complexity of the maritime sector leave its ports and vessels particularly exposed to attack. More specifically, the advent of containerized shipping—which accounts for an estimated 90% of the world's nonbulk cargo and offers high-value goods in closed containers of a standardized size—has offered a tempting target for large-scale theft (McNicholas, 2008, pp. 133, 145).

Mass transit systems provide would-be attackers with open, "fast-paced operations with numerous entry, transfer, and egress points, to transport a high volume of passengers each day.... Multiple stops and interchanges lead to high passenger turnover, which is difficult to monitor effectively" (U.S. Department of Homeland Security, 2010, p. 220).

Passenger rail systems are easily penetrated, have high concentrations of people, and are susceptible to relatively simple means of attack. Freight rail is often used to transport hazardous materials and other dangerous cargo, with an estimated 50% of hazardous materials in the United States transported by this mode. Passenger and freight rail systems each operate across hard-to-patrol rural areas as well as densely populated urban centers, offering a variety of potential points of attack (Riley, 2004, pp. 4–6).

A 2001 *Surface Transportation Vulnerability Assessment* by the U.S. Department of Transportation reported:

> *Highways represent both the most important single surface mode—when looking at the total volume of passengers and freight together—and also the most robust and resilient mode.... The most vulnerable segments of this network appear to be bridges and tunnels, due to their accessibility, the expense and difficulty of replacing them, and their concentration of several routes into a single infrastructure segment.*
>
> *U.S. Department of Transportation, Research and Special Programs Administration, 2001, p. x*

Figure 3.5 shows the entrance to the Holland Tunnel in New York City.

Similar to freight rail, pipelines are vulnerable because of the hazardous materials they transport and the broad range of their networks through both cities and countryside (U.S. Department of Homeland Security, 2010, p. 318).

FIGURE 3.5 Entrance to the Holland Tunnel in New York City. *(Source: National Park Service, http://en.wikipedia.org/wiki/File:Haer_hollandtunnel.jpg.)*

Some of the vulnerabilities in commercial aviation were demonstrated by the tragic events of September 11, 2001. A number of actions have been taken to lessen the security weaknesses revealed then in passenger prescreening and screening and onboard security. However, certain vulnerabilities persist even in those areas, as well as in the remaining security layers of intelligence, airport access control, checked bag screening, and air cargo (Johnstone, 2006, pp. 72–83).

The GAO identified a number of potential weaknesses at general aviation airports with respect to access controls and the screening of passengers and cargo (GAO, 2011, pp. 1–4, 8). (Although the TSA provides security guidance to general aviation operators, private operators are responsible for most security functions.)

Threat Assessment

The threat component in DHS's risk management assessment is "an individual, entity, or action that has the potential to deliberately harm life and/or property." The assessment seeks to ascertain the likelihood that the threat will result in an actual attack. The threats are organized into two broad categories: the originating entities and their targets and tactics (U.S. Department of Homeland Security, 2010, pp. 32, 134).

The primary focus of transportation security threat assessments of the sources of potential attacks is terrorism, in which the threat "is determined by an assessment of terrorist capabilities and intents as derived from intelligence analyses. Terrorism threat assessments must consider the degree of uncertainty associated with estimates of capability and intent." The U.S. State Department maintains a listing of "foreign terrorist organizations, (FTOs)" which are foreign organizations that engage in terrorist activity or retain the capability and intent to engage in such activity and "threaten the security of U.S. nationals or the national security of the United States." As of September 2012, there were 52 organizations designated as FTOs (Table 3.3) (DHS, 2010, p. 38; U.S. Department of State, 2012b).

Table 3.3 U.S. State Department Listing of Foreign Terrorist Organizations (as of September 2012)

Organization Name	Base/Area of Operation
Abdallah Azzam Brigades (AAB)	Lebanon; Arabian Peninsula/Israel; Persian Gulf
Abu Nidal Organization (ANO)	Palestine/Middle East; Europe (inactive)
Abu Sayyaf Group (ASG)	Southern Philippines/Philippines; Malaysia
Al-Aqsa Martyrs Brigade (AAMB)	Palestine/Israel
al Qaeda (AQ)	Middle East/International
al Qaeda in the Arabian Peninsula (AQAP)	Yemen; Saudi Arabia/Yemen; International
al Qaeda in Iraq (AQI)	Iraq/Iraq; Jordan
al Qaeda in the Islamic Maghreb (AQIM)	Algeria/Algeria; Mali; Mauritania; Niger
al-Shabaab	Southern Somalia/Somalia; Uganda
Ansar al-Dine (AAD)	Mali/Mali
Ansar al-Islam (AAI)	Iraq/Iraq
Army of Islam (AOI)	Gaza/Israel; Egypt; Gaza

Table 3.3 U.S. State Department Listing of Foreign Terrorist Organizations (as of September 2012) *(cont.)*

Organization Name	Base/Area of Operation
Asbat al-Ansar (AAA)	Southern Lebanon/Lebanon; Iraq
Aum Shinrikyo (AUM)	Japan/Japan
Basque Fatherland and Liberty (ETA)	Northern Spain/Spain; France
Communist Party of the Philippines/New People's Army (CPP/NPA)	Philippines/Philippines
Continuity Irish Republican Army (CIRA)	Northern Ireland/Northern Ireland; Ireland
Gama'a al-Islamiyya (Islamic Group) (IG)	Southern Egypt/Egypt (inactive)
HAMAS	Palestine/Palestine; Israel
Haqqani Network (HQN)	Afghanistan/Afghanistan
Harakat ul-Jihad-i-Islami (HUJI)	Afghanistan/Afghanistan; India; Pakistan
Harakat ul-Jihad-i-Islami/Bangladesh (HUJI-B)	Bangladesh/Bangladesh; India
Harakat ul-Mujahidin (HUM)	Pakistan/Kashmir; Afghanistan
Hizballah	Lebanon/Lebanon; Israel; Argentina
Indian Mujahedeen (IM)	India/India
Islamic Jihad Union (IJU)	Pakistan/Afghanistan; other Central Asia
Islamic Movement of Uzbekistan (IMU)	Central Asia/Afghanistan; other Central Asia
Jaish-e-Mohammed (JEM)	Pakistan/Kashmir; India; Pakistan
Jemaah Anshorut Tauhid (JAT)	Indonesia/Indonesia
Jemaah Islamiya (JI)	Indonesia/Indonesia; Philippines; Malaysia
Jundallah	Southeastern Iran/Southeastern Iran
Kahane Chai (Kach)	Israel/Israel; Palestine
Kata'ib Hizballah (KH)	Iraq/Iraq
Kurdistan Workers Party (PKK) (Kongra-Gel)	Northern Iraq/Turkey; Iraq; Europe
Lashkar-e Tayyiba (LeT)	Pakistan/Kashmir; India
Lashkar i Jhangvi (LJ)	Pakistan/Pakistan; Afghanistan
Liberation Tigers of Tamil Eelam (LTTE)	Northern Sri Lanka/Sri Lanka; India
Libyan Islamic Fighting Group (LIFG)	Libya/Libya (inactive)
Moroccan Islamic Combatant Group (GICM)	Western Europe/Morocco; Europe (inactive)
National Liberation Army (ELN)	Colombia/Colombia
Palestine Liberation Front (PLF)	Palestine/Palestine; Israel; Europe
Palestinian Islamic Jihad (PIJ)	Gaza/Israel; Palestine
Popular Front for the Liberation of Palestine (PFLP)	Palestine/Israel; Palestine
PFLP-General Command (PFLP-GC)	Palestine/Israel; Europe
Real Irish Republican Army (RIRA)	Northern Ireland/Great Britain; Northern Ireland; Ireland
Revolutionary Armed Forces of Colombia (FARC)	Colombia/Colombia
Revolutionary Organization 17 November (17N)	Greece/Greece
Revolutionary People's Liberation Party/Front (DHKP/C)	Turkey/Turkey
Revolutionary Struggle (RS)	Greece/Greece
Shining Path (SL)	Peru/Peru
Tehrik-e Taliban Pakistan (TTP)	Pakistan/Pakistan; Afghanistan; United States
United Self Defense Forces of Colombia (AUC)	Colombia/Colombia (inactive)

(*Sources*: U.S. Department of State. 2012a, July 31. Country reports on terrorism: Chapter 6. Foreign terrorist organizations. http://www.state.gov/j/ct/rls/crt/2011/195553.htm (accessed 10.21.14.); U.S. Department of State. 2012b, September 28. Foreign terrorist organizations. http://www.state.gov/j/ct/rls/other/des/123085.htm (accessed 10.21.14.))

Nonterrorist threats include individuals and organizations engaging in criminal activities of all kinds on or against transportation systems, including pirates and hijackers; drugs and weapons smugglers; and thieves of vessels, vehicles, passengers, or cargo, among others.

In developing the other components of the threat assessment, possible targets may be found in the transportation system's facilities (e.g., ports, airports, terminals, and stations), means of conveyance (e.g., roads, tracks, waterways, and pipelines), vessels and vehicles (e.g., ships, trucks, trains, and aircraft), supporting infrastructure (e.g., power substations, locks, and dams), control and information systems (e.g., signal and navigation tracking systems), and passengers or cargo. Potential tactics are generally derived from the historical record and expert opinion. The combination of targets and tactics produces an almost limitless set of potential attacks that would seriously complicate the process of developing a meaningful analysis. Therefore, transportation security risk managers frequently use a finite set of illustrative scenarios in their threat assessments. For example, a 1998 Department of Transportation threat and vulnerability assessment considered a number of scenarios for attacks on land transportation systems, including:

- Series of small explosives on highway bridge
- Bomb(s) detonated at pipeline storage facility
- Simultaneous attacks on ports
- Attack on passenger vessel in port
- Shooting in rail station
- Bus bombing
- Bomb detonated on train in rail station
- Anthrax release in transit station
- Physical attack on railcar carrying a toxic chemical
- Cyber attack on pipeline automated control system (National Research Council, 2000, p. 15)

■ ■ Critical Thinking ■

Responding to the terrorist threat has dominated transportation security efforts even before 9/11. Do you think this has been appropriate? Why or why not?

Consequence Assessment

Consequence (or criticality) assessments examine "the effect of an event, incident, or occurrence" and consider "health and human safety, economic impact, national security, and cross-sector effects" (U.S. Department of Homeland Security, 2010, pp. 32, 37).

The evaluation of the potential consequences of a terrorist or other intentional action against transportation systems is complicated by the difficulty in measuring intangible and secondary effects. According to a 2010 report by the National Research Council (NRC), "DHS's consequence analyses tend to limit themselves to deaths; physical damage;

first-order economic effects; and in some cases, injuries and illness. Other effects, such as interdependencies, business interruptions, and social and psychological ramifications, are not always modeled, yet for terrorism events, these could have more impact than consequences that are currently included" (National Research Council, 2010, p. 51).

One example of a simple consequence assessment model is the Coast Guard's risk management and analytical tool for port operators. Its consequence component divides port facilities into one of three categories based on the probable impact of a successful attack. "Facilities that transfer, store, or otherwise handle certain dangerous cargoes" are ranked as the highest consequence level (level 3) followed by those that handle other major cargoes, receive passenger vessels capable of carrying more than 150 passengers, or receive vessels on international voyages (level 2), with all other facilities receiving the lowest consequence assessment (level 1) (McNicholas, 2008, pp. 321–322).

DIGGING DEEPER

EVALUATING RISK ASSESSMENT AT THE DEPARTMENT OF HOMELAND SECURITY

2010 Review by the National Research Council

In 2008, the U.S. Congress directed the NRC to assess risk management within the DHS and its constituent agencies. Of the approximately 60 risk models and processes identified by the DHS, the NRC narrowed its focus to six examples that it would examine in detail. One of these was of particular relevance to transportation security: the risk assessments covering critical infrastructure and key resources (CIKR) (pp. 15, 18–19).

The NRC report acknowledges the considerable difficulties involved in conducting risk analyses for human-made threats, especially those involving terrorism.

> *Risk analysis for natural hazards is based on a foundation of data. For terrorism risk analysis, neither threats nor consequences are well characterized by data. Risk analysis for terrorism involves an open rather than a closed system: virtually anyone can be a participant (ranging from intentionally malevolent actors to bystanders who may respond in ways for better or worse), and parts of systems may be used in ways that are radically different from those for which they were designed (e.g., aircraft as weapons rather than means of transportation). Also, terrorism, unlike natural disasters, involves intentional actors. Not only are many terrorist threats low-likelihood events, but their frequency is evolving rapidly over time as terrorists observe and respond to defenses and to changing political conditions. Thus, it will rarely be possible to develop statistically valid estimates of attack frequencies (threat) or success probabilities (vulnerability) based on historical data.*
>
> *National Resource Council, 2010, pp. 46–47*

In its examination of the CIKR risk analyses, the NRC found that DHS does have in place processes for developing threat, vulnerability, and consequence assessments but noted the following problems:

- *Threat analyses: DHS does strive to get the best and most relevant terrorism experts to assess threats. However, regular, consistent access to terrorism experts is very difficult. Due to competing priorities at other agencies, participation is in reality a function of who is available… Rotation of subject matter experts also puts a premium on documenting, testing, and validating [their] assumptions… To provide the best possible analyses of terrorism threats, DHS has a goal to incorporate more state and local threat information into its risk assessments and has started numerous outreach programs… Despite these efforts, information sharing between the national and local levels and among state and local governments still faces many hurdles. The most serious challenges are security policies and clearances, common standards for reporting and data tagging, numbers and skill levels of analysts at the state and local levels, and resources to mature the information technology architecture.*

- *Vulnerability analyses: DHS's work in support of critical infrastructure protection has surely instigated and enabled more widespread examination of vulnerabilities, and this is a positive move for homeland security. The Department's process for conducting vulnerability analyses appears quite thorough within the constraints of how it has defined "vulnerability…" To date, it seems that vulnerability is heavily weighted toward site-based physical security considerations… However, vulnerability is much more than physical security; it is a complete systems process consisting at least of exposure, coping capability, and longer term accommodation or adaptation. Exposure used to be the only thing people looked at in a vulnerability analysis; now there is consensus that at least these three dimensions have to be considered. The [NRC] committee did not hear these sorts of issues being raised within DHS.*
- *Consequence analyses: The consequence analysis done in support of infrastructure protection… is carried out with skill… The [NRC] committee was concerned that none of DHS's consequence analyses—including, but not limited to, the analyses done in support of infrastructure protection—address all of the major impacts that would come about from a terrorist attack. Consequences of terrorism can range from economic losses to fatalities, injuries, illnesses, infrastructure damage, psychological and emotional strain, disruption to our way of life, and symbolic damage (e.g., an attack on the Statue of Liberty, Washington Monument, or Golden Gate Bridge).*

National Research Council, 2010, pp. 58–65

Considering the challenges to risk assessment efforts identified by the NRC, those cited above from the DHS Transportation Sector Specific Plan, and other potential limitations (political, economic, and so on), what factor or factors pose the greatest limitation to the implementation of effective risk management for transportation security? How can or should those factors be overcome or minimized?

Conclusion

Transportation systems are accessible, extensive, diverse (in kind, ownership, and usage), and integral to the operation of society in general and the economy in particular. Even in the more security-conscious post-9/11 world, such features continue to make these systems the target of attacks by terrorists and others while complicating the task of protecting them.

With the framework used by the DHS, the transportation sector may be divided into six major modes, four of which are components of land transportation: (1) maritime; (2) mass transit and passenger rail; (3) freight rail; (4) highways, bridges, and tunnels; (5) pipelines; and (6) aviation.

The post-2001 terrorist attacks on transportation largely resembled those that had come before, with the land mode experiencing a substantial majority of the incidents and fatalities and the maritime sector the fewest. The major exception has been in aviation, where, no doubt in response to the substantial increase in security attention (including intelligence), the number of successful attacks has been much reduced compared with previous periods. Although none was on the same scale as the 9/11 hijackings in the United States, certain terrorist attacks produced substantial impacts, especially in the countries that experienced them:

- The 2002 attack on an oil tanker off the coast of Yemen that did severe damage to the Yemeni economy
- The 2004 bombing of a Filipino ferry, which killed more than 100 passengers
- The 2004 bombings on the commuter rail system in Madrid, Spain that resulted in almost 200 fatalities and more than 1800 injuries
- The 2005 bombings on the mass transit system in London, England that left more than 50 dead and over 700 wounded
- The 2006 bombings on the Mumbai, India passenger rail system, which produced more than 180 deaths and more than 80 injuries
- The 2008 sea-based attack on targets in Mumbai that killed more than 160 and injured more than 300

As was true before 2001, the United States has experienced few terrorist incidents since then. According to the GTD, just five attacks occurred between 2002 and 2013: the 2002 shootings at the El Al ticket counter at Los Angeles International Airport that left two passengers dead; a minor 2009 attack on a railroad facility in Illinois; the Christmas 2009 attempted suicide bombing of Northwest Flight 253 in which the perpetrator ignited the explosive device but it failed to detonate; the 2013 discovery and defusing of an explosive device at a Nebraska airport; and the 2013 shootings of TSA agents at Los Angeles International Airport. Although not successful, the attempted 2009 attack on the Northwest aircraft was one of several cases involving U.S. aviation in which the attack failed but the attempt produced changes in U.S. security measures.

Transport systems continued to receive far more attacks based on economic and other motives than from terrorism. Although the latter usually generated more casualties, media attention, and security responses, the nonterrorist incidents also had a significant impact, especially economically.

Faced with the need to defend vast and largely open transportation networks in which it is impossible to protect all targets or remove all vulnerabilities, government authorities have sought to use a risk management system, which attempts to focus security attention and resources on the most vulnerable targets and those whose disruption or destruction would produce the most harmful consequences. In implementing this approach, the U.S. DHS has adopted a framework that assesses vulnerabilities (attributes that make a system susceptible to attack), threats (the terrorists and others likely to target transportation systems and capable of causing harm to life or property), and consequences (the likely health, economic, and other effects of a successful attack).

A 2010 review by the NRC found that although DHS has made progress in its use of risk management principles, problems remain in its capability for assessing risks. Part of the difficulty stems from the nature of the terrorist threat that rarely materializes (thus producing limited historical data) and involves thinking adversaries whose methods can and do evolve. The NRC report pointed to further constraints, including DHS's challenges in gaining sufficient access to terrorism experts, facilitating information sharing with the state and local level, and factoring in such important considerations as the coping and adaptive capacity of transportation systems and the full range of physical and psychological impacts that may result from a terrorist strike.

Discussion Questions

1. What key characteristics of transportation systems have an impact on security measures?
2. Name and briefly describe the six major transportation modes as defined by DHS.
3. Why have transportation systems continued to be major targets for terrorists?
4. Describe the major post-9/11 trends in terrorist and other attacks on maritime, land, and aviation transportation systems.
5. What is risk management, and how has it been applied to transportation security in the United States?
6. Describe the features and weaknesses of U.S. transportation vulnerability, threat, and consequence assessments.

References

American Public Transportation Association. 2012. Public transportation fact book. Washington, DC.

American Society of Civil Engineers. 2013. 2013 Report Card for America's Infrastructure. <http://www.infrastructurereportcard.org/a/> (accessed 10.21.14.)

Amtrak, 2012. National fact sheet: FY 2011. Washington, DC.

Archick, K., July, 2006. European approaches to homeland security and counterterrorism. Congressional Research Service, Washington, DC.

Associated Press. October, 2007. The 2004 Madrid bombings. <http://www.guardian.co.uk/world/2007/oct/31/spain.menezes> (accessed 10.21.14.)

Association of American Railroads. 2012. An overview of America's freight railroads. Washington, DC.

BBC News. 2004. Madrid Attacks Timeline. <http://newsvote.bbc.co.uk/mpapps/pagetools/print/news.bbc.co.uk/2/hi/europe/3504912.stm> (accessed 10.21.14.)

Carafano, J., Bucci, S., Zuckerman, J., April, 2012. Fifty terror plots foiled since 9/11: The homegrown threat and the long war on terrorism. The Heritage Foundation, Washington, DC.

Chalk, P., 2008. The maritime dimension of international security: Terrorism, piracy, and challenges for the United States. RAND, Santa Monica, CA.

CNN. June, 2012a. Air rage: Passengers 'quicker to snap.' http://cpf.cleanprint.net/cpf/cpf?action(print&type(filePrint&key(cnn&url(http:%3A%2F (accessed 10.21.14.)

CNN. November, 2012b. India executes last gunman from Mumbai attacks. <http://www.cnn.com/2012/11/20/world/asia/india-mumbai-execution> (accessed 10.21.14.)

FBI. November, 2010. Inside cargo theft: A growing, multi-billion-dollar problem. <http://www.fbi.gov/news/stories/2010/november/cargo_111210> (accessed 10.21.14.)

FBI. September, 2011. Uniform Crime Report: Crime in the United States, 2010. Washington, DC.

Federal Transit Administration. December, 2005. Transit security & safety statistics & analysis: 2003 annual report. Washington, DC.

GAO. October, 2001. Homeland security: A risk management approach can guide preparedness efforts. Washington, DC.

GAO. March, 2010. TSA is increasing its procurement and deployment of the advanced imaging technology, but challenges to this effort and other areas remain. Washington, DC.

GAO. May, 2011. General aviation: Security assessments at selected airports. Washington, DC.

Government of India. January, 2009. Mumbai terrorist attacks (Nov. 26-29, 2008). <http://www.hindu.com/nic/dossier.htm> (accessed 10.21.14.)

Government of the United Kingdom. May, 2006. Report of the official account of the bombings in London on 7th July 2005. <http://www.official-documents.gov.uk/document/hc0506/hc10/1087/1087.pdf> (accessed 10.21.14.)

Hamilos, P., October, 2007. The worst Islamist attack in European history. *The Guardian.* <http://www.guardian.co.uk/world/2007/oct/31/spain> (accessed 10.21.14.)

Hamilos, P., Tran, M., October, 2007. 21 guilty, seven cleared over Madrid train bombings. <http://www.guardian.co.uk/world/2007/oct/31/spain.marktran.> (accessed 10.21.14.)

International Maritime Organization. 2003–2013. Reports on acts of piracy and armed robbery against ships: annual report—2002–2012. London.

Jenkins, B.M., Butterworth, B.R., March, 2010. Explosives and incendiaries used in terrorist attacks on public surface transportation: A preliminary empirical examination. Mineta Transportation Institute, San Jose, CA.

Jenkins, B.M., Trella, J., April, 2012. Carnage interrupted: An analysis of fifteen terrorist plots against public surface transportation. Mineta Transportation Institute, San Jose, CA.

Johnstone, R.W., 2006. 9/11 and the Future of Transportation Security. Praeger, Westport, CT.

Johnstone, R.W., Winter 2007. Not safe enough: Fixing transportation security. Issues in Science and Technology, 51–60.

Kauppila, J., 2011. Ten stylized facts about household spending on transport. Joint Transport Research Centre of the OECD and the International Transport Forum, Paris.

Matthew, T., February, 2009. Policy brief: India's confrontation with terror: Need for bold initiatives. Institute for Defence Studies and Analyses, New Delhi.

McNicholas, M., 2008. Maritime Security. Butterworth-Heinemann/Elsevier, Burlington, MA.

National Consortium for the Study of Terrorism and Responses to Terrorism (START). n.d. Global terrorism database. <http://www.start.umd.edu/gtd.> (accessed 10.21.14.)

National Research Council. 2000. Improving surface transportation: A research and development strategy. The National Academies Press, Washington, DC.

National Research Council. 2010. Review of the Department of Homeland Security's approach to risk analysis. The National Academies Press, Washington, DC.

Price, J.C., Forrest, J.S., 2009. Practical aviation security: Predicting and preventing future threats. Butterworth-Heinemann/Elsevier, Burlington, MA.

Rabasa, A., Blackwill, R., Chalk, P., Cragin, K., Fair, C., Shestak, N., Tellis, A., 2009. The lessons of Mumbai. Rand, Santa Monica, CA.

Reinares, F., October, 2008. After the Madrid bombings: internal security reforms and the prevention of global terrorism in Spain. (Translated from Spanish). <http://www.realinstitutoelcano.org> (accessed 10.21.14)

Riley, J., March, 2004. Statement before the U.S. Senate Committee on commerce, science, and transportation [Presentation]. Washington, DC.

Rodrigue, J., Comtois, C., Slack, B., 2006. The geography of transport systems. Routledge, New York.

Sherman, R.B. n.d. Seaport governance in the United States and Canada. American Association of Port Authorities, Alexandria, VA. Retrieved from http://www.aapa-ports.org/files/PDFs/governance_uscan.pdf. (accessed 10.21.14.)

Taylor, B.D., 2005. Designing and operating safe and secure transit system: Assessing current practices in the United States and abroad. Mineta Transportation Institute, CA, San Jose.

The White House. 2009. Summary of the White House review of the December 25 2009 attempted terrorist attack. Washington, DC.

Transportation Research Board. 2002. Deterrence, protection, and preparation: the new transportation security imperative. Washington, DC.

U.S. Department of Homeland Security. 2010. Transportation systems sector-specific plan: an annex to the National Infrastructure Protection Plan. Washington, DC.

U.S. Department of Homeland Security. February 2011. Budget-in-brief fiscal year 2012. Washington, DC.

U.S. Department of State. July, 2012a. Country reports on terrorism: chapter 6. foreign terrorist organizations. <http://www.state.gov/j/ct/rls/crt/2011/195553.htm> (accessed 10.21.14.)

U.S. Department of State. September, 2012b. Foreign terrorist organizations. <http://www.state.gov/j/ct/rls/other/des/123085.htm> (accessed 10.21.14.)

U.S. Department of Transportation. 2013. Research Innovative Technology Administration. Pocket guide to transportation, Washington, DC.

U.S. Department of Transportation, Research and Special Programs Administration. 2001. Surface transportation vulnerability assessment: General distribution version. Washington, DC.

U.S. Senate Committee on Appropriations. April, 2011. Summary: Homeland Security Subcommittee: FY2011 continuing resolution. Washington, DC.

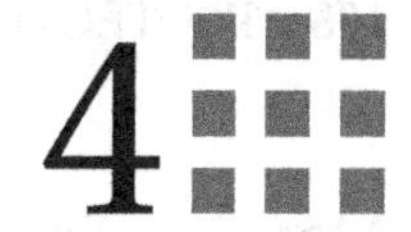

Transportation Security Roles and Responsibilities

CHAPTER OBJECTIVES:

In this chapter, you will learn about the organization of transportation security, including:

- International institutions
- The process of defining transportation security roles in the United States
- The organization of transportation security in the United States at the federal, state, and local levels and within the private sector
- The evolution of the concept of "homeland security"
- The organization of transportation security in Canada, India, and the United Kingdom

Introduction

The extent and diversity of transportation systems is reflected in the wide array of entities involved in their protection. International, national, regional, state or provincial and local institutions all play a part in transportation security, and although the key organizations involved in making and implementing transportation security policy are usually governmental bodies, private sector groups (including representatives of all key stakeholders) have a considerable impact as well.

International

The two most influential international organizations with respect to transportation security—the International Maritime Organization (IMO) and the International Civil Aviation Organization (ICAO)—each operate as a specialized agency of the United Nations. Because of the more localized focus of land transportation systems, outside of Europe there has been less emphasis on transnational arrangements in that sector, although that has begun to change in the post-9/11 world. Within all transportation modes, private sector associations exist and serve to make sure that the interests and concerns of their constituent members are taken into account in the development of transportation security measures.

Maritime Security

The IMO was established by the United Nations in 1948,[1] and its objective is to promote "safe, secure and efficient shipping on clean oceans." The IMO is headquartered in London, England and has 170 member nations (Figure 4.1). Its primary work is carried out by two committees, the Marine Environment Protection Committee (responsible for prevention and control of pollution caused by ships) and the Maritime Safety Committee (responsible for safety and security, including piracy and armed robbery against ships).

The IMO first began to focus on security issues after the *Achille Lauro* hijacking of 1985 and increased its security efforts after 9/11. Today there are three main instruments for promoting maritime security at the international level, all of which were developed through the work of the IMO:

- The Convention for the Suppression of Unlawful Acts against the Safety of Maritime Navigation (SUA Convention), which was first adopted in 1988 and revised in 2005. (The 2005 amendments expanded the Convention's coverage to include carriage or use of weapons of mass destruction and terrorist actions.)
- The 2002 Amendments to the International Convention for the Safety of Life at Sea (SOLAS), which created a new regulatory system for international maritime security.
- The 2002 International Ship and Port Facility Security Code (ISPS), whose purpose is "to provide a standardized, consistent framework for evaluating risk, enabling Governments to offset changes in threat with changes in vulnerability for ships and port facilities through determination of appropriate security levels and corresponding security measures."

In addition, the IMO has undertaken a number of efforts to combat piracy, including facilitating the 2004 "Regional Co-operation Agreement on Combating Piracy and Armed Robbery against ships in Asia" and the 2009 Djibouti "Code of Conduct concerning the repression of piracy and armed robbery against ships in the Western Indian Ocean and the Gulf of Aden" (largely directed against the Somali pirates), as well as providing ongoing guidance for governments, ship operators and crews on preventing and suppressing piracy (International Maritime Organization, 2013, pp. 2, 7–9).

The 2002 SOLAS Amendments and the ISPS together spell out security obligations and responsibilities for member governments, as well as for companies, ships, port facilities, and crews.

- Member governments are required to set security levels for and provide relevant information to all ships flying their flag, port facilities in their territory and foreign-flag shipping entering their territory, and issue and renew International Ship Security Certificates to ships flying their flag after verifying compliance with the ISPS.
- Companies and ships must have their compliance with the ISPS Code verified and certified by their host government, comply with the security requirements of the port they enter if those requirements are more stringent than their flag country's, designate

[1] It was known as the Inter-Governmental Consultative Organization before 1982.

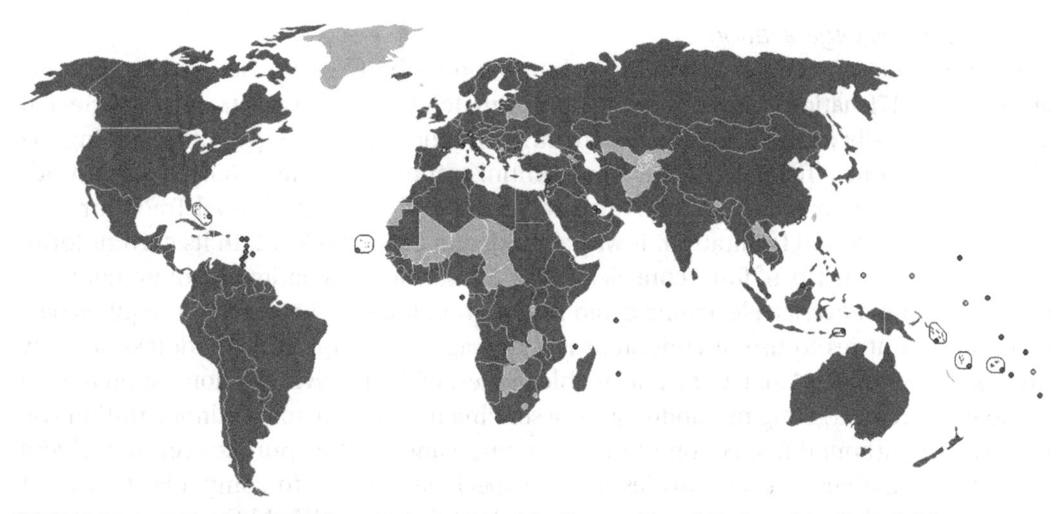

FIGURE 4.1 International Maritime Organization membership. *(Source: Alinor, http://en.wikipedia.org/wiki/File:International_Maritime_Organization.png.)*

at least one Company Security Officer per company and Ship Security Officer per ship, and provide them with the necessary support to fulfill their duties, and each ship must carry a ship's security assessment, a government-approved ship's security plan, and a government-issued certificate of compliance with the ISPS.

- Port facilities are mandated to have a government-approved port security facility plan (based on a port facility security assessment) and a port facility security officer.
- Ship and port personnel with security duties are required to understand their security duties, receive relevant security training, and participate in required periodic security drills.
- Other ship and port personnel must understand their role in security plans and participate in periodic security drills (Bennett, 2008, pp. 162–169).

Although representing a significant expansion and improvement over the pre-2001 international maritime security program, the existing IMO-supported system still shares its predecessor's reliance on member nations for enforcement, a reliance that imposes significant limitations and produces substantial variations in the quality of maritime security. For example, a 2012 Government Accountability Office (GAO) report stated: "DHS and its component agencies face inherent challenges and limitations working with international partners because of sovereignty issues," citing difficulties encountered by the Coast Guard in assessing foreign port security (because some countries insisted on visiting and assessing a sample of U.S. ports before permitting the Coast Guard assessments) and Customs and Border Protection in fully staffing foreign ports under the Container Security Initiative (because of host government requirements) (GAO, 2012, p. 21).

In addition to the IMO, a number of other entities play significant roles in international maritime security.

World Customs Organization

The World Customs Organization is an international body composed of the customs authorities of 179 nations who account for approximately 99% of all world trade. It was created in 1952 (originally as the Customs Co-operation Council) with the goal of improving the effectiveness and efficiency of customs administration. In June 2005, the group adopted the SAFE Framework of Standards to Secure and Facilitate Global Trade for supply chain security and facilitation. It was amended in 2007 and 2012. In its current form, the framework consists of four "core elements:" (1) harmonizes national requirements for the provision of advance electronic cargo information for all shipments; (2) requires participating countries to use "a consistent risk management approach to address security threats;" (3) stipulates "that at the reasonable request of the receiving nation, based upon a comparable risk targeting methodology, the sending nation's Customs administration will perform an outbound inspection of high-risk cargo and/or transport conveyances;" and (4) suggests that customs authorities provide specified benefits to companies that meet minimum supply chain security standards and best practices (World Customs Organization, 2012, pp. 1–4).

International Maritime Bureau

The International Maritime Bureau (IMB) was established in 1981 as a specialized, nonprofit division of the International Chamber of Commerce. Its mission is "to protect the integrity of international trade by seeking out fraud and malpractice." It authenticates trade finance documents and investigates and reports on credit fraud, cargo theft, and ship finance fraud, but perhaps its most important function in recent years has been the 1992 establishment of the *IMB Piracy Reporting Centre* in Kuala Lumpur, Malaysia. The Centre has become a leading source of information about maritime piracy and "maintains a round-the-clock watch on the world's shipping lanes, reporting pirate attacks to local law enforcement and issuing warnings about piracy hotspots to shipping" (ICC Commercial Crime Services, n.d.).

International Chamber of Shipping

The International Chamber of Shipping "is the principal international trade association for the shipping industry, representing all sectors and trades." Its membership accounts for two-thirds of merchant shipping tonnage. Among the organization's objectives is to "promote properly considered international regulation of shipping and oppose unilateral and regional action by governments" (International Chamber of Shipping, n.d.).

World Shipping Council

The World Shipping Council was created in 2000 by the CEOs of the world's major containership companies to work with international organizations and national governments "to develop new laws, regulations and programs designed to better secure international maritime commerce and the thousands of supply chains that importers and exporters around the world depend upon" (World Shipping Council, n.d.).

International Transport Workers' Federation: Seafarers, Dockers, and Inland Navigation Workers

The International Transport Workers' Federation (ITF) is an umbrella organization representing labor unions around the world whose members work in various transportation sectors. In the maritime mode, ITF-affiliated unions have more than 600,000 members who are seafarers (mainly ship crews), more than 350,000 who are dock workers, and more than 46,000 who are inland navigation workers. The role of the ITF is to "improve conditions for seafarers [and dock and inland navigation workers] of all nationalities and to ensure adequate regulation of the shipping industry." With regard to security, the ITF commissioned a survey of its member unions in 2005 about the then newly established International Ship and Port Facility Security Code. Although a majority of the respondents indicated that the code had improved security, 86% thought its implementation had resulted in extra workloads and impaired crew performance, and 58% reported a denial of adequate shore leave, especially for those entering U.S. ports. A 2006 ITF background paper commented, "Whilst the need to improve security in the maritime sector is recognized and not disputed, it must be implemented in such a way as to safeguard the human rights of seafarers" (International Transport Workers' Federation, n.d.c.; 2006, pp. 33-34).

Land Transportation Security

There is no international governmental body, nor set of treaties or other international obligations, for land transportation security comparable to those for the maritime and aviation modes. Part of the reason for this is the lesser role of land transportation in international trade and travel, and part stems from the greater fragmentation of ownership and regulation. A partial exception is in rail transportation (both freight and passenger) in which an early need for coordination across national borders arose (with respect to tracks, vehicles, and schedules), primarily in Europe. More recently, the significant terrorist threat to mass transit and passenger rail systems promoted greater transnational cooperation.

International Union of Railways

The International Union of Railways (UIC, from the French name) was founded in Europe in 1922 and is now a worldwide organization with 202 members from the railroad industry. Its mission is to promote rail transport at the international level by, among other things, facilitating the sharing of best practices, creating new international standards for railways, and developing centers of competence (including for security). In 2006, the UIC established its Security Platform as the sole UIC entity "empowered to develop and formulate analyses and policy positions on behalf of the rail sector in matters relating to the security of persons, property and installations." It seeks to "defend the common interests of UIC members in the security field vis-à-vis European and international institutions" (International Union of Railways, n.d.).

International Association of Public Transport

The International Association of Public Transport (UITP, from the French name) is composed of 1300 member companies from 92 countries, including public transit operators, the public transport supply and service industry, policymakers, and research institutes. It serves as the global advocate for public transport systems. In 2004, the UITP created a Security Group, which was renamed as the Security Commission in 2006. The Commission is responsible for all aspects of public transit security, including the assessment and promotion of innovative operations and technology and implementation of UITP's security strategy and objectives. In November 2010, the Security Commission published a position paper on "Secure Public Transport in a Changeable World," which recommended that UITP members (1) make security a corporate priority, (2) consider security as an investment in helping increase ridership, rather than as a financial burden, (3) conduct a security risk assessment, (4) emphasize preparedness, (5) focus on the human factor (i.e., training), (6) make security an integral part of customer service, and (7) foster relationships with governmental and nongovernmental partners to help clearly define roles and relationships and to facilitate joint exercises (International Association of Public Transport, n.d.; 2010, p. 4).

Aviation Security

The international framework for addressing aviation security issues that evolved during the 1960s and 1970s and was strengthened after 9/11 continues to function under the direction of the International Civil Aviation Organization (ICAO). The ICAO, which was founded in 1944 to promote the safe and orderly development of global civil aviation, currently has 191 member nations, with security concerns becoming an increasing part of its mission. Annex 17 to the Convention on International Civil Aviation of 1974, which sets forth a baseline security program of standards and recommended practices for national aviation authorities, continues to be a central component of the ICAO's security activities and is now in its ninth edition (issued in 2011). However, maintaining and updating Annex 17 is just one of three major ICAO security efforts at present, the other two being audits of the capabilities of member states to adequately oversee aviation security measures and assistance to countries unable to address serious deficiencies identified by the audits. In addition, the ICAO undertakes to improve the security of travel documents, enhance the training of security personnel, and support regional security initiatives (International Civil Aviation Organization, n.d.).

The audit and assistance efforts together represent the Universal Security Audit Program (USAP). The program acknowledges that each member state has complete sovereignty over its own airspace and is fully responsible for aviation security, including any corrective actions arising from the audits. The audits are designed to comprehensively cover all recommended standards under Annex 17, as well as any other ICAO security guidelines. The audit process involves multiple stages to allow the audited entity to monitor, comment, and respond to the findings. The ICAO security training program is designed to further assist members in meeting the Annex 17 standards and includes guidelines for instruction in a wide range of aircraft and airport security measures (Price and Forrest, 2009, pp. 88–90).

The ICAO's 2013 annual report provided the results of 178 USAP audits, which indicated that 86% of member states had adopted aviation security legislation, 78% had in place aviation security programs and regulations, 67% had established security personnel qualifications and training standards, 59% had in place measures for the resolution of security concerns, and 52% were meeting quality control guidelines (International Civil Aviation Organization, 2014).

In achieving a global standard of security, however, the ICAO and others involved in international aviation security face problems similar to those encountered by their maritime counterparts. In 2010 testimony to a Congressional committee, the GAO detailed some of the limitations.

> *The framework for developing and adhering to international aviation standards is based on voluntary efforts from individual states.... Foreign countries, as sovereign nations, generally cannot be compelled to implement specific aviation security standards or mutually accept other countries' security measures.... Some foreign governments do not share the United States government's position that terrorism is an immediate threat to the security of their aviation systems, and therefore may not view international aviation security as a priority.... In contrast to more developed countries, many less developed countries do not have the infrastructure or financial or human resources necessary to enhance their aviation security programs.... Legal and cultural differences among nations may hamper efforts to harmonize aviation security standards. For example, some nations, including the United States, limit or even prohibit the sharing of sensitive or classified information on aviation security procedures with other countries.... Cultural differences also serve as a challenge in achieving harmonization because aviation security standards and procedures that are acceptable in one country may not be in another. For example, international aviation officials explain that the nature of aviation security oversight varies by country—some countries rely more on trust and established working relationships to facilitate security standard compliance than direct government enforcement.*
>
> *GAO, 2010, pp. 13–16*

International organizations representing important stakeholder groups have a considerable impact on global aviation security.

International Air Transport Association

The International Air Transport Association (IATA), which was founded in 1945, represents 240 airlines from 118 countries. Its mission is "to represent, lead, and serve the airline industry" by helping airlines "to operate safely, securely, efficiently and economically under clearly defined rules." The IATA has taken positions on a number of key security questions, including advocating increased use of risk management in allocating resources, development of "one-stop security" (wherein passengers would only be subjected to security screening once, when they begin their trip), and greater government sharing of security costs (IATA, n.d.).

Airports Council International

The Airports Council International (ACI) consists of 591 members operating 1861 airports in 177 nations and territories around the world. Founded in 1991 to represent the common interests of global airports, it "defends airports' positions and develops standards and recommended practices in the areas of safety, security and environment initiatives. It also advances and protects airport interests in important policy changes on airport charges and regulation, strengthening the hand of airports in dealing with airlines." On security issues, it has taken positions similar to those of IATA in supporting greater government funding of security measures and increased usage of risk management in which "measures to protect and prevent an attack should be commensurate with the risk" (Airports Council International, n.d.; 2010).

International Federation of Air Line Pilots' Associations

The International Federation of Air Line Pilots' Associations (IFALPA) has more than 100 member associations from around the world, which represent more than 100,000 airline pilots. Its objective is "to be the global voice of professional pilots by providing representation, services and support in order to promote the highest level of aviation safety worldwide." IFALPA's Security Committee: prepares working papers on IFALPA security policy "consistent with international baseline security measures;" assists member associations in maintaining "in a harmonized way a level of security in their region;" contributes to the development of relevant ICAO standards and recommended practices; and reviews new aviation security technologies and other developments. (International Federation of Air Line Pilots' Associations, n.d.).

International Transport Workers' Federation: Civil Aviation

The ITF Civil Aviation Section represents civil aviation unions and is particularly involved in working with the ICAO on international aviation safety standards. In the aftermath of 9/11, the ITF developed a set of aviation security principles, which included government and industry sharing of security costs; the need for a "a global minimum framework" of security standards; and improved training for cabin crew, counter personnel, and security screeners. In addition, it called for governments and industry to "see aviation workers and our unions as partners for security" (International Transport Workers' Federation, n.d.a.; n.d.b.).

■ ■ Critical Thinking ■

Most transportation security specialists back greater international standardization and coordination in the belief that a security measure is only as strong as its weakest link. Against this is the long-standing concern for national sovereignty, espoused by most national governments and their political leaders and publics, as well as the other limiting factors cited by the GAO in its testimony on international aviation security. How have these conflicting objectives (international standardization vs. national sovereignty) been addressed in transportation security? In your opinion, what, if anything, should be done to change the current arrangements? Why?

United States

The foundations of current transportation security organizational arrangements in the United States were contained in four instruments, all developed in the aftermath of 9/11: the Aviation and Transportation Security Act (ATSA), which created the TSA and gave it overall responsibility for transportation security; the Maritime Transportation Security Act (MTSA), which established a national maritime security system; the Homeland Security Act, which created the DHS from 22 existing federal departments and agencies and assigned the lead role in maritime security to the Coast Guard; and Homeland Security Presidential Directive-7 (HSPD-7), which was issued by President Bush on December 17, 2003.

HSPD-7 established "a national policy for federal departments and agencies to identify and prioritize United States critical infrastructure and key resources and to protect them from terrorist attacks," and directed DHS to "produce a comprehensive integrated National Plan for Critical Infrastructure and Key Resources Protection to outline national goals, objectives, milestones, and key initiatives." In carrying out this responsibility, DHS developed Sector-Specific Plans (SSP) covering the various infrastructure sectors and designated the TSA and the Coast Guard as lead agencies for the Transportation Systems SSP. That document was to identify sector participants (including their roles and relationships), assess vulnerabilities, prioritize assets, identify protective programs, measure performance, and prioritize research and development (Johnstone, 2006, p. 49).

Defining Roles and Responsibilities

The 9/11 Commission believed that defining roles for the various entities involved in transportation security was an important undertaking, and recommended that the transportation security plan it called for "should assign roles and missions to the relevant authorities (federal, state, regional, and local) and to private stakeholders." The Intelligence Reform and Terrorism Prevention Act of 2004 retained most elements of the Commission's recommended transportation security plan but modified the role definition language so that the plan "sets forth the agreed upon roles and missions of federal, state, regional, and local authorities and establishes mechanisms for encouraging private sector cooperation in the implementation of such plan." DHS's 2005 National Strategy for Transportation Security made little headway with respect to these requirements, neither indicating how federal roles were to be "agreed upon" nor how private sector participation was to be accomplished (Johnstone, 2006, p. 111).

In 2010, DHS issued three documents that could have clarified the question of homeland security roles and responsibilities: the Quadrennial Homeland Security Review (QHSR) Report, the Bottom-Up Review Report, and the Transportation Systems Sector-Specific Plan (SSP). The 2010 QHSR includes an Appendix that "reflects the current alignment of roles and responsibilities across the [homeland security] enterprise." The "key current roles and responsibilities of the many actors" were "derived largely from statutes, Presidential directives and other authorities, as well as from the National Infrastructure

Protection Plan (NIPP)[2] and National Response Framework (NRF)[3]." The resulting descriptions of organizational roles are thus more an amalgamation of individually defined, broad mission statements than a clearly differentiated articulation of functions and relationships. Indeed, its preface states explicitly that the QHSR does not "detail the roles and responsibilities of Federal or other institutions for each mission area." The second QHSR, which was released in June 2014, essentially reiterates the 2010 role definitions (Table 4.1) (U.S. Department of Homeland Security, 2010b, pp. vi, 13, A-1; 2014b, pp. 83–89).

The Bottom-Up Review Report goes further than the QHSR in providing an assessment of how DHS's organizational structure aligns with the homeland security missions defined by the QHSR and a listing that matches individual homeland security programs with the responsible agency. However, the review is confined to DHS and its components and does not cover any shared or separate responsibilities held by other federal or nonfederal entities. The absence of a fuller consideration of the roles of other federal departments and agencies is a potentially serious limitation in both the QHSR and the Bottom-Up Review considering that DHS itself accounts for only slightly more than half of annual federal homeland security funding, with the remainder distributed among 30 other federal entities (U.S. Department of Homeland Security, 2010c, pp. vii, Annex D; Reese, 2013, p. 1).

The initial Transportation Systems SSP was released in 2007, with a revised version published by DHS in 2010. As the designated lead agencies for the transportation sector, the TSA and the Coast Guard are directed to, in collaboration with the Department of Transportation (DOT), "coordinate the preparedness activities among the sector partners to prevent, protect against, respond to, and recover from all hazards that could have a debilitating effect on homeland security, public health and safety, or economic well-being" (U.S. Department of Homeland Security, 2010a, p. 3).

More specifically, TSA is given "a lead role for security of the aviation and surface [land] modes and supports the Coast Guard as the lead for maritime security." In this capacity, TSA's responsibilities include "assessing intelligence, issuing and enforcing security directives, ensuring the adequacy of security measures at transportation facilities, and assuring effective and timely distribution of intelligence to sector partners." Also, TSA is to collaborate with the U.S. DOT "in its capacity as the lead for transportation safety, response, and recovery—to manage protection and resiliency programs for all hazards." Along with its multiple non–homeland security missions, the Coast Guard "has the primary responsibility for the security of the maritime domain, including coordinating mitigation

[2]The NIPP of 2009 seeks to provide a unifying framework for infrastructure protection, with a goal of building a safer, more secure, and more resilient national system by preventing, deterring, neutralizing, or mitigating the effects of a terrorist attack or natural disaster and strengthening national preparedness, response, and recovery in the event of an emergency. The NIPP assigns Sector-Specific Agencies for each of 18 critical infrastructure sectors, including transportation (U.S. Department of Homeland Security, 2009).

[3]The 2013 edition of the NRF "covers the capabilities necessary to save lives, protect property and the environment and meet basic human needs after an incident has occurred…Core capabilities are the distinct elements needed to achieve the National Preparedness Goal." There are 14 such capabilities contained in the National Response Framework, one of which is Critical Transportation (FEMA, 2013).

Table 4.1 National Infrastructure Protection Plan Sector-Specific Agencies

Sector-Specific Agency	Critical Infrastructure and Key Resources Sector
Department of Agriculture*	Agriculture and food
Department of Health and Human Services†	
Department of Defense	Defense industrial base
Department of Energy	Energy
Department of Health and Human Services	Healthcare and public health
Department of the Interior	National monuments and icons
Department of the Treasury	Banking and finance
Environmental Protection Agency	Water
DHS Office of Infrastructure Protection	Chemical
	Commercial facilities
	Critical manufacturing
	Dams
	Emergency services
	Nuclear reactors, materials, and waste
DHS Office of Cybersecurity and Communications	Information Technology
	Communications
DHS Transportation Security Administration	Postal and shipping
DHS Transportation Security Admin‡	Transportation systems¶
DHS United States Coast Guard§	
DHS Immigration and Customs Enforcement, Federal Protective Service	Government facilities

*Responsible for agriculture and meat, poultry, and egg products.
†Responsible for food other than meat, poultry, and egg products.
‡Responsible for nonmaritime transportation modes.
§Responsible for maritime transportation mode.
¶As stated in Homeland Security Presidential Directive-7, the Department of Transportation and the Department of Homeland Security (DHS) will collaborate on all matters relating to transportation security and transportation infrastructure protection.
(*Source*: U.S. Department of Homeland Security. 2009. National infrastructure protection plan. p. 3. http://www.dhs.gov/national-infrastructure-protection-plan (accessed 10.24.14.))

measures to expedite the recovery of maritime infrastructure and transportation systems and to support incident response in coordination with the Department of Defense" (U.S. Department of Homeland Security, 2010a, p. 18).

The Transportation Systems SSP provides role and responsibility descriptions of the various parts of the transportation sector that are similar to those included in the QHSR but with somewhat greater elaboration in the individual modal annexes. The mass transit and passenger rail annex in particular goes into greater detail in delineating the division of responsibilities between TSA, the Federal Transit Administration (FTA), and the Federal Railroad Administration. It includes a discussion of the 2005 memorandum of understanding executed by the DHS and DOT, along with their TSA and FTA components, that aimed "to ensure that all gaps, fragmented efforts and unnecessary redundancies and overlaps in security roles and responsibilities are identified and addressed.... Through these types of cooperative efforts, respective roles and responsibilities are now more clearly defined and security partners work in a collaborative environment to ensure that security gaps are mitigated and a high level of security is achieved and maintained in mass transit and passenger rail systems" (U.S. Department of Homeland Security, 2010b, pp. 217–220).

Department of Homeland Security

When DHS opened on March 1, 2003, it and its components were faced with a number of daunting challenges. Its agencies brought with them different histories and organizational cultures and in some cases (notably the Coast Guard, Federal Emergency Management Agency [FEMA], Immigration and Naturalization Service, and Customs Service) substantial nonsecurity missions, all of which had to be harmonized, at least to some degree.

Although assigned the lead role, not all homeland security functions were given to the new department. In transportation, for example, the various modal administrations within the DOT retained roles ranging from significant (in the case of the FTA) to narrow (the Federal Aviation Administration [FAA], whose security component was removed and served as the basis for the TSA). Furthermore, limited federal ownership or control of most aspects of homeland security—which was certainly the case in transportation—imposed constraints on DHS's authority and influence.

In aviation security, TSA underwent rapid changes in organizational status, from being a relatively small component within the FAA to being an independent agency within the DOT to becoming part of DHS, all within a 15-month period and faced a number of congressionally mandated deadlines (under ATSA). The Coast Guard and Customs had to cope with MTSA mandates (Johnstone, 2006, p. 55).

In an assessment of the department's progress after 10 years in existence, a February 2013 GAO report on DHS indicated, "it has implemented key homeland security operations and achieved important goals and milestones in many areas to create and strengthen a foundation to reach its potential. As it continues to mature, however, more work remains for DHS to address gaps and weaknesses in its current operational and implementation efforts, and to strengthen the efficiency and effectiveness of those efforts." (GAO, 2013, Highlights).

As of 2014, DHS consists of more than 240,000 employees distributed across a number of entities (Figure 4.2), including the following with significant roles in transportation security.

The *Science and Technology Directorate* is the primary research and development arm of DHS and manages the Transportation Security Laboratory.

The *Directorate for National Protection and Programs* advances DHS's risk reduction mission through an integrated approach encompassing both physical and cyber threats and their associated human elements. Its *Office of Infrastructure Protection* division has overall responsibility for coordinating implementation of the National Infrastructure Protection Plan, including the Transportation Systems SSP; providing security training and plans for owners and operators of critical infrastructure; and providing assistance to state, local, tribal, territorial, and private sector entities to develop measures to mitigate critical infrastructure vulnerabilities.

The *Office of Intelligence and Analysis* is responsible for using information and intelligence from multiple sources to assess current and future security threats and for disseminating such information to appropriate federal and nonfederal entities. The *Homeland Infrastructure*

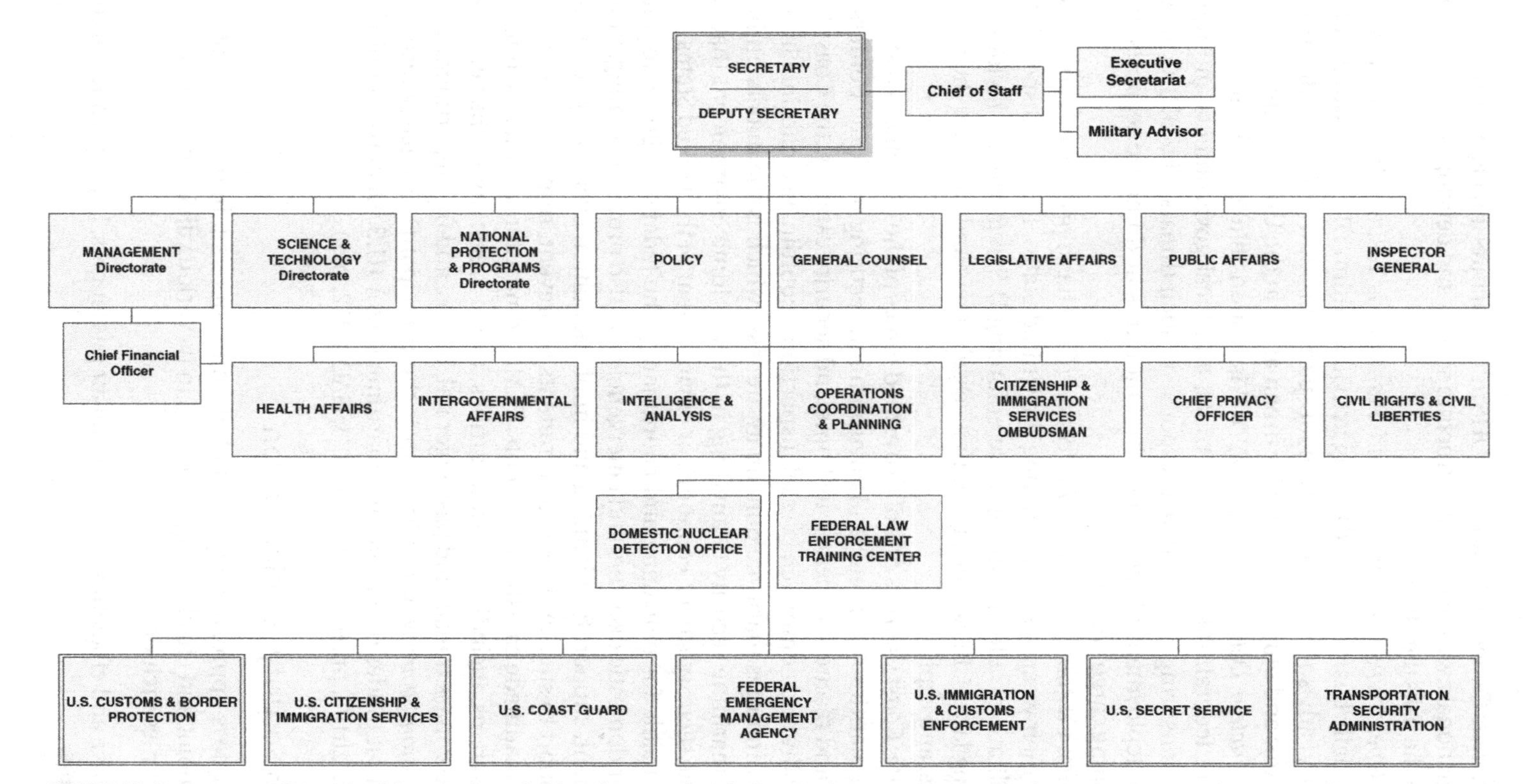

FIGURE 4.2 Department of Homeland Security Organization Chart as of April 10, 2013. *(Source: Department of Homeland Security, http://www.dhs.gov/xlibrary/assets/dhs-orgchart.pdf.)*

Threat and Risk Analysis Center (HITRAC) is the DHS infrastructure–intelligence fusion center that maintains situational awareness of infrastructure sectors and develops long-term strategic assessments of the risks they face.

The *Office of Operations Coordination and Planning* is responsible for monitoring U.S. security on a daily basis and coordinating activities within DHS and with governors, homeland security advisors, law enforcement partners, and critical infrastructure operators in all states and more than 50 major urban areas in the United States.

The *Domestic Nuclear Detection Office (DNDO)* is responsible for enhancing nuclear detection efforts of federal, state, territorial, tribal, and local governments and the private sector and for ensuring a coordinated response to such threats. The DNDO acquires and supports the deployment of radiation portal monitors at domestic seaports for the scanning of incoming cargo containers.

The *United States Customs and Border Protection (CBP)* has a priority mission of keeping terrorists and their weapons from entering the United States. It is also responsible for securing and facilitating trade and travel while enforcing U.S. regulations, including immigration and drug laws. CBP screens incoming vessels' crew and cargo for the presence of weapons, drugs, and explosives.

The *United States Coast Guard* is one of five armed forces of the U.S., and is responsible for protecting the maritime economy and environment, securing maritime borders, and providing search and rescue services. In its homeland security capacity, the Coast Guard conducts port facility and commercial vessel inspections, coordinates maritime information sharing, and promotes maritime domain awareness (which is the understanding by those involved in maritime security of anything in the maritime environment that could adversely affect the security, safety, economy, or environment of the United States).

FEMA is responsible for managing and coordinating the federal response to and recovery from major domestic disasters and emergencies. It also provides training, funds to purchase equipment, support for planning and implementation exercises, and technical and financial support to state and local governments to prevent, respond to, and recover from natural and manmade catastrophic events. FEMA administers a variety of homeland security grant programs, including for states; tribes; urban areas; nonprofits; emergency management; ports; mass transit; and passenger rail, ferries, and fire departments.

The *Transportation Security Administration* is the principal DHS agency responsible for all transportation modes other than the maritime sector (U.S. Department of Homeland Security, n.d.; 2010a, pp. 91–92; 2013, p. 175; GAO, 2012, p. 2).

Transportation Security Administration

TSA currently employs approximately 53,600 full-time workers (almost 47,000 of whom are airport checkpoint and checked bag screeners) in its various divisions (Figure 4.3).

TSA's specific duties include:

- Ensuring effective and efficient screening of all air passengers, baggage, and cargo on passenger planes

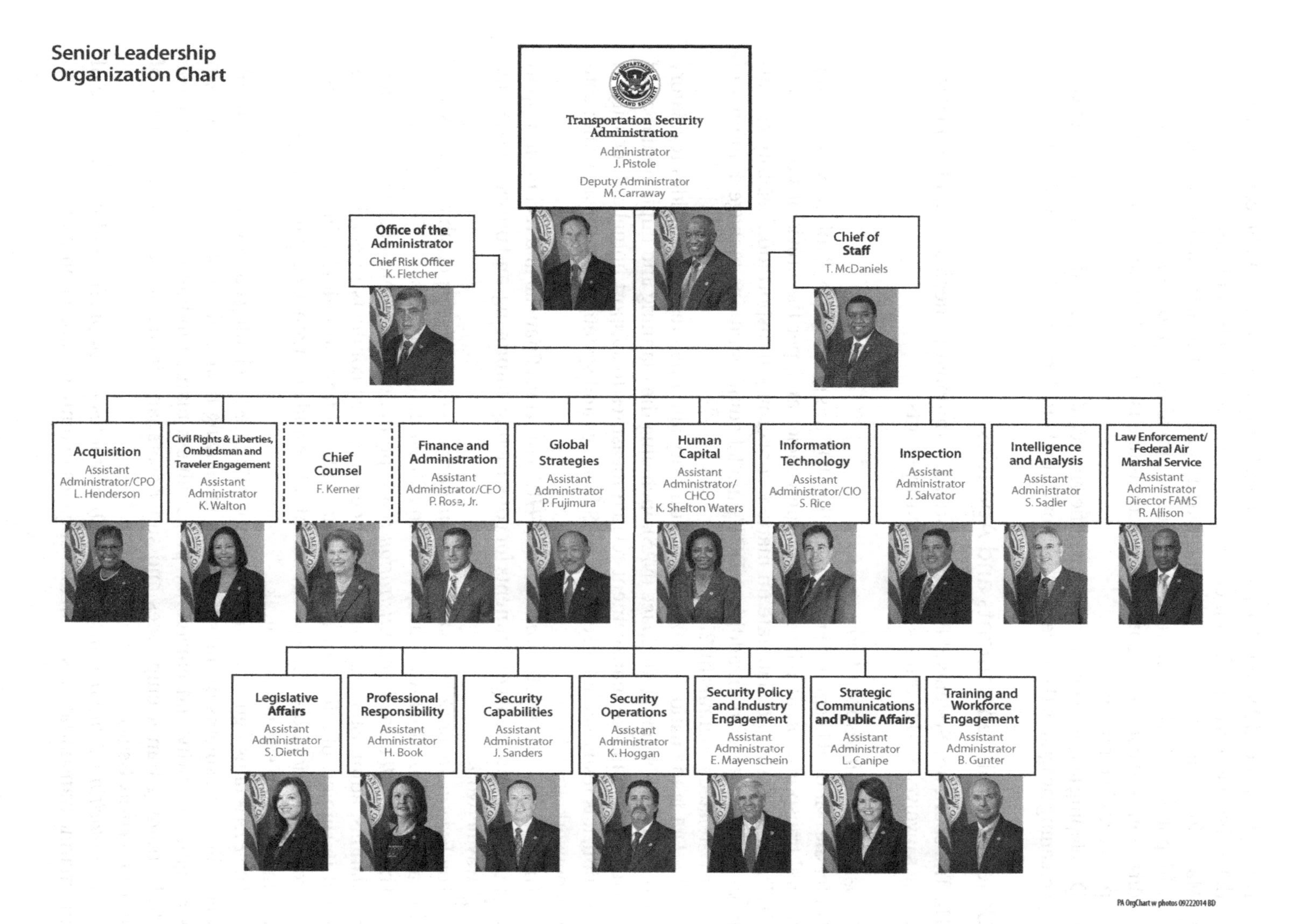

FIGURE 4.3 Transportation Security Administration organization chart as of October 2014. *(Source: Department of Homeland Security, http://www.tsa.gov/sites/default/files/assets/pdf/tsa_orgchart.)*

- Deploying Federal Air Marshals internationally and domestically to detect, deter, and defeat hostile acts targeting air carriers, airports, passengers, and crews
- Managing security risks of the surface [land] transportation systems by working with public and private sector stakeholders, providing support and programmatic direction, and conducting onsite inspections to ensure the freedom of movement of people and commerce
- Developing and implementing more efficient, reliable, integrated, and cost-effective screening programs (U.S. Department of Homeland Security, 2014a, pp. 67–68).

Other Federal Departments and Agencies

The Transportation Systems SSP provides brief descriptions of other federal and nonfederal entities involved in U.S. transportation security efforts.

Department of Transportation

Under the National Response Framework, the DOT was given the lead role in coordinating federal transportation activities during emergencies and, per the provisions of HSPD-7, DOT and DHS are to collaborate on matters concerning transportation security and infrastructure protection. In addition, DOT's modal administrations manage a number of security-related programs (p. 22).

The *FAA* is the nation's civil aviation authority and operates and provides regulatory oversight of the National Air Space (NAS), including the planning and implementation of air traffic and airspace management-related measures to support homeland security. In addition, it is responsible for securing NAS facilities and systems (e.g., the air traffic control network) (p. 132).

The *Federal Highway Administration (FHWA)* provides financial and technical support to state, local, and tribal governments for constructing, improving, and preserving the U.S. highway network.

The *Federal Motor Carrier Safety Administration (FMCSA)* is responsible for overseeing the safe and secure highway transportation of hazardous materials.

The *Federal Railroad Administration (FRA)* promulgates and enforces railroad safety regulations; administers railroad assistance programs; conducts research and development in support of rail safety and national railroad transportation policy; and reviews, approves, and monitors implementation of safety and emergency preparedness plans for commuter rail systems and Amtrak.

The *FTA* retains significant security responsibilities, including conducting risk and vulnerability assessments of transit systems, deploying technical assistance teams to help strengthen security and emergency preparedness plans, funding emergency response drills, training transit employees and supervisors, and increasing public awareness of transit security issues.

The *Maritime Administration (MARAD)* promotes development and maintenance of the marine transportation system and engages in outreach and coordination activities to assist the maritime industry in emergency preparedness, response, and recovery efforts for maritime transportation security incidents and natural disasters.

The *Office of Intelligence, Security, and Emergency Response (S-60)* is the focal point within DOT for intelligence, security, and emergency management matters, including lead responsibility for implementation of the DOT's NRF obligations. It issues notifications of threats to transportation systems, develops and maintains DOT's emergency management strategy, policy, and plans and operates the DOT's Crisis Management Center.

The *Pipeline and Hazardous Materials Safety Administration (PHMSA)* oversees the safety of the 1.2 million daily shipments of hazardous materials within the United States and the 2.3 million miles of pipeline in the country.

The *Research and Innovative Technologies Administration (RITA)* coordinates DOT's research and development programs; performs transportation statistics research, analysis, and reporting; and provides transportation education and training.

The *Saint Lawrence Seaway Development Corporation (SLSDC)* is a federally owned corporation and operating administration within DOT that is responsible for the operations and maintenance of the U.S. portion of the St. Lawrence Seaway (pp. 93–94).

Department of Commerce

The Department of Commerce's involvement in transportation security is largely limited to services it provides to the transportation industry with respect to the supply chain. (p. 21).

Department of Defense

The Department of Defense (DOD) is involved in transportation security matters both as a user of private sector transportation systems (for movement of personnel and equipment) and as a provider of security services during natural or human-made disasters. The *North American Aerospace Defense Command (NORAD)* provides detection and warning of attacks against North America by aircraft, missiles, and space vehicles. The *Defense Joint Intelligence Operations Center (DJIOC)* integrates and synchronizes military and other national intelligence capabilities. The *U.S. Army Corps of Engineers* maintains the nation's commercial waterways, including levees, and operates the locks and dams that facilitate commerce on U.S. inland waterways. *U.S. Northern Command* conducts operations to deter, prevent, and defeat threats and aggression against the United States and its territories and interests within the continental United States, Alaska, Canada, Mexico, and the surrounding waters out to approximately 500 nautical miles (including the Gulf of Mexico and the Straits of Florida) (pp. 21, 92).

Department of Energy

The Department of Energy (DOE) is responsible for securing the nation's electricity, petroleum, and natural gas energy resources. Because of the interdependencies between the energy and transportation sectors (including the former's reliance on hazardous liquids and natural gas pipelines), the DOE and TSA established a cross-sector partnership to coordinate security in the oil and natural gas industries (p. 21).

Department of Justice

In collaboration with DHS, the Department of Justice and its components investigate and prosecute actual or attempted attacks on, sabotage of, or disruptions of critical infrastructure (including transportation systems). More specifically, the TSA works closely in this regard with the *Federal Bureau of Investigation (FBI)*, which has the lead responsibility for investigation of terrorist acts or threats by individuals or groups within the United States (pp. 21–22).

Department of State

The State Department leads the representation of the United States overseas and advocates U.S. policies with foreign governments and international organizations, including transportation security matters such as the protection and security of pipelines crossing international borders, transportation-related concerns over the use of international waterways, and the transportation of passengers and freight across international boundaries via aviation (p. 22).

National Counterterrorism Center

The National Counterterrorism Center (NCTC), which is part of the Office of the Director of National Intelligence and draws its personnel from throughout the U.S. intelligence community, is the primary federal organization in integrating and analyzing all intelligence pertaining to terrorism and counterterrorism (p. 95).

Amtrak

The National Rail Passenger Corporation, most commonly known as Amtrak (a blending of "America" and "track"), is a quasi-governmental agency established by the Rail Passenger Service Act of 1970 as a federally funded but independently operated and managed for-profit corporation. It was created in an attempt to preserve a national passenger rail system by taking over the intercity passenger rail service previously operated by private railroads that had been experiencing severe drops in ridership and coverage over the previous decade. Currently, Amtrak operates more than 21,000 miles of routes, serving more than 500 destinations in 46 states, the District of Columbia, and three Canadian provinces. In terms of security, it has its own police department, requires photo identification for ticket purchasers, provides some security training for its 19,000 employees, and implements relevant security directives issued by TSA (Amtrak, n.d.).

State, Territorial, Tribal, and Local Governments

In the U.S. system of federalism, sovereignty is constitutionally divided between the national and subnational governments, with states, certain Native American tribes, and U.S. territories retaining substantial authority within their jurisdictions. Although homeland and transportation security has largely been instigated and directed by the federal government, these subnational authorities also play important roles. For example, most of those likely to be the "first responders" to acts of terrorism and other major disasters are employees of local police, fire, and other emergency management departments. Indeed, a 2009 analysis by the Heritage Foundation indicated that state and local governments accounted for a majority of national homeland security spending, and when including firefighters, emergency management personnel and law enforcement officials, "the state and local personnel advantage is roughly 2,200,000 to 50,000" (Mayer and Baca, 2010, pp. 5–6).

Federal agencies provide support to these governments in meeting their emergency response and recovery needs, while the latter in turn assist DHS and DOT in collecting information about critical infrastructure and providing impact assessments about any security incidents. The State, Local, Tribal, and Territorial Government Coordinating Council (SLTTGCC) was formed in 2007 to represent these entities within the National Infrastructure Protection Plan partnership framework (U.S. Department of Homeland Security, 2010a, p. 22).

State and Territorial Governments

Governments of the 50 U.S. states and five territories (American Samoa, Guam, Northern Marianas, Puerto Rico, and Virgin Islands) coordinate the transportation security activities of cities, counties, and intrastate regions within their jurisdictions. Furthermore (along with the District of Columbia), they administer State Homeland Security Grants from FEMA designed to address planning, organization, equipment, training, and exercise needs to prevent, protect against, mitigate, respond to, and recover from acts of terrorism and other catastrophic events. State and territorial agencies perform law enforcement, security, and disaster response and recovery activities (U.S. Department of Homeland Security, 2010b, p. A-6; FEMA, 2012).

Some states have created port authorities, which may be run as either state or local entities, to manage U.S. coastal or Great Lakes ports. The *American Association of Port Authorities (AAPA)* was formed in 1912 to represent the interests of these organizations and currently has more than 130 members, mostly in the United States but also in Canada, the Caribbean, and Latin America. The AAPA has supported maintaining funding and the eligibility of all U.S. seaports for FEMA's Port Security grant program (American Association of Port Authorities, n.d.).

State governments are particularly influential in certain land transportation sectors, where they own or operate one-fifth of the nation's roads, almost half of its bridges, and a portion of its passenger rail system. For example, state agencies are responsible for overseeing compliance with the FTA's regulations on transit safety and security, and the *American Association of State Highway and Transportation Officials (AASHTO)* is active in all transportation modes (but with particular emphasis on the highway sector), with the aim of representing states' (and the District of Columbia's and Puerto Rico's) interests in developing, operating, and maintaining an integrated national transportation system. Its Special Committee on Transportation Security and Emergency Management provides support to state transportation agencies in developing security and emergency management plans and policies (U.S. Department of Homeland Security, 2010a, p. 219; American Association of State Highway and Transportation Officials, n.d.).

Tribal Governments

There are 566 federally recognized Native American tribes, and depending on location, land base, and resources, their governments can provide law enforcement, fire, emergency, and other public safety services within their boundaries. In carrying out their responsibilities, tribal governments may coordinate resources and capabilities with neighboring jurisdictions and establish mutual aid agreements with other tribal governments, local jurisdictions, and state governments. They are eligible for FEMA's Tribal Homeland Security Grants that support homeland security preparedness and response in their territories (U.S. Department of Homeland Security, 2010b, p. A-6).

Local Governments

As mentioned previously, local governments serve as the front-line authorities for the provision of local law enforcement, fire, public safety, environmental response, public health, and emergency medical services for all types of hazards (natural and human-made) and emergencies. They generally coordinate resources and capabilities with other entities (including neighboring local governments, nongovernmental organizations, their state government, and the private sector) during such events. At present, 36 "high-threat, high-density" cities are eligible for FEMA's Urban Area Security Initiative (UASI) grants, which are to be used "in building an enhanced and sustainable capacity to prevent, protect against, mitigate, respond to and recover from acts of terrorism." As with the states, local governments occupy a prominent place in the highway mode (owning and operating three-quarters of U.S. roads, more than half of its bridges, and 70% of school buses) (U.S. Department of Homeland Security, 2010b, p. A-7).

Local governments play an especially large part in mass transit and passenger rail, where local agencies are typically responsible for operating the systems, including providing for their security, whether directly through a dedicated law enforcement or separate security component or through contract with outside law enforcement or private security agencies. Furthermore, although other levels of government are involved, primary responsibility for implementing transit and passenger rail security measures falls on these local operators (U.S. Department of Homeland Security, 2010a, pp. 219–220).

The *American Public Transportation Association (APTA)*, which was founded in 1882, represents almost 1500 members, including the public authorities that operate transit systems and commuter, intercity, and high-speed rail systems, as well as private companies that design, construct, supply, and sometimes operate those systems. More than 90% of transit passengers in the United States and Canada ride on APTA member systems. As part of its advocacy efforts, APTA has regularly lobbied the U.S. Congress to increase funding for transit security, primarily through FEMA's Transit Security Grant Program (American Public Transportation Association, 2013).

Cities and counties also own and operate a majority of commercial airports in the United States (except in Alaska, Hawaii, and Rhode Island, where the state governments own all airports within their boundaries). Management of these municipal airports can be directly through a local governmental agency or through a separate "airport authority" in which the local government (or governments in the case of authorities that encompass more than one jurisdiction) retains ownership but the authority is responsible for a large part of management and planning. Commercial airport operators are required to have a TSA-approved Airport Security Program specifying how the airport will comply with federal security regulations. The interests of airport operators in the United States are represented by the North American branch of the *Airports Council International* and the *American Association of Airport Executives* (Price and Forrest, 2009, pp. 18–21).

Private Sector

Although private companies own little of the transportation infrastructure in the United States (with the notable exceptions of freight railroads and pipelines), they control most of the vehicles and containers that carry passengers and cargo across those networks and operate most marine and inland waterway terminals. The Transportation Systems SSP describes the role of private (as well as public) owners and operators of U.S. transportation systems.

> *Owners and operators participate in a variety of ways to protect the sector's infrastructure and to assure its resiliency through business continuity planning and risk mitigation activities. In the wake of the attacks of September 11, 2001, many trade associations developed and encouraged participation in security best practices, planning, training, and exercises. Numerous owners and operators of transportation infrastructure as well as members of representative associations provide technical expertise during the development of voluntary standards and regulations.*
>
> *U.S. Department of Homeland Security, 2010a, p. 23*

Although this description of largely voluntary, privately operated transportation security programs applies across most modes, there are exceptions. In the maritime sector, MTSA requires foreign and U.S.-flagged commercial vessels of greater than 100 gross tons, passenger vessels capable of carrying 12 or more, tugs and barges carrying hazardous materials, and most offshore drilling platforms to comply with the provisions of the International Ship and Port Facility Security Code. In addition, MTSA mandates that most marine terminals have approved facility security assessments and security plans. In commercial aviation, scheduled airlines and public charter operators flying within, to or from the United States must manage a TSA-developed Aircraft Operators Standard Security Program (AOSSP) and specify how they will comply with its requirements. TSA deploys inspectors and conducts oversight to enforce compliance. Private charters and cargo aircraft operators are subject to less extensive federal security requirements (McNicholas, 2008, pp. 117–118; Price and Forrest, 2009, pp. 256–258).

Numerous trade and other associations represent the interests of the various sectors of the U.S. transportation industry. Below are listed some of those that have been most active in security matters.

American Waterways Operators (AWO) represents the owners and operators of tugboats, towboats, and barges involved in U.S. waterborne commerce. After consulting with the Coast Guard and the U.S. Army Corps of Engineers, AWO issued a Model Vessel Security Plan for its members in April 2002, and after MTSA was enacted in November 2002, the organization worked with the Coast Guard to modify the plan so that it was an acceptable Alternative Security Plan under the new law (American Waterways Operators, n.d.).

Chamber of Shipping of America (CSA) represents U.S.-based companies that own, operate, or charter oceangoing tank, container, or dry bulk vessels (Chamber of Shipping of America, n.d.).

Inland Rivers Ports & Terminals (IRPT) promotes the interests of the owners and operators of U.S. inland waterways, ports and terminals (Inland Rivers Ports & Terminals, n.d.).

The *National Association of Waterfront Employers (NAWE)* represents private sector U.S. marine terminal operators and the associated stevedoring industry, which is responsible for the loading and unloading of cargo. The NAWE "has partnered with Congress and federal agencies to resolve key facility security and cargo screening issues in order to insure a secure and efficient U.S. marine transportation system" (National Association of Waterfront Employers, n.d.).

The *Passenger Vessel Association (PVA)* represents the owners of ferries, tour vessels, charter boats, and other types of passenger vessels operating in the United States. Its "PVA Industry Standard for Security of Passenger Vessels and Small Passenger Vessels and their Facilities" has been approved by the Coast Guard as an Alternative Security Plan under MTSA. The Association has also successfully lobbied the federal government to exempt smaller passenger vessels—those carrying less than 150 passengers—from security plan requirements and reduce the costs to the industry of other security mandates, such as the installation of Automatic Identification Systems (Passenger Vessel Association, n.d.).

The *American Bus Association (ABA)* represents approximately 1000 intercity bus and tour companies in the United States and Canada. The ABA supports full funding for FEMA's intercity bus security grant program and the setting by DHS of funding priorities "that address the greatest threat to bus passengers and others" (American Bus Association, n.d.).

The *American Trucking Association (ATA)* is a federation of 50 affiliated state organizations and is the largest trade association representing the trucking industry. Its primary security goal "is to consolidate, harmonize and better coordinate multiple security requirements so that commercial drivers are not required to undergo and pay for multiple background checks, and carriers do not have to develop multiple security plans and training requirements when transporting certain types of cargo or operating in higher risk environments" (American Trucking Association, n.d.).

The *Association of American Railroads (AAR)* represents Amtrak and the major freight railroads of the United States, Canada, and Mexico. Its members account for more than 43% of intercity freight volume and almost 100% of intercity passenger service in the United States. After 9/11, the AAR developed a classified analysis of freight rail risks and potential countermeasures, and its Security and Emergency Response Training Center (SERTC) provides free training to state and local emergency responders for handling hazardous material accidents (Association of American Railroads, n.d.).

The *Association of Oil Pipe Lines (AOPL)* represents the interests of owners and operators of U.S. pipelines that transport liquid materials (primarily crude oil and refined petroleum products), including before DHS on facility security programs and rulemakings (Association of Oil Pipe Lines, n.d.).

The *Interstate Natural Gas Association of America (INGAA)* has 25 members in the United States and Canada that own most of the interstate natural gas pipelines in the two countries. Working with DHS, its members "are focusing their [security-related] prevention and recovery planning on the potential disruption of natural gas supply as the primary issue to be addressed in connection with any natural gas pipeline incident" (Interstate Natural Gas Association of America, n.d.).

Airlines for America (A4A; formerly known as the Air Transport Association) is the sole trade organization of the principal U.S. airlines. Its members and affiliates handle more than 90% of U.S. airline passenger and cargo traffic. A4A has long been an active participant in aviation security matters, with a goal of implementing security measures "effectively and efficiently to maximize the security benefits and minimize passenger/shipper inconvenience." In recent years, A4A has focused on the need to expand risk-based security to target security measures on the greatest threats and has opposed any increase in airline passenger security fees (Airlines for America, 2013).

Labor unions, including those listed below, have also played a significant role in representing private sector interests in transportation security.

The *Air Line Pilots Association, International (ALPA)* is the largest airline pilot union, representing more than 50,000 pilots in the United States and Canada. ALPA has also been an active participant in the making of aviation security policy. Among other actions, the organization conceived and successfully lobbied for the creation of the Federal Flight Deck Officer program under which eligible flight crew members receive training by the Federal Air Marshal Service and are then authorized to have access to firearms to defend against acts of criminal violence and hijackings.

The *American Federation of Government Employees (AFGE)* is the largest federal employee union, representing 650,000 federal and D.C. government workers. In transportation security, AFGE has been involved in an ongoing effort to obtain collective bargaining rights and win representation for TSA's checkpoint screeners. In 2011, the screeners were granted limited collective bargaining rights and, as AFGE members, ratified their first collective bargaining agreement with TSA (American Federation of Government Employees, 2012).

The *Association of Flight Attendants-CWA (AFA-CWA)* is the world's largest labor union organized by and for flight attendants, with almost 60,000 members. As part of its efforts to promote safety and security for its members and passengers, the organization has been actively involved in seeking to improve security training for flight crews and in opposing TSA's proposal to allow smaller knives to be carried onboard aircraft (Association of Flight Attendants-CWA, 2013).

The *International Longshoremen's Association (ILA)* represents approximately 65,000 maritime workers in North America (International Longshoremen's Association, n.d.).

The *Transport Workers Union (TWU)* represents more than 200,000 members in the United States, including 130,000 transit workers and 45,000 workers in the aviation industry (Transport Workers Union, n.d.).

The *United Transportation Union (UTU)* represents 125,000 active and retired railroad, bus, and mass transit workers in the U.S. and Canada (United Transportation Union, n.d.).

■ ■ Critical Thinking ■

Do you think more needs to be done to clarify transportation security roles in the United States? If so, what should be done? If not, why not?

DIGGING DEEPER
WHAT IS HOMELAND SECURITY?
An Evolving Concept

In its 2007 "National Strategy for Homeland Security," President Bush's Homeland Security Council wrote:

> *Homeland security before September 11 existed as a patchwork of efforts undertaken by disparate departments and agencies across all levels of government. While segments of our law enforcement and intelligence communities, along with our armed forces, assessed and prepared to act against terrorism and other significant threats to the United States, we lacked a unifying vision, a cohesive strategic approach, and the necessary institutions within government to secure the Homeland against terrorism. The shock of September 11 transformed our thinking. In the immediate aftermath of history's deadliest international terrorist attack, we developed a homeland security strategy [in which homeland security is defined as] a concerted national effort to prevent terrorist attacks within the United States, reduce America's vulnerability to terrorism, and minimize the damage and recover from attacks that do occur.*
>
> *Homeland Security Council, 2007, p. 3*

Three years later, the Obama Administration issued the "Quadrennial Homeland Security Review Report," which offered the following on the same subject:

> Homeland security *is a relatively new concept. Yet it is one that can trace its roots to traditional functions such as civil defense, emergency response, law enforcement, customs, border control, and immigration. Homeland security captures the effort to adapt these traditional functions to confront new threats and evolving hazards.... The question "What is homeland security?" recognizes that, in fact, securing the United States and its people represents an overarching national objective. Equally important, and aside from obviously identifying a Cabinet-level department of the federal government, homeland security is a widely distributed and diverse—but unmistakable—national enterprise. The term "enterprise" refers to the collective efforts and shared responsibilities of Federal, State, local, tribal, territorial, nongovernmental, and private-sector partners—as well as individuals, families, and*

> *communities—to maintain critical homeland security capabilities.... Homeland security is a concerted national effort to ensure a homeland that is safe, secure, and resilient against terrorism and other hazards where American interests, aspirations, and way of life can thrive.*
>
> *U.S. Department of Homeland Security, 2010b, pp. 11–13*

A 2013 report by the Congressional Research Service traces the evolution of the homeland security concept after 9/11 with particular attention to the most recent attempts by the federal government to define the term.

> *The competing and varied definitions in these documents may indicate that there is no succinct homeland security concept. Without [such] a . . . concept, policymakers and entities with homeland security responsibilities may not successfully focus on the highest prioritized or most necessary activities. Coordination is especially essential to homeland security because of the multiple federal agencies and the state and local partners with whom they interact. Coordination may be difficult if these entities do not operate with the same understanding of the homeland security concept.*
>
> *Reese, 2013, pp. 9–10*

Table 4.2 provides a listing of some of the evolving definitions of "homeland security" from 2007 to 2012.

Compare and contrast the 2007 and 2010 descriptions of U.S. homeland security. What factors (including chronological, political, and economic circumstances) account for the continuities and changes? Do you agree that the differences in definitions of homeland security within various federal strategic documents represent a serious obstacle to successful homeland security implementation? Why or why not?

Other Nations

According to Morag (2011), the combined experiences of 9/11 and the 2005 devastation caused by Hurricane Katrina led to the development of a unique "homeland security" program in the United States. Lacking these "dual shocks," other countries did not "view [preparedness for terrorism, natural disasters, public health emergencies, threats to critical infrastructure and the like] as interlinked and part of a common effort designed to head off, and failing that, cope with and recover from events that could produce massive social and economic disruption." Thus, there has been little adoption by other nations of a separate department devoted to "homeland security" (p. 1).

Many other countries, however, had extensive pre-9/11 experience with terrorism and other acts of violence against transportation, and in general, they have modified existing institutions to deal with the evolving security threats of the 21st century (pp. 1–3).

Canada

Transport Canada (TC), a division of the Ministry of Transportation, Infrastructure and Communities, is the primary Canadian government agency responsible for transportation policy and programs, including security. Maritime security is handled by the department's *Marine*

Table 4.2 Homeland Security Definitions, 2007 to 2012

Source Document	Definition
2007 National Strategy for Homeland Security (Homeland Security Council)	A concerted effort to prevent terrorist attacks within the United States, reduce America's vulnerability to terrorism, and minimize the damage and recover from attacks that do occur.
2008 U.S. Department of Homeland Security Strategic Plan Fiscal Year 2008-2013 (DHS)	A unified national effort to prevent and deter terrorist attacks, protect and respond to hazards, and to secure the national borders.
2010 National Security Strategy (Office of the President)	A seamless coordination among federal, state, and local governments to prevent, protect against and respond to threats and natural disasters.
2010 Quadrennial Homeland Security Review (DHS)	A concerted national effort to ensure a homeland that is safe, secure, and resilient against terrorism and other hazards where American interests, aspirations, and ways of life can thrive.
2010 Bottom-Up Review (DHS)	Preventing terrorism, responding to and recovering from natural disasters, customs enforcement and collection of customs revenue, administration of legal immigration services, safety and stewardship of the Nation's waterways and marine transportation systems, as well as other legacy missions of the various components of DHS.
2011 National Strategy for Counterterrorism (Office of the President)	Defensive efforts to counter terrorist threats.
2012 Strategic Plan (DHS)	Efforts to ensure a homeland that is safe, secure and resilient against terrorism and other hazards.

DHS, Department of Homeland Security.
(*Source*: Reese, S., January 8, 2013. Defining homeland security: Analysis and congressional considerations. Congressional Research Service, Washington, DC, p. 8.)

Security Regulatory Affairs branch, which "develops regulations, security measures and other legal tools to safeguard the marine transportation industry," and the *Marine Security Operations branch*, which "provides [security] oversight for Canadian vessels on international voyages, domestic ferries on certain routes, and foreign vessels in Canadian waters, as well as Canadian marine facilities and ports that interface with these vessels." The Operations branch also "acts as the functional authority for security related matters for the regional Marine Safety and Security Offices, including Marine Security personnel working in the interdepartmental *Marine Security Operations Centres* (MSOCs)" (Government of Canada, Transport Canada, n.d.a).

Land transportation security is handled by TC's *Surface and Intermodal Security (SIMS) Directorate*. For mass transit and commuter rail, the Directorate provides codes of best practices, security tools, and stakeholder guidelines, and between 2006 and 2009, provided financial assistance to operators in major urban areas for the accelerated implementation of enhanced security measures. Freight rail security is implemented through a 2007 memorandum of understanding between Transport Canada and the Railway Association of Canada, which represents the freight rail industry. The agreement reflects best security

FIGURE 4.4 Ambassador Bridge connecting Windsor, Ontario with Detroit, Michigan, viewed from the Canadian side. *(Source: Patr1ck, http://upload.wikimedia.org/wikipedia/commons/9/9d/Ambassador_bridge_evening.jpg.)*

practices of the industry and commits its members to prepare a risk-based security plan, perform security exercises, provide security training for workers, and expeditiously report security incidents to the government. The SIMS Directorate also manages the highway security programs created by the International Bridges and Tunnels Act of 2007 (which seeks to secure all international bridges and tunnels in Canada while not impeding traffic flow) and the Transportation of Dangerous Goods Act (which authorizes the creation of security regulations and the establishment of mandatory security plans and training standards) (Figure 4.4) (Government of Canada, Transport Canada, n.d.b; n.d.c).

In aviation security, Transport Canada develops policy and regulations, conducts oversight of the Canadian Air Transport Security Authority and the aviation industry, and verifies Canada's compliance with international security obligations, such as ICAO Annex 17 (Government of Canada, Transport Canada, 2013, p. 4).

Other government entities also have significant transportation security roles.

The *Canada Border Services Agency (CBSA)* is responsible for a number of border control functions, including trade security programs, such as the multinational Container Security Initiative, the Advance Commercial Information Program for obtaining advance reporting of cargo shipments, and the Integrated Cargo Security Strategy (ICSS) with the

United States. It also operates radiation detection equipment at various points of entry (Government of Canada, Canada Border Services Agency, n.d.).

The *Canadian Air Transport Security Authority (CATSA)* was established in 2002 as a Crown Corporation[4] reporting through TC. It screens air passengers and their baggage and certain nonpassengers (including airport workers) and administers the Restricted Area Identity Card Program for controlling access to certain areas of airports (Government of Canada, Transport Canada, 2013, p. 19).

Public Safety Canada (PS) is the lead Canadian department of public safety and coordinates federal security and emergency management programs. It has the lead role in administering (with TC) the Passenger Protect program for prescreening individuals to identify those who pose a threat to civil aviation and for taking action to respond to such threats (Government of Canada, Public Safety Canada, n.d.; Government of Canada, Transport Canada, 2013, p. 19).

The *Royal Canadian Mounted Police (RCMP)* is the Canadian national police service and an agency of the Ministry of Public Safety Canada. It conducts national security investigations within the various transportation modes and law enforcement records checks in support of the Transportation Security Clearance Program. Its National Ports Strategy is designed "to prevent, deter and detect any illicit and/or terrorist activity, cargo or people at Canada's major marine ports that may pose a threat to national, US and global safety and security." In aviation, it places specially trained RCMP covert In Flight Security Officers on board certain Canadian-registered flights on the basis of risk (Government of Canada, Transport Canada, 2013, p. 19; Government of Canada, RCMP, n.d.).

India

In India, transportation security roles have been evolving rapidly in recent years and are divided among a number of government agencies, with the *Ministry of Home Affairs*, which is responsible for a wide range of domestic policy matters, taking the lead in several areas. Its *Department of Border Management* was established in 2004 to provide more focused attention to the management of India's international coastal and land borders. Among its initiatives is the development of Integrated Check Posts at 13 major entry points on India's land borders to address inadequacies in existing customs, immigration, cargo inspection, and other border control functions. The goal is to house adequately resourced and supported border control agencies in one facility. The *Department of Internal Security* deals with internal law enforcement and security matters, including terrorism. It manages India's Central Armed Police Forces, one of which is the *Central Industrial Security Force (CISF)*. CISF was formed in 1969 to provide security for critical infrastructure across the country, and it currently covers 307 key installations, including 12 major seaports, the Delhi Metro Rail system, and 59 domestic and international airports. It also provides technical

[4]Crown Corporations are wholly owned by the government but operate at "'arm's length" from it. They are established to further certain public needs that the government believes are not being achieved by the private sector (Stastna, 2012).

Table 4.3 India's International Borders

Bordering Country or Zone	Length (in km)
Coastline	7516.6
Bangladesh	4096.7
China	3488.0
Pakistan	3323.0
Nepal	1751.0
Myanmar	1643.0
Bhutan	699.0
Afghanistan	106.0
Total	22,623.3

(*Source*: Government of India, Ministry of Home Affairs. 2013. Annual Report 2012–13. Delhi, p. 28.)

security consulting services to both public and private sector industries (Table 4.3) (Government of India, Ministry of Home Affairs, 2013, pp. 1, 26–29, 37–39, 46, 101, 104).

In 2009, the *Indian Navy* was designated as the lead authority for overall maritime security, with the *Indian Coast Guard* made responsible for coastal security in territorial waters. Coordination is facilitated through four Joint Operations Centres located at major ports (including Mumbai) and the conduct of multi-agency coastal security exercises (Government of India, Ministry of Home Affairs, 2013, pp. 46–47).

Maritime ports and vessels are under the purview of the *Ministry of Shipping*. Among its security functions is to monitor India's compliance with the International Ship and Port Facility Security Code. Although 12 major ports, 53 other ports, and five shipyards have been made ISPS compliant according to security experts, the remaining facilities have not, and the Ministry established a Working Group in 2009 to develop security standards for all Indian ports. The Ministry of Shipping is also responsible for the registration of all boats (including fishing boats) operating in Indian waters, as well as for the installation of identification and tracking transponders on all vessels except for smaller fishing boats (Government of India, Ministry of Home Affairs, 2013, pp. 44–46; Kurup, 2012, p. 50).

Indian Railways is the government-owned Indian national intercity passenger and freight rail system operated through the *Ministry of Railways*. Its Security Directorate seeks to "protect and safeguard railway passengers, passenger areas and railway property," primarily through two security organizations:

- The *Government Railway Police* are generally responsible for the prevention and detection of serious crimes on passenger trains, including track patrolling and the investigation of acts of sabotage. Along with the local police, the Railway Police provide security for the Kolkata Metro rail system.
- The *Railway Protection Force* escorts passenger trains in "vulnerable areas" and provides access control, regulation, and general security on passenger platforms and terminals. In addition, the Force is responsible for freight rail security (Government of India, Indian Railways, n.d.b; Government of India, Ministry of Home Affairs, 2013, p. 27).

The *Bureau of Civil Aviation Security* was made an independent agency under the Ministry of Civil Aviation in 1987 and regulates civil aviation security in India. Its functions include:

- Promulgating aviation security standards in accordance with ICAO Annex 17 for airport operators and airlines
- Monitoring the implementation of security rules and regulations
- Ensuring that persons implementing security measures receive appropriate training and possess all competencies necessary to perform their duties
- Planning and coordinating all aviation security matters
- Conducting unannounced tests to evaluate the performance of security personnel
- Conducting training exercises to evaluate the efficacy of contingency plans and the operational preparedness of all agencies involved in aviation emergencies (Government of India, Bureau of Civil Aviation Security, n.d.a)

United Kingdom

After undergoing major changes in response to the Pan Am Lockerbie and 9/11 disasters, transportation security in the United Kingdom has experienced another significant shift, this time as the result of the May 2010 election of a new government. Under the latest plans, the *Department for Transport* (DfT) retains overall responsibility for transportation policy, including security matters, but its Transportation Security division (TRANSEC) has been abolished, with its responsibilities divided up within the department, and a *Transport Security Strategy Division* was created to coordinate departmental security efforts. The rationale for the changes is outlined in the "Transport Security Annual Report: April 2010-March 2011":

> *In order to develop effective, sensible and fully integrated policies that tackle these constantly evolving threats [to the transportation network], close working with the industry is vital. With this in mind, we have taken the decision to move away from a stand-alone security Directorate to a distributed transport security organization, which will bring our modal security teams (e.g. for aviation, rail and maritime) alongside their non-security counterparts, to enable them to work more closely together and with the industry concerned.*
>
> *Government of the United Kingdom, Department for Transport, 2011, pp. 4, 6*

The United Kingdom's current approach to maritime security was outlined in "The UK National Strategy for Maritime Security," which was issued in May 2014 and explains how the government organizes and utilizes its various capabilities to identify, assess, and address maritime security challenges. Pursuant to the Strategy, a ministerial working group, chaired by the Foreign and Commonwealth Office, was formed and charged with considering maritime security issues and making policy decisions. The Department for Transport was designated as the regulator responsible for the security of U.K. ports and port facilities and for U.K.-flagged vessels. Its security duties include the following:

- Ensuring the U.K. balances commercial interests with safety, security, and environmental considerations;
- Encouraging shipping companies to register in the United Kingdom;
- Implementing and influencing international and European Union safety and security legislation; and
- Representing the United Kingdom in the IMO.

The *Maritime and Coastguard Agency* (*MCA*) supports the DfT security role by being responsible for cargo ship security compliance activity (Government of the United Kingdom, 2014, pp. 12, 48–49).

The DfT's *Land Transport Security Division* has the lead for security in this sector. The Division:

- Sets and enforces security standards for the national rail network, the London Underground, certain light rail systems, and the U.K. portion of the Channel Tunnel;
- Develops and issues recommendations on best security practices for buses, coaches, and certain light rail systems; and
- Inspects and enforces compliance with the U.K.'s secure carriage of dangerous goods by road and rail program (Government of the United Kingdom, n.d.).

Aviation security has undergone the biggest changes, with the passage by Parliament of the Civil Aviation Act 2012. The law makes a number of revisions in the powers and responsibilities of the independent *Civil Aviation Authority (CAA*; heretofore responsible for the regulation of safety since its creation in 1972) "in order to enable the sector to make a full contribution to economic growth without compromising standards." Effective April 2014, the Act transfers a number of security functions previously carried out within the DfT to the CAA, including regulation, inspection, and enforcement and vetting of security personnel, although ultimate responsibility for aviation security policy—including directing industry compliance with national and international security standards—is to remain with DfT. Another key provision of the law transfers the costs of security regulation from the government to industry. In justifying the changes, the U.K. government stated it "believes that industry will benefit from the efficiencies that could be gained through having aviation security and safety regulation in one place. The CAA also has valuable experience of safety management systems designed to manage risks as effectively as possible. The move would also mean that the 'user pays' principle is applied to aviation security as it is currently applied to aviation safety" (Government of the United Kingdom, Department for Transport, 2012, pp. 3, 15–16).

■ ■ Critical Thinking ■

The United Kingdom's Civil Aviation Act 2012 and its transportation security reorganization begun in 2010 represent, in a number of respects, a move toward the type of transportation security organization used in the U.S. before 9/11. Under that approach, security was a component of transportation policy handled along with safety and economic regulation within the main transportation department and its modal divisions rather than within a "stand-alone" security agency, and industry (and its customers) bore most of the cost. Discuss the pros and cons of this approach versus the current U.S. transportation security model.

Conclusion

The institutions of transportation security have continued to evolve in the second decade of the 21st century. At the international level, IMO and the security standards established under the 2002 SOLAS Amendments and the International Ship and Port Facility Security Code have become increasingly important in defining a global security baseline for the maritime sector. ICAO, which has long performed a similar function in aviation, has taken the next step by seeking to promote implementation of its standards through its security audits and training and technical assistance programs. The land mode still lacks international governmental institutions and instruments, but through the recent efforts of such groups as the UIC and UITP, the sharing of best practices and the creation of multinational security guidelines has become more prevalent. However, although these and other international efforts have undoubtedly improved the world's transportation security systems, they remain—as previously—dependent on national governments for their enforcement, and given the national differences in outlook, resources, and circumstances, actual security standards continue to display considerable variation.

In the United States, which has pioneered the very concept of "homeland security," the definition of transportation security roles and responsibilities remains a challenge. At the federal level, DHS houses the key agencies responsible for transportation security: TSA for aviation and land transportation and the Coast Guard for maritime transportation. Yet even within DHS, other agencies (notably Customs and Border Protection) have major transportation security roles, as do a number of non-DHS agencies (especially the modal divisions within the DOT). Furthermore, in the U.S. federalist system, state and local governments retain substantial authority within their jurisdictions, including for transportation security. Progress is being made in describing institutional security roles through such documents as the 2010 Transportation Systems Sector-Specific Plan.

Although largely operating through existing institutions rather than creating a new "homeland security" structure, other nations have also elaborated their institutional framework to address the threat to their transportation systems. Canada has continued to use its transportation department as the focal point for security, but it has created separate entities within that department for maritime and land transportation security and an independent agency to be responsible for screening commercial aviation passengers and their baggage. India's situation in facing a number of external and internal armed threats has produced greater reliance on the Indian military (with the Navy the designated lead for maritime security, for example) and armed police forces (e.g., the Central Industrial Security Force, which provides security for major ports and airports) in its transportation security program. In the United Kingdom, the election of a new government—rather than the traditional response to a security incident—has produced significant institutional changes since 2010, with the new system to be somewhat akin to the pre-9/11 U.S. organization of transportation security (although at a much greater level of resources and requirements).

At all governmental levels, the role of private trade associations and labor unions continues to be substantial in seeking to shape transportation security policies to meet the interests and needs of their constituents.

Discussion Questions

1. What are the major international organizations involved in transportation security, and what are their roles?
2. Describe how the U.S. DHS's Quadrennial Homeland Security Review, Bottom-Up Review, and Transportation Systems SSP define transportation security roles.
3. Briefly describe the transportation security roles of the TSA, Coast Guard, Customs and Border Protection, FEMA, and Office of Infrastructure Protection.
4. What role do state, tribal, and local governments and the private sector (including trade associations and labor unions) play in U.S. transportation security? Give examples.
5. How has the U.S. government defined "homeland security," and why has this concept not been more generally applied around the world?
6. Compare and contrast the organization of national transportation security efforts in the United States, Canada, and the United Kingdom.

References

Airlines for America. 2013. A4A testimony by Sharon Pinkerton before the U.S. House of Representatives Committee on Homeland Security, Subcommittee on Transportation Security, April 11, 2013. <http://www.airlines.org/Pages/A4A-Testimony-by-Sharon-Pinkerton> (accessed 10.22.14.)

American Association of State Highway and Transportation Officials. n.d. AASHTO governing documents. <http://www.transportation.org/Pages/Organization.aspx> (accessed 10.24.14.)

American Association of Port Authorities. n.d. About AAPA. <http://www.aapa-ports.org/About/?navItemNumber=495> (accessed 10.24.14.)

American Waterways Operators. n.d. <http://www.americanwaterways.com/security/awovesselsec.html> (accessed 10.24.14.)

Amtrak. n.d. Our commitment to safety & security. <http://www.amtrak.com> (accessed 10.24.14.)

Airports Council International. n.d. ACI overview: The community of airports. <http://www.aci.aero/About-ACI/Overview> (accessed 10.24.14.)

Airports Council International. 2010. Aviation security. <http://www.aci.aero> (accessed 10.24.14.)

American Bus Association. n.d. Motorcoach security. <http://www.buses.org/Government-Affairs/Current-Legislation/Policy-Papers/Security> (accessed 10.24.14.)

American Federation of Government Employees. 2012. Transportation Security Administration (TSA) and Transportation Security Officers (TSOs). <http://www.afge.org> (accessed 10.24.14.)

American Public Transportation Association. 2013. Testimony submitted to the House Appropriations Subcommittee on Homeland Security on fiscal year 2014 appropriations for the Federal Emergency Management Agency. Washington, DC.

American Trucking Association. n.d. Related to: Security. <http://www.truckline.com/Trucking_Issues_Security.aspx> (accessed 10.24.14.)

Association of American Railroads. n.d. Hazardous materials transportation. <https://www.aar.org/safety/Pages/Hazardous-Materials-Transportation.aspx> (accessed 10.24.14.)

Association of Flight Attendants-CWA. June, 2013. Association of Flight Attendants applauds TSA policy reversal and seeks permanent ban on knives on planes. <http://www.afanet.org/> (accessed 10.24.14.)

Association of Oil Pipe Lines. n.d. About AOPL. <http://www.aopl.org/aboutAOPL/> (accessed 10.24.14.)

Bennett, J.C., 2008. Maritime security. In: Bragdon, C.R. (Ed.), Transportation Security. Butterworth-Heinemann/Elsevier, Burlington, MA, pp. 149–181.

Chamber of Shipping of America. n.d. Chamber of Shipping of America. <http://www.knowships.org> (accessed 10.24.14.)

FEMA. 2012. State Homeland Security Grant Program. <http://www.fema.gov/fy-2013-homeland-security-grant-program-hsgp-0> (accessed 10.24.14.)

FEMA. 2013. National response framework. <http://www.fema.gov/national-response-framework> (accessed 10.24.14.)

GAO. December, 2010. Aviation security: DHS has taken steps to enhance international aviation security and facilitate compliance with international standards, but challenges remain. Washington, DC.

GAO. September, 2012. Maritime security: Progress and challenges 10 years after the Maritime Security Act. Washington, DC.

GAO. February, 2013. Department of Homeland Security: Progress made and work remaining after nearly 10 years in operation. Washington, DC.

Government of Canada, Canada Border Services Agency. n.d. Facilitating trade. <http://www.cbsa-asfc.gc.ca/trade-commerce/facil-eng.html> (accessed 10.30.14.)

Government of Canada, Public Safety Canada. n.d. About Public Safety Canada <http://www.publicsafety.gc.ca/cnt/bt/index-eng.aspx> (accessed 10.30.14.)

Government of Canada, RCMP. n.d. About the RCMP. <http://www.rcmp-grc.gc.ca/about-ausjet/index-eng.htm> (accessed 10.24.14.)

Government of Canada, Transport Canada. n.d.a. Marine security. <http://www.tc.gc.ca/eng/marinesecurity> (accessed 10.29.14.)

Government of Canada, Transport Canada. n.d.b. Rail security. <http://www.tc.gc.ca/eng/railsecurity/menu.htm> (accessed 10.24.14.)

Government of Canada, Transport Canada. n.d.c. Road security. <http://www.tc.gc.ca/eng/roadsecurity/menu.htm> (accessed 10.24.14.)

Government of Canada, Transport Canada. 2013. Canada's National Civil Aviation Security Program. Ottawa.

Government of India, Bureau of Civil Aviation Security. n.d.a. About us. <http://bcasindia.nic.in/aboutus/aboutus.html> (accessed 10.24.14.)

Government of India, Indian Railways. n.d.b. About Indian Railways. <http://www.indianrailways.gov.in/railway board/> (accessed 10.24.14.)

Government of India, Ministry of Home Affairs. 2013. Annual Report 2012-13. Delhi.

Government of the United Kingdom. n.d. Land Transport Security Division. <http://www.gov.uk/government/groups/land-transport-security-division> (accessed 10.30.14.)

Government of the United Kingdom. May, 2014. The UK National Strategy for Maritime Security. London.

Government of the United Kingdom, Department for Transport. November, 2011. Transport security annual report: April 2010–March 2011. London.

Government of the United Kingdom, Department for Transport. December, 2012. Civil Aviation Act 2012 questions & answers. London.

IATA. n.d. <http://www.iata.org> (accessed 10.24.14.)

Inland Rivers Ports & Terminals. n.d. Our mission. <http://www.irpt.net/organization/> (accessed 10.24.14.)

International Association of Public Transport. n.d. What is UITP? <http://www.uitp.org/about/What_is_UITP.cfm> (accessed 10.24.14.)

International Association of Public Transport. 2010. Secure public transport in a changeable world. <http://www.uitp.org> (accessed 10.24.14.)

ICC Commercial Crime Services. n.d. International Maritime Bureau. <http://www.icc-ccs.org/icc/imb> (accessed 10.24.14.)

International Chamber of Shipping. n.d. What is ICS? <http://www.ics-shipping.org> (accessed 10.24.14.)

International Civil Aviation Organization. n.d. About ICAO. <http://www.icao.int/about-icao/Pages/default.aspx> (accessed 10.24.14.)

International Civil Aviation Organization. 2014. Annual report of the Council—2013: Universal Security Audit Programme. <http://www.icao.int/annual-report-2013/Pages/progress-on-icaos-strategic-objectives-strategic-objective-b-security-universal-security-audit-programme.aspx> (accessed 10.24.14.)

International Federation of Air Line Pilots' Associations. n.d. <http://www.ifalpa.org> (accessed 10.24.14.)

International Longshoremen's Association. n.d. Welcome. <http://www.ilaunion.org/index.html> (accessed 10.24.14.)

International Maritime Organization. 2013. IMO—What It Is. London.

International Transport Workers' Federation. n.d.a. About the civil aviation section. <http://www.itfglobal.org/civil-aviation/about.cfm> (accessed 10.24.14.)

International Transport Workers' Federation. n.d.b. Aviation security principles. <https://www.itfglobal.org/files/extranet/-1/769/Security%20Policy.pdf> (accessed 10.24.14.)

International Transport Workers' Federation. n.d.c. Sefarers, dockers, inland navigation section. <http://www.itfglobal.org/> (accessed 10.24.14.)

International Transport Workers' Federation. 2006. Out of sight, out of mind. <http://www.itfglobal.org/seafarers/index.cfm> (accessed 10.24.14.)

International Union of Railways. n.d. Introduction. <http://www.uic.org/spip.php?article528&lang=en> (accessed 10.24.14.)

Interstate Natural Gas Association of America. n.d. Security. <http://www.ingaa.org/Topics/Security.aspx> (accessed 10.24.14.)

Johnstone, R.W., 2006. 9/11 and the Future of Transportation Security. Praeger, Westport, CT.

Kurup, V., June, 2012. Shielding seas. Maritime Gateway, 50–51.

Mayer, M.A., Baca, L., September, 2010. Want real homeland security? Give state and local governments a real voice. The Heritage Foundation, Washington, DC.

McNicholas, M., 2008. Maritime Security. Butterworth-Heinemann/Elsevier, Burlington, MA.

Morag, N., 2011. Does homeland security exist outside the United States? Homeland Security Affairs Journal 7, the 9/11 Essays.

National Association of Waterfront Employers. n.d. Industry overview. <http://www.nawe.us/information/> (accessed 10.24.14.)

Passenger Vessel Association. n.d. About PVA. <http://www.passengervessel.com/about.aspx> (accessed 10.24.14)

Price, J.C., Forrest, J.S., January, 2009. Practical Aviation Security: Predicting and Preventing Future Threats. Butterworth-Heinemann/Elsevier, Burlington, MA.

Reese, S., 2013. Defining homeland security: Analysis and Congressional considerations. Congressional Research Service, Washington, DC.

Stastna, K., April, 2012. What are crown corporations and why do they exist? CBC News.

Transport Workers Union. n.d. Our union. <http://www.twu.org/OurUnion/> (accessed 10.24.14.)

United Transportation Union. n.d. About UTU. <http://www.utu.org> (accessed 10.24.14.)

U.S. Department of Homeland Security. n.d. Department components. <http://www.dhs.gov/department-components> (accessed 10.24.14.)

U.S. Department of Homeland Security. 2009. National infrastructure protection plan. <http://www.dhs.gov/national-infrastructure-protection-plan> (accessed 10.24.14.)

U.S. Department of Homeland Security. 2010a. Transportation systems sector-specific plan: an annex to the National Infrastructure Protection Plan. Washington, DC.

U.S. Department of Homeland Security. 2010b. Quadrennial homeland security review report: a strategic framework for a secure homeland. Washington, DC.

U.S. Department of Homeland Security. 2010c. Bottom-Up Review Report. Washington, DC.

U.S. Department of Homeland Security. 2013. Budget-in-brief fiscal year 2014. Washington, DC.

U.S. Department of Homeland Security. 2014a. Budget-in-brief fiscal year 2015. Washington, DC.

U.S. Department of Homeland Security. 2014b. 2014 quadrennial homeland security review. Washington, DC.

World Customs Organization. 2012. WCO SAFE framework of standards. <http://www.wcoomd.org/en> (accessed 10.24.14.)

World Shipping Council. n.d. History. <http://www.worldshipping.org/about-the-council/history> (accessed 10.24.14.)

Transportation Security Policymaking

CHAPTER OBJECTIVES:

In this chapter, you will learn about how transportation security policy is made, including:

- An overview of public policy and how "policy goals and means" are adopted
- The authorizing, budget, and appropriations processes in the United States and how they operate with respect to transportation security
- The role of presidential directives and rulemaking in setting U.S. transportation security policy
- Specific examples of the policymaking process at work through case studies of authorization, appropriations, and rulemaking proceedings
- An outline of the policymaking process in parliamentary systems

Introduction

Much of the attention to transportation security in news accounts, as well as scholarly assessments, is devoted to examining the performance of front-line personnel—including checkpoint screeners at airports, customs inspectors at ports of entry, and transit systems' operational and security personnel, among others—in implementing security measures. Yet all of these activities are grounded in legislative acts that authorize and fund programs, and directives and regulations promulgated by high-level officials within the executive branch. This is the realm of policy.

Public policy may be defined as "a course of government action or inaction in response to public problems. It is associated with formally approved policy goals and means, as well as the regulations and practices of agencies that implement programs." In this chapter, the focus is on the processes involved in adopting "policy goals and means." Implementation and the additional policy stage of evaluation are addressed in subsequent chapters (Kraft and Furlong, 2007, p. 5).

Before the formal adoption of a policy, a problem must come to the attention of policymakers. In the words of Kraft and Furlong (2007):

> *The mere existence of a problem is no guarantee that it will attract government attention or be acted on...When policymakers begin active discussions about a problem and potential solutions, the issue is said to be "on the agenda." .... Some issues make it to the agenda automatically. They are mandated, or required, actions with which government must deal. Examples include passing the annual budget, legislating to*

reauthorize existing programs, and acting on a president's or governor's nominees for executive appointments .... A focusing event, such as a crisis, usually improves an issue's chance of getting on the agenda, in part because of the exceptional media coverage it receives. The terrorist attacks of September 11, 2001, clearly altered the agenda status of airport and airline security in extraordinary ways.

Kraft and Furlong, 2007, pp. 74–77

In the case of transportation security issues, the initial impetus for getting on the policy agenda has almost always come from the reactions of the public, the news media, and policymakers themselves to specific incidents when transportation systems have been attacked. However, as the institutions of transportation security have grown and matured, increasingly more of the policy considerations fall into the "automatic" categories of appropriations and presidential nominations.

In the United States, the policymaking function is shared by Congress, composed of the House of Representatives and the Senate, and the executive branch, led by the president.

Under the U.S. Constitution's separation of powers doctrine, the two branches operate distinctly from each other. Indeed, the independence of the Congress from the president is one of the distinguishing characteristics of the American system, in contrast to most parliamentary democracies (including the United Kingdom, Canada, and India) where the executive and legislative branches are interconnected, with the chief executive elected by, and accountable to, the members of the legislative body. On a related note, the U.S. system also features "checks and balances" in which no branch (legislative, executive, judicial) can exercise authority in any policymaking sphere without being subject to some form of constraint by one or both of the other branches. An example is the presidential power to veto most forms of legislation. Figure 5.1 shows the U.S. Capitol.

FIGURE 5.1 U.S. Capitol. *(From Wikimedia Commons, http://upload.wikimedia.org/wikipedia/commons/b/b2/United_States_Capitol_-_west_front.jpg.)*

Beyond these formal governmental institutions, interest groups can have a major impact on the making of public policy. Transportation security policy is significantly affected by the actions of such entities, including trade associations and labor unions, with a direct economic stake in policy decisions. They may be "involved in direct lobbying of policymakers, indirect or grassroots lobbying aimed at mobilizing the public or the group's supporters, and public education campaigns." In addition, "because of the complexity of public problems and policies, and the often detailed knowledge required to understand them," interest groups with special expertise may work with officials in the Congress or the executive branch (or both) to develop or modify policy proposals. According to Kraft and Furlong:

> *These networks or subsystems are … important. To varying degrees, their participants remain preoccupied with narrow economic interests; they may afford limited participation beyond their core members; and they may be able to resist external influences. If nothing else, it is clear that much U.S. policymaking involves informal networks of communication in which prevailing policy ideas and the evaluation of new studies and information shape what is likely to be acceptable to the major policy actors.*
>
> *Kraft and Furlong, 2007, pp. 54–58*

A number of different instruments are used in the United States (with similar counterparts in other nations) to set transportation security policy. The most important of these are:

- *Authorizing legislation*, which may "establish, continue, or modify an agency, program, or activity for a fixed or indefinite period of time … set forth the duties and functions of an agency or program, its organizational structure, and the responsibilities of agency or program officials … and authorize, implicitly or explicitly, the enactment of appropriations for an agency or program"
- *Federal budget process*, which begins with the submission of the president's budget and continues with action in the House and Senate on a Congressional budget resolution
- *Appropriations legislation*, which provide funds for federal agencies, programs, or activities
- *Presidential directives and nominations*
- *Federal regulation*, which is the means by which laws and certain presidential orders are carried out (Office of Management and Budget, 2013b, pp. 117–120; Heniff, 2012a, pp. 1–2; Carey, 2013, p. 1).

Authorizing Legislation

The legislative process in the U.S. Congress has multiple stages (Figure 5.2), with possibilities for delay at each stage. After a bill is introduced by an individual member of the House or Senate (sometimes after being developed with the help of outside parties, such as trade

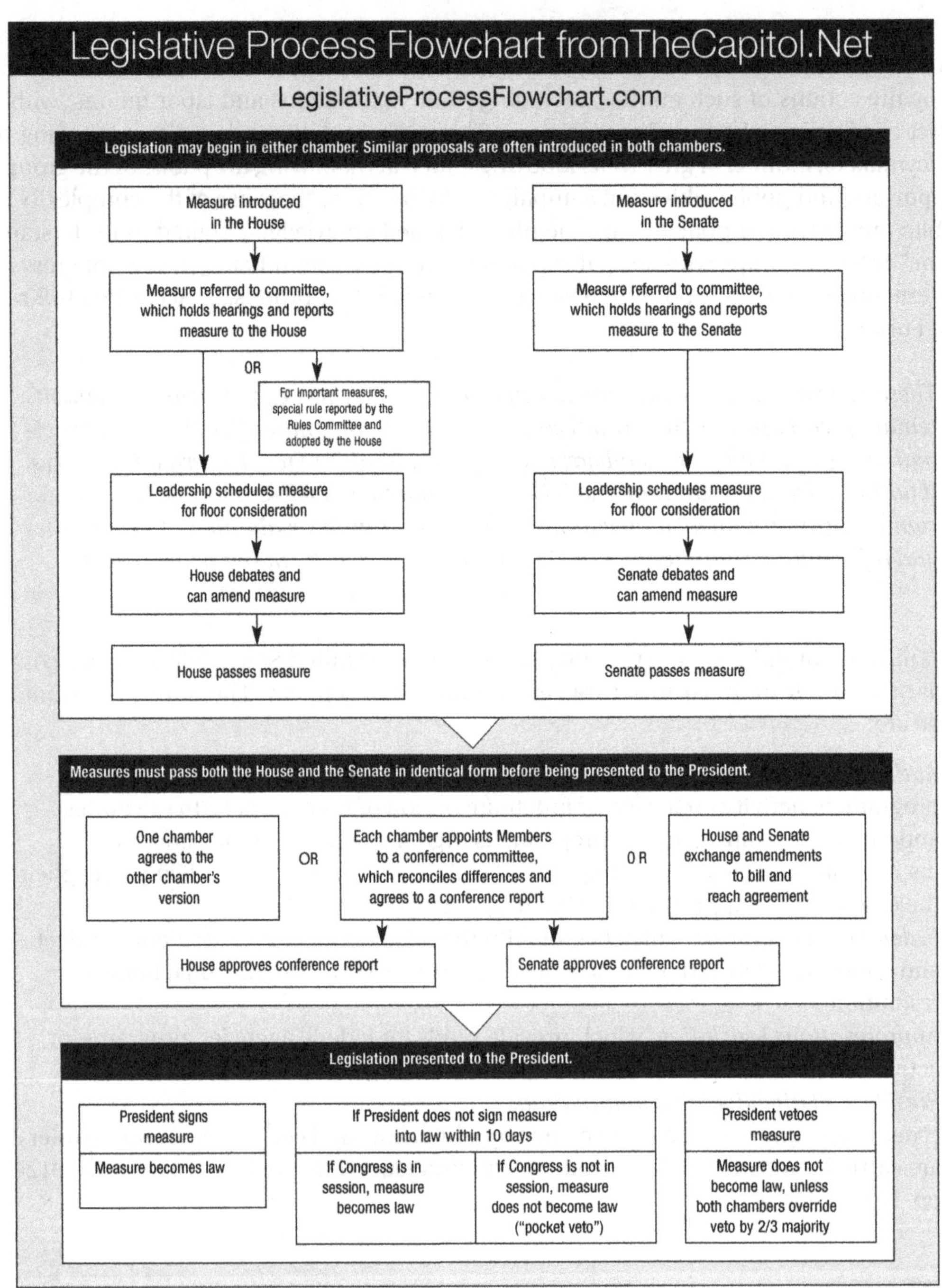
Legislative Process Flowchart fromTheCapitol.Net
LegislativeProcessFlowchart.com
Legislation may begin in either chamber. Similar proposals are often introduced in both chambers.
Measure introduced in the House
Measure introduced in the Senate
Measure referred to committee, which holds hearings and reports measure to the House
Measure referred to committee, which holds hearings and reports measure to the Senate
OR
For important measures, special rule reported by the Rules Committee and adopted by the House
Leadership schedules measure for floor consideration
Leadership schedules measure for floor consideration
House debates and can amend measure
Senate debates and can amend measure
House passes measure
Senate passes measure
Measures must pass both the House and the Senate in identical form before being presented to the President.
One chamber agrees to the other chamber's version
OR
Each chamber appoints Members to a conference committee, which reconciles differences and agrees to a conference report
OR
House and Senate exchange amendments to bill and reach agreement
House approves conference report
Senate approves conference report
Legislation presented to the President.
President signs measure
Measure becomes law
If President does not sign measure into law within 10 days
If Congress is in session, measure becomes law
If Congress is not in session, measure does not become law ("pocket veto")
President vetoes measure
Measure does not become law, unless both chambers override veto by 2/3 majority

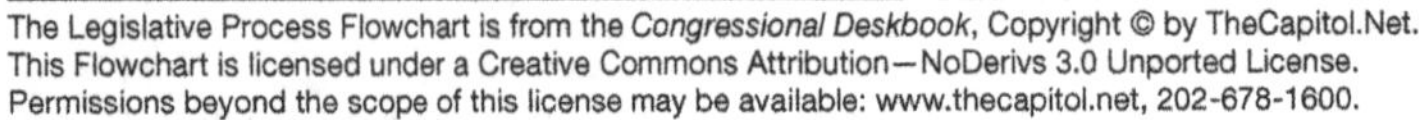

BY
ND

FIGURE 5.2 Congressional legislative process.

associations), it is referred to the committee (sometimes multiple committees) that has jurisdiction over its subject matter. For homeland and transportation security, the primary (but by no means only) committees that handle authorizing legislation are the House Committee on Homeland Security (which was established in 2002) and the Senate Committee on Homeland Security and Governmental Affairs (which had the homeland security mission added to its jurisdiction and title in the aftermath of the 9/11 attacks). The jurisdiction of both committees is essentially confined to the Department of Homeland Security (DHS), so they have limited roles in non-DHS homeland security matters.

The key authorization measures for transportation security were all adopted within the first 14 months after the 9/11 hijackings:

- The *Aviation and Transportation Security Act of 2001 (ATSA)*, which created the TSA and assigned it responsibility for transportation security, established a number of deadlines for the deployment of specific commercial aviation security measures, and authorized a passenger fee to help pay for the expanded aviation security program
- The *Maritime Transportation Security Act of 2002 (MTSA)*, which established a national maritime security system, with specific security requirements for federal agencies, local port authorities, and maritime vessel owners
- The *Homeland Security Act of 2002 (HSA)*, which created the DHS via the merger of 22 existing federal departments and agencies, including TSA, the Coast Guard (given lead responsibility for port, waterway and coastal security), and the Customs Service

After the 9/11 Commission presented its recommendations to Congress and the president, the *Intelligence Reform and Terrorism Prevention Act* of 2004 was enacted. This measure was designed to address the Commission's recommendations and other perceived shortcomings in homeland security. It adopted most of the 9/11 Commission's transportation security–related proposals, including those calling for a TSA takeover of the administration of the "no fly" list, prioritization of improved detection of explosives on passengers or in baggage, and deployment of advanced "in-line" baggage screening systems at airports.

Since that time, only two major transportation security authorization measures have been enacted: the Security and Accountability for Every Port Act of 2006 (SAFE Port Act) and the Implementing the Recommendations of the 9/11 Commission Act of 2007.

Security and Accountability for Every Port Act of 2006 (SAFE Port Act)

The SAFE Port Act, which was signed into law by President Bush on October 13, 2006 (PL 109-347), made amendments to the original MTSA framework for port security, codified certain programs that had been developed outside MTSA, and created new programs. Its major provisions included (1) the codification of Customs and Border Protection's (CBP's) Container Security Initiative (CSI) and Customs-Trade Partnership Against Terrorism (C-TPAT) for enhancing container security; (2) the establishment within DHS of the Domestic Nuclear Detection Office (DNDO) to conduct research, development, testing, and evaluation of radiation detection equipment; (3) a requirement for the creation of interagency operational centers to organize security at selected major ports within 3 years; (4) the establishment of an implementation schedule and fee restrictions for the Transportation Worker Identification Credential (TWIC);[1] (5) a mandate that all shipping containers entering high-volume U.S. ports be scanned for radiation by no later than December 31, 2007; (6) a requirement that shippers make additional information available to CBP for its use in targeting cargo containers for inspection; and 7) a mandate for the Coast Guard to track all large commercial vessels operating within U.S. waters by April 1, 2007 (McNicholas, 2008, pp. 128–129; Government Accountability Office [GAO], 2007, p. 2).

IMPLEMENTING RECOMMENDATIONS OF THE 9/11 COMMISSION ACT OF 2007
How a Bill Became a Law

Two major factors placed further consideration of the 9/11 Commission recommendations back on the congressional agenda at the start of 2007. First, the Commission formed a successor group, the 9/11 Public Discourse Project (9/11 PDP), to monitor the status of its proposals. On December 5, 2005, the group issued its Final Report, which provided letter grades for the federal government's implementation efforts and found that full implementation of the Commission recommendations was lacking in many areas. In transportation security, the PDP assigned grades of "C" to improving airline passenger explosives screening, "C-" to development of a National Strategy for Transportation Security, "D" to improving checked bag and cargo screening, and "F" to the TSA assumption of full responsibility for airline passenger prescreening (i.e., "no fly" list). PDP members, as well as organizations representing the families of 9/11 victims, used this document to push for further congressional action throughout 2006 (Johnstone, 2006, p. 53).

The other factor was political. In the 2002 and 2004 elections, President Bush and Congressional Republicans had generally benefitted from the issue of homeland security, with the president winning reelection and the Republicans obtaining significant majorities in both houses of Congress as a result of the 2004 contests. As the 2006 elections approached, Congressional Democrats (especially in the House) made full implementation of the 9/11 Commission's recommendations one of their top priorities. After Democratic victories in those elections that turned control of the House and Senate over to the Democrats, newly installed House Speaker Nancy Pelosi (D-CA) announced that legislation to implement the 9/11 Commission recommendations would be taken and passed within the "first 100 hours" of the new Congress (SourceWatch, 2008).

Fulfilling Pelosi's pledge meant that a highly unusual and expedited process would be used. First, the bill was introduced on the second day of Congress (January 5) and, pursuant to standard practice, was immediately referred to the committees of jurisdiction. However, the normal procedure of committee consideration (hearings) and deliberation (mark-up and reporting out) was bypassed, and the measure was brought to the House floor just 4 days later under terms of a special rule that limited floor action to 3 hours of debate, with only one amendment to be in order.

Despite congressional Republican and White House objections to the manner in which the legislation was developed and considered, as well as to provisions providing collective bargaining rights for TSA's checkpoint screeners and mandating the physical screening for radiation of all container cargo entering the United States from more than 600 international ports, the House approved the Implementing Recommendations of the 9/11 Commission Act by a 299 to 128 margin (with 231 Democrats and 68 Republicans voting yes and 128 Republicans voting no) on January 9.

The Senate followed its regular procedures, with its version (Improving America's Security Act) introduced on January 4 and referred to the Committee on Homeland Security and Governmental Affairs, which considered and modified the original bill and reported it out to the full Senate on February 22 (still representing a much faster than usual process). The Senate considered the bill with no limits on time for debate or amendment (189 amendments were drafted, but most of these were either not offered or ruled out of order for technical reasons). Senate floor consideration commenced on February 27 and continued into March, with final approval (by a vote of 60 to 38, with 48 Democrats, 10 Republicans and 2 Independents voting yes and 38 Republicans voting no) coming on March 13. The Senate bill was similar to the House version in most respects but eased the requirements for container screening.

After Senate approval, however, there was a longer than normal delay in appointment of a conference committee to reconcile the competing versions caused largely by White House concerns (and a threatened veto) over certain provisions in both bills (including collective bargaining for airport screeners). After lengthy negotiations, agreement was reached between Congressional Democrats and the president (which included dropping the screener bargaining rights), clearing the way for appointment of Senate conferees (on July 9) and House conferees (on July 17). The conference report containing the final bill language agreed upon by the conferees was filed on July 25, approved by the Senate on July 26 (by a vote of 85 to 8, with 46 Democrats, 37 Republicans, and 2 Independents voting yes and 8 Republicans voting no) and the House on July 27 (by a vote of 371 to 40, with 221 Democrats and 150 Republicans voting yes and 1 Democrat and 39 Republicans voting no). Although still voicing concerns about the final version's container screening provisions, the president signed the bill into law on August 3, 2007 (PL 110-53) (Library of Congress, 2007; NBCNews.com, 2007).

Implementing the Recommendations of the 9/11 Commission Act of 2007

The Implementing Recommendations of the 9/11 Commission Act covered the whole range of subject matter addressed in the Commission's proposals, including emergency response, intelligence, border control, civil liberties, and foreign policy. One notable omission was the absence of further action in implementing the Commission's recommendation that Congress streamline its system of oversight of homeland security by creating, in each house, "a single, principal point of oversight and review for homeland security." (Although both the House and Senate possess homeland security committees, these share

[1]A program administered jointly by TSA and the Coast Guard aiming to provide a tamper-resistant, biometric identification credential to be used in granting maritime workers unescorted access to secure areas of ports as well as maritime vessels (McNicholas, 2008, p. 128).

jurisdiction with a number of other congressional panels, including the homeland security appropriations subcommittees in each chamber.) The new law contained a number of transportation security–related provisions derived from the 9/11 Commission recommendations and in some cases (notably, in land transportation) going beyond them.

Maritime Security Provisions

- Established a deadline of July 1, 2012, by which all containers loaded on vessels in foreign ports must be scanned by nonintrusive imaging and radiation detection equipment at a foreign port before being permitted entry to the United States, but authorized DHS to extend the deadline by 2 years initially and to renew the extension in 2-year intervals, provided that the DHS certified to Congress that certain conditions could not be met. The conference report also called on DHS to work with the State Department and the United States Trade Representative in pressing for the establishment of an international framework for scanning and securing containers.

Land Transportation Security Provisions

- Established the National Strategy for Public Transportation Security, which included conduct of security assessments of transit systems, security grants to public transportation agencies, security training and exercise programs for public transit workers, and identification checks against the consolidated terrorist watchlist and on the immigration status for public transportation employees. The Act authorized up to $650 million in fiscal year (FY) 2008, $750 million in FY 2009, $900 million in FY 2010, and $1.1 billion in FY 2011 for the security grant program.
- Created similar security programs for the railroad, bus, and trucking transportation modes. The award of railroad security assistance grants was made discretionary for DHS, and the authorization levels for bus security assistance were set at $12 million for FY 2008 and $25 million per year for fiscal years 2009 to 2011.
- Increased the number of surface (land) transportation security inspectors (who were primarily assigned to the rail and transit modes) from 100 in FY 2007 to 200 by FY 2010.

Aviation Provisions

- Required that the level of screening for cargo placed on passenger aircraft must be "commensurate" with that of passenger checked bag screening and mandated that 50% of all cargo on passenger aircraft be screened within 18 months of the bill's enactment and 100% be screened within 3 years.
- Authorized up to $450 million in discretionary funds per year for fiscal years 2008 to 2011 "to fund the installation of in-line Explosives Detection Systems at U.S. airports at a level approximate to the TSA's strategic plan for the deployment of such systems." These amounts were to be used in combination with the $250 million per year in funding from the Aviation Security Capital Fund dedicated to this purpose.

- Mandated that DHS submit a plan to Congress within 120 days of enactment that included timelines for testing and implementation of DHS's advanced airline passenger prescreening system (now called "Secure Flight")
- Required that a standardized threat and vulnerability assessment program be established for General Aviation airports and that TSA conduct a feasibility study of a security grant program for such airports and devise a means by which general aviation aircraft entering the United States from a foreign location submit passenger information to TSA for checking against appropriate watchlists

Multimodal Provisions

- Increased the use of canine explosives detection teams for land transportation and aviation systems (U.S. House of Representatives, 2007, pp. 289–290, 330–371)

Department of Homeland Security Authorization

One of the key methods through which congressional authorizing committees exert influence over both federal agencies and the congressional appropriations process is through development of legislation to regularly reauthorize a department or agency and its programs. The rationale for such action in the case of the DHS was laid out in the committee report filed by the House Committee on Homeland Security to accompany the DHS Authorization Act for Fiscal Year 2006 developed by the Committee:

> *DHS is the third largest Cabinet agency, and its challenges are surely magnified by the fact that it is the result of a recent merger of 22 legacy agencies, each of which brought with it its own policies, systems, processes, and culture. The complexity of the Department's missions, coupled with the enormity of its management and operational challenges, requires the close and continuing oversight that an annual Congressional re-authorization provides. Like the Department of Defense and the Intelligence Community agencies, DHS is—first and foremost—a national security agency. And like those other national security agencies, DHS should be subject to an annual authorization process through which the evolving needs of the Department can be met, and through which Congressional direction, oversight, and prioritization can take place. An annual authorization will help the Department improve the overall management and integration of its various legacy agencies, to guide resource allocation and prioritization, to set clear and achievable benchmarks for progress and success, and to enhance the Department's implementation of its critical mission.*
>
> *U.S. House of Representatives, Committee on Homeland Security, 2005, pp. 23–24*

Despite those sentiments, since the initial creation of DHS in the Homeland Security Act of 2002, Congress has never enacted an authorization bill for the department. On two occasions (in 2005 and 2007), such a measure was passed by the House but received no further action in the Senate. The Senate Committee on Homeland Security and Governmental

Affairs finally succeeded in approving a DHS authorization bill in 2011, but the legislation was not brought to the Senate floor until the closing days of the 112th Congress in December 2012 and was never voted on. In contrast to this record, the House and Senate Armed Services Committees annually develop authorization bills for the Department of Defense, and these measures have been enacted into law in every single year throughout the existence of DHS (2003-present).

In the absence of regular authorizing legislation, congressional policymaking with respect to homeland security has generally been exercised through the annual appropriations process or through periodic specialized bills (e.g., the Implementing Recommendations of the 9/11 Commission Act).

■ ■ Critical Thinking ■

In your opinion, why has Congress never adopted an authorization bill for DHS since the initial creation of the Department? Consider the circumstances usually required to produce action on homeland security issues, congressional and presidential politics, the relative influence of the homeland security authorizing committees, and other factors as you see fit.

Federal Budget Process

The budget for the U.S. government "provides the means for the President and the Congress to decide how much money to spend, what to spend it on, and how to raise the money they have decided to spend." In these terms, it would include all individual authorization, appropriations and revenue legislation through which policy decisions are made. However, as the size and scope of the federal government grew over the course of the 20th century and deficits became a recurring problem, a distinct process was developed to coordinate and enforce overall budgetary decision making.

The budgetary process in the United States derives largely from two statutes, the Budget and Accounting Act of 1921, which established the system for executive branch budgeting, and the Congressional Budget and Impoundment Control Act of 1974, which created a formal congressional budget process. Although there have been modifications to each, these remain the basis for the current system (Office of Management and Budget, 2013b, p. 117; Heniff, 2012b, p. 1).

Presidential Budget Process

The U.S. Constitution did not lay out a clear role for the president and the executive branch with respect to spending and taxation, which were defined explicitly as Congressional authorities, and for more than 130 years, no system was put in place to provide for "a coordinated set of actions covering all federal spending and revenues." The Budgeting and Accounting Act of 1921 changed that by requiring the president to submit an annual budget to Congress, creating the Bureau of the Budget (since renamed the Office of Management and Budget [OMB]) within the executive branch to assist in preparing the president's budget, and establishing GAO (originally the General Accounting Office but later renamed the Government Accountability Office) as the primary auditing agency within the federal government.

In its current form, the presidential budget process begins with the issuance of general budgetary guidelines by the president and OMB to federal agencies (including DHS and its components) in the spring, at least 9 months before the submission of the budget to Congress (required by law to occur not later than the first Monday in February) and 17 months before the beginning of the fiscal year (October 1) to which it pertains. The agencies then formulate their own recommendations in consultation with OMB and by early fall submit their formal budget requests to that office. Final decision making by the president and OMB is usually completed by the end of December to allow time for the preparation and presentation of the budget to Congress by early February. The president must submit a budget update by July 15 (reflecting changing economic conditions and congressional actions, among other things) and is permitted to offer revised budget proposals at any point (Heniff, 2012b, pp. 9–11).

FEDERAL BUDGETING

Key Concepts and Terms

Budget authority: Amount of money agencies are authorized to spend in current or future years. It may be provided by an annual appropriations bill (also called discretionary spending) or by an authorization bill (mandatory or direct spending). Budget authority is the primary means used by Congress in measuring federal spending obligations in a given fiscal year.

Functional classification: Categorical division of the federal budget based on the major purpose being served (such as agriculture or national defense) rather than by agency. Currently, there are 20 major functions, 17 of which represent broad national needs and are further divided into subfunctions.

Outlays: Disbursements from the Federal Treasury in the form of checks or cash that reflect actual federal spending in a given fiscal year. Outlays result from both prior and current fiscal year budget authority.

Rescission: Cancellation of budget authority before the time that authority would otherwise cease to be available for obligation (spending).

Sequestration: Presidential order permanently cancelling budget authority in nonexempt programs by a uniform percentage to achieve a required amount of outlay savings.

User fees: Charges levied on select individuals or organizations directly benefitting from or subject to regulation by a government program or activity and generally treated (for budgetary and appropriations purposes) as an offset against outlays. User fees are classified as either discretionary (when controlled through annual appropriations acts) or mandatory (when controlled by authorizations).

Source: Office of Management and Budget. April 10, 2013b. pp. 124, 127, 129, 221

Release of the president's budget for FY 2014 (beginning October 1, 2013) was delayed until April 14, 2013, because of a continuing budgetary stalemate between President Obama and congressional Republicans. Strong disagreements over federal spending and taxation had roiled the federal budgetary process since Republicans regained control of the House of Representatives in the 2010 elections, producing lengthy delays in action on both the budget and appropriations bills. In an attempt to resolve some of the differences and address a widening budget deficit, the Budget Control Act (BCA) of 2011 (PL 112-25) was enacted on August 2, 2011. The BCA:

- Established separate discretionary spending limits for security and nonsecurity categories for FY 2012 to FY 2021 (with DHS included under the security category)
- Created a Congressional Joint Select Committee on Deficit Reduction, which was instructed to develop a proposal to reduce the federal budget deficit by at least an additional $1.5 trillion over the FY 2012 to FY 2021 period
- Provided that, in the event the Joint Committee did not succeed by January 15, 2012 in developing and getting enacted legislation to reduce the deficit by at least $1.2 trillion during the 10-year period, an automatic process (called sequestration) was to be used to achieve the $1.2 trillion deficit reduction target beginning on January 2, 2013, with the cuts equally divided between the Department of Defense and all other federal agencies

The Joint Committee was unsuccessful in meeting its goal, but before the sequester was triggered, on January 2, 2013, the American Taxpayer Relief Act of 2012 was signed into law (PL 112-240). Among other provisions, the measure postponed the effective date for imposition of sequestration until March 1, 2013, restored the previous spending caps and further reduced those limits by $4 billion in FY 2013 and $8 billion in FY 2014 (split evenly between defense and nondefense) (Office of Management and Budget, 2013b, pp. 120–123; Library of Congress, 2011).

On March 1, 2013, the OMB issued its sequestration order for FY 2013. To achieve the required $85 billion in savings, the OMB determined that nonexempt defense discretionary spending would have to be cut by 7.8% and nonexempt nondefense discretionary funding (including most transportation security programs) would have to be reduced by 5% across-the-board. For example, at TSA the sequestration order resulted in the following reductions in previously appropriated spending levels: $49 million for Federal Air Marshals, $276 million for aviation security programs, $7 million for land transportation security, $52 million for transportation security support, and $12 million for transportation threat assessment and credentialing (Office of Management and Budget, 2013a, pp. 1, 27).

Two key documents describe the President's budget for homeland and transportation security: the "Homeland Security Funding Analysis" included within the *Analytical Perspectives* portion of OMB's annual *Budget of the United States Government* and DHS's annual *Budget-in-Brief.* The former was required by a provision in the Homeland Security Act of 2002 and is particularly important in attempting to assess overall federal homeland security efforts given that (1) a total of 31 departments and independent agencies would receive homeland security funding under the president's FY 2014 budget, with DHS accounting for just 49% of the total (followed by the Defense Department at 24% and the Department of Health and Human Services at 6%); (2) for budgetary purposes, homeland security is not classified as a separate functional category but rather is spread across all 17 of the major functional categories; and (3) the agencies within DHS perform a number of non–homeland security functions (including immigration services, Coast Guard rescue operations, and many others), which represented more than 41% of the total amount requested by the president in the FY 2014 DHS budget. Under these circumstances, it is difficult enough to simply account for homeland security spending, let alone to subject it to analytical and accountability measures. The OMB report helps to fill the gap, but it is limited in size (a total of eight pages in the FY 2014 analysis) and scope. As required by the Homeland Security Act, it divides homeland security spending into three broad categories: "prevent and disrupt terrorist attacks," "protect the American people, our critical infrastructure, and key resources," and "respond to and recover from incidents," with limited detail supplied beyond calculation of departmental-level spending in each category. Transportation security falls mostly into the first category, but some elements are included in the other two as well (Table 5.1) (Office of Management and Budget, 2013b, pp. 415–422).

Table 5.1 President's Fiscal Year 2014 Budget Request for the Department of Homeland Security and the Transportation Security Administration (in $thousands)

	FY 2012 Final[†]	FY 2013 Estimate[†,¶]	FY 2014 Proposal	FY 2014 +/– FY 2012
Department of Homeland Security				
Net discretionary	46,381,144	46,560,550	44,672,346	–1,708,798
+ Discretionary fees	3,515,166	3,639,720	3,785,021	269,855
– Rescissions[‡]	196,468	131,412	—	—
= Gross discretionary	49,699,841	50,068,858	48,457,367	–1,438,943
+ Mandatory[§]	10,271,646	10,616,486	11,501,970	1,230,324
= Total budget authority*	59,971,487	60,685,344	59,959,337	–208,619
Transportation Security Administration (included within DHS totals)				
Net discretionary	7,598,957	7,669,463	7,140,988	–457,969
+ Discretionary fees	2091	2167	2307	216
– Rescissions[‡]	71,596	16,296	—	–71,596
= Gross discretionary	7,529,452	7,655,334	7,143,295	–386,157
+ Mandatory[§]	254,890	255,000	255,000	110
= Total budget authority*	7,784,342	7,910,334	7,398,295	–386,047

*Totals may not add due to rounding.

[†]Fiscal year (FY) 2012 final and FY 2013 estimate include revisions made by the Consolidated Appropriations Act of 2012 (PL 112-74) and the BCA of 2011 (PL 112-25).

[‡]Rescissions are from prior year unobligated balances.

[§]Mandatory includes mandatory spending and user fees and trust funds.

[¶]For comparability purposes, the FY 2013 estimate for Department of Homeland Security (DHS) excludes $12.1 billion provided by the Disaster Relief Appropriations Act of 2013 (PL 113-2).

(*Source*: U.S. Department of Homeland Security. April, 2013. Budget-in-brief: Fiscal Year 2014. Department of Homeland Security, Washington, DC, p. 3, 136, 211.)

The DHS *Budget-in-Brief* is released annually along with the president's budget. It provides an overview of DHS and its components, descriptions of select DHS initiatives, and budget summaries by agency and account (U.S. Department of Homeland Security, 2013).

Most of the funding for transportation security programs in the president's budget is provided through requests to Congress for new spending authority in the annual appropriations process. An important exception is the budgetary treatment of passenger aviation security fees. These charges were established by the ATSA to help cover the costs of the increased aviation security measures mandated by the legislation and consist of the passenger security fee[2] and the Aviation Security Infrastructure Fee imposed on air carriers.[3] The assessments were originally considered as offsetting collections and added to the direct discretionary appropriations amounts for TSA aviation security provided through the regular appropriations process, with the sum equaling the net discretionary appropriations total. Subsequently, the Vision 100 Act of 2003 created the Aviation Security Capital Fund to finance security-related capital improvements at airports and funded, in part, from the first $250 million collected annually in passenger aviation security fees. These latter funds are treated as mandatory spending and not included in TSA appropriations; the remainder of the security fees is still accounted for as offsetting collections. To illustrate, in the president's budget proposal for FY 2014, a total of $5.218 billion was requested in discretionary and mandatory resources for TSA aviation security, broken down as follows:

- $2.722 billion in direct discretionary appropriations
- $2.246 billion in offsetting collections (mainly from passenger aviation security fees, including a portion of the Administration's proposed increase in the passenger security fee)
- $250 million in mandatory spending from the Aviation Security Capital Fund (Office of Management and Budget, 2013c, p. 492).

Congressional Budget Process

Until 1974, Congress lacked a comprehensive counterpart to the president's budget, with its own budgetary framework being simply the sum of all of its actions on appropriations, revenue, and mandatory spending bills. However, in the wake of the Watergate scandal and the resulting assertion of congressional authority vis-à-vis the executive branch, the Congressional Budget and Impoundment Control Act of 1974 was adopted:

> *The congressional budget process initiated in the 1970s did not replace the preexisting revenue and spending processes. Instead, it provided an overall legislative framework within which the many separate measures affecting the budget would be considered. The central purpose of the budget process established by the 1974 act is to coordinate*

[2]Originally, a uniform fee of $2.50 per boarding, capped at $5.00 per one way trip, for passengers on U.S. and foreign air carrier flights originating at U.S. airports.

[3]Collected if the passenger fees are insufficient to pay for all aviation security costs and capped at the amount paid by the airlines for passenger screening in FY 2000.

the various revenue and spending decisions which are made in separate revenue, appropriations, and other budgetary measures.

Heniff, 2012b, p. 1

The centerpiece of this process is a concurrent resolution on the budget that is to be adopted by both houses before consideration of spending or revenue bills.[4] The 1974 law called for the adoption of two budget resolutions per year, but the Balanced Budget and Emergency Deficit Control Act of 1985 removed the requirement for a second budget resolution. As the process now stands, the Congressional Budget Office (established by the act to assist Congress with its budgetary responsibilities) is to submit its analysis of the economic and budgetary outlook to the House and Senate Budget Committees (also created pursuant to the 1974 act) by February 15 of each year, and the various other congressional committees are to provide the Budget Committees with estimates of their anticipated spending and revenue actions for the upcoming fiscal year within 6 weeks of the submission of the president's budget.

After these submissions, each Budget Committee must develop a budget resolution that includes:

- Budget aggregates, including total revenues and the amount by which the total is to be changed by legislative action; total new budget authority and outlays; the surplus or deficit; and the debt limit
- The amounts of new budget authority and outlays for each of the 20 functional categories of the budget, which must add up to the corresponding aggregate total

After being reported by the budget committees, the budget resolutions are considered in each house and are subject to amendment, with differences in the two versions reconciled by a conference committee and final approval of the resolution to occur by April 15. Subsequent action on spending and revenue bills is to be consistent with the aggregates contained in the budget resolution (Heniff, 2012b, pp. 11–14).

In practice, both the timetable and the use of the budgetary aggregates as the primary enforcement tool proved problematic. Over the 39 years between when the budget process was first put into effect (FY 1975) and FY 2014, Congress met the deadline for completion of action on a budget resolution only six times (most recently for FY 2004) and on eight other occasions (for FYs 1999, 2003, 2005, 2007, 2011, 2012, 2013 and 2014) failed to complete action at any point. These delays, combined with the recurring inability of Congress to finish action on appropriations bills before the start of the fiscal year, compromised the effectiveness of the budget resolution's spending and revenue aggregates as compliance mechanisms (Heniff, 2014, pp. 29–30).

[4]A concurrent resolution is subject only to approval by the House and Senate and is not submitted to the president for signature into law or veto. It thus lacks the force of law and cannot authorize spending or raise revenue. In the case of the budget resolution, it is used to set guidelines within which budget-related legislation is to be considered (Heniff, 2012b, p. 11).

Starting in the 1980s, Congress turned for enforcement to the use of the spending allocations to committees provided for under Section 302 of the Budget Act. Section 302(a) allocations are normally made in the statement of the Budget Committee managers accompanying the conference report on the final version of the budget resolution and set forth the amounts of new budget authority and outlays allocated to each committee with jurisdiction over spending. Special provision is made for the House and Senate Appropriations Committees in Section 302(b), which authorizes and requires these panels to subdivide their overall allocation among their 12 subcommittees. When Congress fails to approve a budget resolution by May 15, the House Appropriations Committee is authorized to proceed with development of its spending bills, and Senate appropriators may do so upon agreement of the leadership of both parties. In recent years, when final budget resolutions have been late or absent altogether, the 302(b) allocations—which have become the primary means of enforcing discretionary spending limits—have been established through a variety of legislative means (including the BCA of 2011) (Heniff, 2012b, pp. 14–15; Tollestrup, 2012, pp. 3–4).

With most transportation security funding provided through the appropriations process, the 302(b) allocation to the House and Senate Appropriations Subcommittees on Homeland Security have become the primary means through which the Congressional budget process impacts transportation security.

Appropriations Legislation

Appropriations measures provide agencies with budgetary authority for specified purposes and that "budget authority allows federal agencies to incur obligations and authorizes payments to be made out of the [U.S.] Treasury." Although not required by the Constitution, since the first Congress, appropriations bills have been limited to a single year, and the precedent that these measures should originate in the House of Representatives was also established early on.

There are three basic types of appropriations bills:

- *Regular appropriations bills* are the separate measures reported out of each of the 12 appropriations subcommittees in both the House and Senate (including the Homeland Security subcommittees) that provide budget authority for the coming fiscal year beginning October 1. Traditionally, these have accounted for most of the funding in a given year. *Omnibus appropriations bills* are a special type of regular appropriations, which result when more than one bill are combined together, typically at the conference committee stage of the appropriations process after the component bills have been approved by one or both of the houses.
- *Continuing resolutions* are used to provide for a continuation of funding for a specified time period for those agencies whose regular appropriations bills have not been enacted by October 1. Since FY 1977, Congress has completed action on all appropriations bills by that date on only four occasions (FYs 1977, 1989, 1995, and 1997), and at least one continuing resolution has been required in all other years.

- *Supplemental appropriations bills* may be considered at any time and generally provide additional funding to meet unforeseen needs (often for disaster response and recovery) or to increase resources for select activities previously funded in the regular appropriations bills. Supplemental appropriations may be included in regular appropriations bills or continuing resolutions, rather than through stand-alone legislation (Heniff, 2012a, pp. 1–2, 2012b, pp. 20–21; Tollestrup, 2012, pp. 10–13).

In recent years, partisan divisions over federal spending have resulted in greater delays in the enactment of regular appropriations bills and have necessitated increased use of both omnibus appropriations measures and continuing resolutions. The DHS has not escaped this trend. In its first 4 years of operations (FYs 2004–2007), DHS was funded via a regular appropriations bill enacted on or shortly after October 1. However, since that time, appropriations for DHS have been provided through a separate DHS Appropriations bill only once (for FY 2010 when the DHS Appropriations measure was enacted in late October 2009). DHS did receive its funding authorization on time for FY 2009 but in an omnibus appropriations measure that also included emergency supplemental funding for disaster recovery, regular appropriations for the Department of Defense and for military construction and veterans affairs, and a continuing resolution that funded the rest of the federal government. In all other years, DHS (similar to most other federal agencies) has received its spending authority well after the beginning of the fiscal year:

- For FY 2008 (beginning on October 1, 2007), DHS appropriations were folded into an omnibus appropriations bill that was signed into law on December 26, 2007.
- For FY 2011 (beginning on October 1, 2010), DHS appropriations were ultimately provided for through an eighth continuing resolution for that year, signed into law on April 15, 2011.
- For FY 2012, (beginning on October 1, 2011), DHS appropriations were again folded into an omnibus bill that was signed into law on December 23, 2011.
- For FY 2013 (beginning on October 1, 2012), DHS appropriations were again included in an omnibus measure that was signed into law on March 26, 2013.
- For FY 2014 (beginning October 1, 2013), DHS appropriations were yet again provided in an omnibus bill, which was signed into law on January 17, 2014. (Painter, 2013, pp. 5–6; 2014, pp. 3, 12).

Determining precise funding levels for transportation security programs (and the modal division of such funding) has continued to be difficult despite the creation of DHS and the elaboration of security-related programs within its components. Some of the previous challenges—such as separating security from non-security spending within such agencies as Customs and Border Protection—have persisted, and new ones have arisen, including:

- Changes in agency accounting, with the Department of Transportation moving from having a separate category for its agencies' security-related spending through FY 2007, to merging those accounts into a larger security, preparedness and response category from FY 2008 to FY 2010, to reinstating the security line in FY 2011, to eliminating it altogether in its FY 2012 and FY 2013 budget presentations;

DEPARTMENT OF HOMELAND SECURITY APPROPRIATIONS FOR FISCAL YEAR 2013

How an Appropriations Measure Became a Law

On February 13, 2012, President Obama submitted his proposed budget for FY 2013, which included a request for $59.032 billion in total budget authority for DHS (including $44.942 billion in net discretionary spending, $3.757 billion in discretionary fees, and $10.334 billion in mandatory spending and fees, plus trust funds). This represented a reduction of $681 million (1.1%) from the total enacted for FY 2012. The president's DHS budget reflected a total of $39.510 billion in adjusted net discretionary budget authority subject to the appropriations process.

Congress failed to adopt a budget resolution for FY 2013, so the Section 302 allocations for appropriations were subject to the BCA of 2012, which divided all spending into two categories: defense (capped at a total of $546 billion in discretionary spending) and everything else, including DHS (capped at a total of $501 billion). Whereas the Senate used these totals in arriving at its 302(a) allocation to the Appropriations Committee, the House used the somewhat lower levels included in the House-passed budget resolution. The resulting initial 302(b) allocations to the Senate Homeland Security Subcommittee (made on April 19) totaled $39.514 billion in discretionary budget authority compared with the $39.117 billion allocated to the House Homeland Security Subcommittee (on May 8).

In a break from customary procedure, the Senate Appropriations Committee took up the DHS appropriations measure first, reporting out its version by a bipartisan vote of 27 to 3 on May 22. The Committee bill met its 302(b) target of $39.514 billion in net discretionary spending for DHS, which was $4 million above the president's request. Because the fate of all appropriations measures was dependent on ongoing negotiations between the president and congressional Republicans, the Democratic-controlled Senate took no further action on the DHS (or any other) appropriations bill beyond the Committee approval stage. However, the Committee-passed bill served as the basis for Senate negotiators in subsequent deliberations on DHS appropriations.

The House Appropriations Committee reported out the FY 2013 DHS Appropriations bill by a party-line vote of 28 to 21 (with Republicans in favor and Democrats opposed) on May 16, and the House amended and passed the measure on June 7 by a vote of 234 to 182 (with 217 Republicans and 17 Democrats voting yes and 16 Republicans and 166 Democrats voting no). The House bill conformed to its 302(b) target of $39.114 in net discretionary appropriations, $393 million below the president's request. Democratic opposition in the House to the DHS and other appropriations bills centered on the fact that, through the budget resolution, House Republicans had lowered the caps on discretionary spending contained in the BCA.

As the impasse between the White House and Congressional Republicans over federal fiscal policy continued, no progress was made in resolving appropriations matters before the start of FY 2013. Consequently, a continuing resolution was developed to provide for continued funding for federal agencies through March 27, 2013. This measure passed the House on September 13 by a bipartisan margin of 329 to 91 (with 165 Republicans and 164 Democrats voting yes and 70 Republicans and 21 Democrats voting no) and the Senate on September 22 on a 62 to 30 vote (with 48 Democrats, 12 Republicans, and 2 Independents voting yes and 1 Democrat and 29 Republicans voting no). The "no" votes in the Senate were in protest over the failure to reach an overall agreement on fiscal policy. Under the terms of the continuing resolution, most DHS programs were funded at the FY 2012 spending rate plus 0.612%. The most significant exception was for DHS's cybersecurity activities, which received a $282 million increase.

In March 2013, another continuing resolution was developed to fund federal agencies for the remainder of the fiscal year. This legislation passed the Senate on March 20 (73 to 26; yes: 50 Democrats, 21 Republicans, two Independents; no: 1 Democrat and 25 Republicans) and the House on March 21 (318 to 109; yes: 203 Republicans and 115 Democrats; no: 27 Republicans and 82 Democrats). The president signed the bill into law on March 26, 2013 (PL 113-6). DHS received $39.646 billion in adjusted net discretionary budget authority, which was subject to an across-the-board cut of $54 million in order to comply with agreed-upon discretionary budget caps, resulting in a total figure of $39.592 billion. This amount was subject to a further reduction under terms of the sequester required by the BCA and announced by OMB on March 1, but the exact impact on the funding levels provided for under PL 113-6 was not immediately clear (Painter, 2012, pp. 1–2; 2013, pp. 1–3; Library of Congress, 2013).

- Reorganizations within DHS, such as the shifting in FY 2006 of transportation-specific research and development from TSA to the DHS Science and Technology division, where determining funding for transportation-related projects has grown increasingly problematic through changes in the division's appropriations accounts; and
- Growth in the TSA budget accounts for Transportation Threat Assessment and Credentialing (TTAC)[5] and Security Support, both of which contain nonmodal specific programs and activities that fall into more than one modal category.

Despite the above caveats, an examination of the available evidence on transportation security spending by the U.S. government over the first 10 years of DHS's existence reveals the following trends:

1. Funding for transportation security continued to rise, but at a slower rate of increase than in the immediate aftermath of 9/11, before beginning to level off in FY 2009. The Christmas 2009 attempted bombing of Northwest Flight 253 produced a temporary surge in spending (especially for aviation security) in the FY 2010 budget, but increasing overall budgetary concerns lead to actual reductions since then.
2. Aviation security remains the predominant recipient of federal support. Although the growth in the TSA's multimodal TTAC and Security Support categories has produced an apparent decline in the aviation share of overall transportation security funding, a substantial portion of both of these accounts is used for the aviation sector, indicating that, in fact, this mode's proportion of all transportation security spending is little reduced from the earlier post-9/11 period.
3. Maritime security funding has risen slightly, although it has continued the previous trend of claiming just over one-fifth of total spending.
4. The land mode still trails far behind the other two in terms of funding levels. Most of this support has been derived from DHS security grants, which are particularly susceptible to changes in congressional sentiment in the appropriations process (Table 5.2).

[5]Now called Intelligence and Vetting.

Table 5.2 Transportation Security Appropriations FY2004-FY2013 (amounts in $millions)

	FY 2004	FY 2005	FY 2006	FY 2007	FY 2008	FY 2009	ARRA	FY 2010	FY 2011	FY 2012	FY 2013
Aviation Security											
TSA aviation security	3,724	4,324	4,561	5,129	4,809	4,755	1,000	5,214	5,213	5,254	5,048
Avsec Capital Fund	—	250	250	250	250	250	—	250	250	250	250
Federal Air Marshals	623	663	679	719	770	819	—	860	928	966	907
Secure Flight	—	—	56	15	50	82	—	84	84	92	107
FAA security	231	172	168	173	247	219	—	250	213	245	210
Sub-total	*4,578*	*5,409*	*5,714*	*6,286*	*6,126*	*6,125*	*1,000*	*6,658*	*6,688*	*6,807*	*6,522*
As % of total	*60.0*	*61.9*	*60.3*	*60.7*	*59.0*	*56.1*	*71.4*	*58.6*	*60.1*	*61.6*	*60.7*
Maritime Security											
Coast Guard security	1,853	1,638	1,760	1,362	1,554	1,641	—	1,598	1,651	1,918	1,826
Port security grants	141	150	173	320	400	400	150	300	250	98	97
CBP container security	215	309	274	447	335	360	100	378	294	268	232
DOE Megaports	13	44	74	111	132	108	—	174	125	133	133
Sub-total	*2,222*	*2,141*	*2,281*	*2,240*	*2,421*	*2,509*	*250*	*2,450*	*2,320*	*2,417*	*2,288*
As % of total	*29.1*	*24.5*	*24.1*	*21.6*	*23.3*	*23.0*	*17.9*	*21.6*	*20.8*	*21.9*	*21.3*
Land Transportation Security											
TSA surface transport	35	55	36	37	47	50	—	111	106	135	124
FTA security	38	38	42	42	47	48	—	50	50	23	47
FRA security	1	1	1	1	1	1	—	1	4	1	1
DHS rail/transit grants	50	150	149	275	400	400	150	300	225	88	87
Amtrak security	—	—	—	—	—	—	—	—	20	10	10
Highway security	—	21	23	25	22	22	—	22	22	22	22
Trucking security	22	5	5	12	16	8	—	—	—	—	—
Intercity bus security	10	10	10	12	12	12	—	12	5	—	—
Sub-total	*156*	*280*	*266*	*404*	*545*	*541*	*150*	*496*	*432*	*279*	*291*
As % of total	*2.0*	*3.2*	*2.8*	*3.9*	*5.3*	*5.0*	*10.7*	*4.4*	*3.9*	*2.5*	*2.7*

Multi-modal Security											
TSA TTAC	50	87	198	99	115	74	—	135	120	112	165
TSA security support	438	534	505	525	524	948	—	1,002	987	1,032	953
R&D	154	239	152	139	103	129	—	165	144	—	—
CBP targeting	28	46	44	51	52	68	—	70	90	100	193
DNDO	—	—	315	616	485	514	—	383	342	290	318
OST security	—	2	2	4	8	9	—	9	9	11	11
Sub-total	*670*	*908*	*1,216*	*1,434*	*1,287*	*1,742*	—	*1,764*	*1,692*	*1,545*	*1,640*
As % of total	*8.8*	*10.4*	*12.8*	*13.8*	*12.4*	*16.0*	—	*15.5*	*15.2*	*14.0*	*15.3*
Grand total	**7,626**	**8,738**	**9,477**	**10,364**	**10,379**	**10,917**	**1,400**	**11,368**	**11,132**	**11,048**	**10,741**

Amounts include supplemental appropriations and fee-funded programs. ARRA=American Recovery and Reinvestment Act of 2009 ("Stimulus bill") (PL 111-5). FY 2012 and FY 2013 amounts for FAA, FTA, FRA and Highway security are estimates only. FY 2013 totals are pre-sequester.

Coast Guard security=Ports, Waterways and Coastal Security.

CBP container security includes Container Security Initiative, C-TPAT, CBP inspection and detection technology, Safe Commerce in FY04 and Secure Freight beginning in FY2009.

TSA surface transport=surface transportation security.

DHS rail/transit grants=Rail and transit security grants, including Amtrak FY2005-FY2010.

Highway security includes FHWA and FMCSA security.

TSA TTAC=Transportation Threat Assessment and Credentialing, excluding Secure Flight.

R&D=TSA R&D in FY2004-2005; DHS Science and Technology counter-MANPADS in FY2005-2006; Science and Technology explosives research FY2006-2010; and Science and Technology borders and maritime research FY2007-2008. Account re-organized in FY2011 and relevant figures unavailable for FY2011-2013.

CBP targeting includes Automated Targeting Systems, National Targeting Center and (beginning in FY2009) trusted traveler programs.

OST security=Office of Intelligence, Security and Emergency Response within Office of Secretary of Transportation.

(*Sources*: U.S. House of Representatives, Committee on Appropriations, *Department of Homeland Security Appropriations Bill, 2005* (H. Report 108-541), *2006* (H. Report 109-79), *2007* (H. Report 109-476), *2008* (H. Report 110-181), *2009* (H. Report110-862), *2010* (H. Report 111-157), *2013* (H. Report 112-492) and *2014* (H. Report 113-91);
U.S. Senate, Committee on Appropriations, *Department of Homeland Security Appropriations Bill, 2011* (S. Report 111-222), *2012* (S. Report 112-74) and *2013* (S. Report 112-169);
U.S. Department of Transportation, *Budget in Brief* for *FY2005-FY 2010, Budget Estimates Fiscal Year 2011, 2012* and *2013;*
GAO, 2012, *Maritime Security: Progress and Challenges 10 Years after the Maritime Transportation Security Act;*
GAO, 2012, *Megaports Initiative faces funding and sustainability challenges;*
U.S. Department of Energy, *FY 2014 Congressional Budget Request, volume 1.*)

■ ■ Critical Thinking ■

Name three features in the operation of the legislative process (including authorizations, budgeting, and appropriations) with respect to homeland security that have impeded effective policymaking and explain why.

Presidential Directives and Nominations

Presidents have a definite role in the legislative process (including proposing budgets and legislative language and holding the power to veto authorizing and appropriations legislation), but Congress generally has the last word in that arena (including the ability to override vetoes). There are other policy instruments in which the president's authority is paramount.

Presidential Directives

Although not explicitly provided for in the U.S. Constitution, U.S. presidents have used implicit Constitutional and statutory authority to issue directives "to achieve policy goals, set uniform standards for managing the executive branch, or outline a policy view intended to influence the behavior of private citizens." Over the course of American history, the two best known of these presidential pronouncements have been executive orders, which are directed at government agencies, and proclamations, which concern the activities of private individuals. Both must be published in the *Federal Register*, but whereas executive orders generally have the force of law, proclamations are generally not legally binding (Burrows, 2010, summary and p. 1).

In response to the events of 9/11, on October 29, 2001, President George W. Bush initiated a new type of executive order, termed the Homeland Security Presidential Directive (HSPD) "that shall record and communicate presidential decisions about the homeland security policies of the United States." Although HSPDs were not to be published in the *Federal Register*, they were to be made available to the public in written form on the White House website unless classified. Between that date and the end of his second term, President Bush issued 24 such directives (Relyea, 2008, pp. 6–7).

When he took office, President Barack Obama designated the directives he intended to use to promulgate his decisions on national security matters (including homeland security) as Presidential Policy Directives (PPDs). As of January 2014, President Obama had issued 28 PPDs, but few concerned homeland security (Federation of American Scientists, n.d.).

The presidential directives most relevant to transportation security are summarized below:

- HSPD-6 (September 16, 2003), "Integration and Use of Screening Information," established the policies and procedures for developing and using a database of known or suspected terrorists. This "watchlist" has become the basis for the terrorist identity checks performed by TSA's Secure Flight and other programs.

- HSPD-11 (August 27, 2004), "Comprehensive Terrorist-related Screening Procedures," built upon HSPD-6, declaring "It is the policy of the United States to enhance terrorist-related screening through comprehensive, coordinated procedures that detect, identify, track, and interdict people, cargo, conveyances, and other entities and objects that pose a threat to homeland security, and to do so in a manner that safeguards legal rights, including freedoms, civil liberties, and information privacy guaranteed by Federal law, and builds upon existing risk assessment capabilities while facilitating the efficient movement of people, cargo, conveyances, and other potentially affected activities in commerce." DHS was directed to develop a report outlining a strategy for achieving these goals.
- HSPD-13 (December 21, 2004), "Maritime Security Policy," "establishes U.S. policy, guidelines, and implementation actions to enhance U.S. national security and homeland security by protecting U.S. maritime interests." It created the Maritime Security Policy Coordinating Committee to "act as the primary forum for interagency coordination of the implementation of this directive"; called upon DHS and the Department of Defense to jointly develop a recommended "National Strategy for Maritime Security"; directed DHS, the Department of Defense, and the CIA to better integrate global maritime intelligence; directed the State Department to develop a plan "to solicit international support for an improved global maritime security network"; and required DHS to develop a comprehensive plan for securing the international maritime supply chain.
- HSPD-14 (April 15, 2005), "Domestic Nuclear Detection Office," created the office to "provide a single accountable organization with dedicated responsibilities to develop the global nuclear detection architecture, and acquire, and support the deployment of the domestic detection system to detect and report attempts to import or transport a nuclear device or fissile or radiological material intended for illicit use."
- HSPD-16 (March 26, 2007), "National Strategy for Aviation Security," identified threats to aviation security and targets and tactics of terrorists and criminals; set forth strategic objectives and implementing actions for the U.S. aviation security system; outlined roles and responsibilities within the aviation security system; and directed DHS and other relevant agencies to develop an Aviation Transportation System Security Plan, Aviation Operational Threat Response Plan, Aviation Transportation System Recovery Plan, Air Domain Surveillance and Intelligence Integration Plan, International Aviation Threat Reduction Plan, Domestic Outreach Plan, and International Outreach Plan in support of the strategy.
- HSPD-19 (February 12, 2007), "Combating Terrorist Use of Explosives in the United States," directed the Justice Department, in coordination with DHS and other relevant agencies, to develop "a report, including a national strategy and recommendations, on how more effectively to deter, prevent, detect, protect against, and respond to explosive attacks" to be followed by the development of an implementation plan.
- HSPD-24 (June 5, 2008), "Biometrics for Identification and Screening to Enhance National Security," provided "a Federal framework for applying existing and emerging

biometric technologies to the collection, storage, use, analysis, and sharing of data in identification and screening processes employed by agencies to enhance national security, consistent with applicable law, including information privacy and other legal rights under Unites States law."

- PPD-17 (2012), "Countering Improvised Explosive Devices," replaced HSPD-19 but was not made public. However, on February 26, 2013, a policy statement on this topic was released and outlined approaches for strengthening previous efforts, including increasing domestic and international awareness and information sharing, improving intelligence and information analysis, developing and maintaining counter-IED (Improvised Explosive Devices) resources, improving the performance of explosives screening, detection and protection technologies, and safeguarding explosives and select precursor materials.
- PPD-18 (2012), "National Strategy for Maritime Security," replaced HSPD-14 but has not yet been released (U.S. House of Representatives, Committee on Homeland Security, 2008, pp. 31–32, 67–70, 73-80, 81–84, 87–114, 127–132; White House, 2008; 2013, pp. 1–3; Federation of American Scientists, n.d.; AllGov.com, 2013).

Presidential Nominations

An important presidential power is the authority to make appointments to executive branch agencies, commissions, and other federal entities. In general, these political appointments "are for individuals who make or advocate administration policy or support those positions. Individuals serving in political appointments generally serve at the pleasure of the appointing authority and do not have the job protections afforded to those in career-type appointments." (As of October 2012, there were a total of 3719 such positions.) The politically appointed positions may be divided into the following categories:

- Presidential appointments requiring Senate confirmation, which represent the top positions in the executive branch in terms of authority (e.g., Departmental Secretaries and Deputy Secretaries), and account for 33% of all political appointees.
- Presidential appointments not requiring Senate confirmation (9% of political appointees), which typically are for positions on federal advisory boards and commissions or in the Executive Office of the President.
- Other political appointments (the remaining 58%) are generally subordinate to presidential appointees in the other categories, and often appointed by them rather than directly by the president (GAO, 2013, pp. 4, 35).

From a policy standpoint, the most important appointments are those that require the approval of the U.S. Senate. The process in such cases is a multi-stage one. First, the president selects and officially nominates the candidate. Then the nomination is referred to one or more Senate committees, which hold hearings and consider whether or not to report the nomination to the full Senate. If the committee or committees report out the nomination or if the Senate votes to "discharge" the candidate from further consideration, then the nomination is placed on the Senate's Executive Calendar and may be called up by the majority leader for a vote. Only a majority of Senators present and voting is required for approval of the nomination, but it may be subjected to a filibuster, in which case 60 votes are needed to end the debate and proceed to a vote (Rybicki, 2013, Summary).

The Senate Committee on Homeland Security and Governmental Affairs has jurisdiction over most, but not all, high-level nominations within DHS, but reflective of the divided nature of congressional oversight of homeland security, that panel shares responsibility for the TSA administrator position with the Committee on Commerce, Science, and Transportation, and the latter has exclusive jurisdiction over the nomination for commandant of the Coast Guard. Another key position with respect to transportation security, the commissioner of U.S. Customs and Border Protection, is subject to review by the Finance Committee (Table 5.3) (Davis and Mansfield, 2012, pp. 13, 21, 33).

Table 5.3 Transportation Security–Related Positions Requiring Senate Confirmation

Position	Senate Committee(s) of jurisdiction
DHS Secretary	Homeland Security and Governmental Affairs
DHS Deputy Secretary	Homeland Security and Governmental Affairs
DHS Assistant Secretary—National Protection and Programs	Homeland Security and Governmental Affairs
DHS Assistant Secretary—Policy	Homeland Security and Governmental Affairs
DHS Assistant Secretary/Administrator—TSA	Homeland Security and Governmental Affairs Commerce, Science, and Transportation
Administrator—FEMA	Homeland Security and Governmental Affairs
Deputy Administrator—FEMA	Homeland Security and Governmental Affairs
Deputy Administrator—Protection and National Preparedness (FEMA)	Homeland Security and Governmental Affairs
DHS Under Secretary—Science and Technology	Commerce, Science, and Transportation
DHS Under Secretary for Intelligence and Analysis	Intelligence
Commissioner—U.S. Customs and Border Protection	Finance
US Coast Guard—Commandant	Commerce, Science, and Transportation
Privacy and Civil Liberties Oversight Board	Judiciary
DOT Secretary	Commerce, Science, and Transportation
DOT Deputy Secretary	Commerce, Science, and Transportation
Federal Aviation Administration—Administrator	Commerce, Science, and Transportation
Federal Motor Carrier Safety Administration—Administrator	Commerce, Science, and Transportation
Federal Railroad Administration—Administrator	Commerce, Science, and Transportation
Pipeline and Hazardous Materials Safety Administration—Administrator	Commerce, Science, and Transportation
Amtrak—Board of Directors	Commerce, Science, and Transportation

DHS, Department of Homeland Security; DOT, Department of Transportation; FEMA, Federal Emergency Management Agency.
(*Source*: Davis, C., Mansfield, J., November 15, 2012. Presidential appointee positions requiring senate confirmation and committees handling nominations. Congressional Research Service, Washington, DC, pp. 13, 15, 21, 33, 37, 39.)

POLICYMAKING IN PARLIAMENTARY SYSTEMS
A Comparison with the United States

The policymaking process in the United States is unusual among national governments. Many democracies (including those in the United Kingdom, Canada, and India) operate under a parliamentary system, whose distinguishing characteristics include:

- A "fusion of powers" in which "the executive branch is 'fused with' and 'dependent upon' the legislative branch."
- "Parliamentary supremacy," which makes the parliament (legislative branch) "the supreme legal authority. … which can create or end any law," with, in most cases, the separate judicial system unable to overrule its acts and future parliaments enabled to alter any previously passed laws.
- Governing authority vested exclusively in the executive branch, which is specifically termed the "government" and "has responsibility for developing and implementing policy and for drafting laws."

The interrelationship between the parliament and the executive is especially crucial in understanding the workings of parliamentary systems. First, in most cases, the chief executive (usually termed the prime minister) is elected not by the general electorate but by a vote of the members of the lower house of parliament (the House of Commons in the United Kingdom). In practice, the prime minister is thus the leader of the majority party in the lower house. (In the elections for the lower house, voters tend to cast their ballot based on which party or party leader they prefer to form the government rather than on the merits of the individual lower house candidates.) After a prime minister is chosen, he or she forms the new government, with the prime minister and the heads of government agencies (Cabinet ministers) usually drawn from the majority party membership of the legislative branch (predominantly from the lower house but occasionally from the appointive upper chamber, known as the House of Lords in the United Kingdom).

After it is installed, the government proceeds to develop and implement policy, with the legislative role limited to debating and passing all laws submitted by the executive, examining the work of government agencies, and authorizing the government to raise taxes. Because the government is selected by the majority party in the lower house and, to a large extent, is able to control the parliamentary schedule, it has a high probability of success in getting its proposed legislation adopted. Indeed, although referred to as a "parliamentary system," the legislative bodies in these governments typically play a lesser role in the policymaking process than their U.S. counterparts. However, one crucial power vested in parliaments is the ability to unseat the government in power, either by defeating a major government policy proposal or by voting against a motion expressing "confidence" in the government. In either case, "typically the government will fall, and elections will be held to select a new executive." If the government's majority in the lower house is a slim one, the defection of even a few of its members can thus bring down the government.

A further check on the power of the leading party is the potential for no single party to obtain an absolute majority in the lower house because of more than two parties winning seats. Such a result occurred in the 2010 U.K. elections, when the Conservative Party won a 59-seat advantage in the House of Commons over the incumbent Labour Party (307 to 248) but failed to gain a majority of the 650 seats. This necessitated the Conservatives forming a coalition with the third-place Liberal Democrats (who won 57) to form the new government. Such coalition governments have been frequent in continental Europe and require the partners to reach an accommodation on policy proposals.

The "fusion of powers" in parliamentary systems can enable a new government to move relatively quickly in changing policies, as was the case in the United Kingdom's adoption of substantial organizational changes in transportation security in 2011 and enactment of the Civil Aviation Act of 2012, both of which elevated the importance of economic factors and coordination with industry in the development and implementation of transportation security policies.

What are some of the pros and cons of parliamentary systems versus the U.S. model with respect to transportation security policymaking? Consider such factors as ability to act, accountability, and "checks and balances." Which system do you believe will prove more effective in the long run in adapting transportation security policies to changing circumstances and threats?

Sources: U.K. Parliament, n.d.; Mapleleafweb.com, n.d.; BBC News, 2010

Federal Regulation

In the United States, the regulatory, or rulemaking,[6] process is a principal means by which public policy is implemented. The process generally begins with an act of Congress that either requires or authorizes a federal agency to issue a rule to carry out its provisions. The agency then proceeds to develop a draft regulation that, in the case of "significant" rules that would have a major economic impact or raise important policy issues, is submitted to the Office of Information and Regulatory Affairs (OIRA) within the Office of Management and Budget in the Executive Office of the president for review. The next step is the publication by the agency of a Notice of Proposed Rulemaking (NPRM) in the *Federal Register*[7] that includes (1) a summary of the issues involved in the policy under consideration, (2) the timing of the proposed rulemaking (including provision for a public comment period generally lasting from 30 to 60 days from the date of the NPRM), (3) citation of the specific legal authority under which the rule is being proposed, (4) a discussion of the merits of the proposal, and (5) the full text of the proposed rule.

As described in the *Federal Register's* "Guide to the Rulemaking Process,"

> *The notice-and-comment process enables anyone to submit a comment on any part of the proposed rule.... If the rulemaking record contains persuasive new data or policy arguments, or poses difficult questions or criticisms, the agency may decide to terminate the rulemaking. Or the agency may decide to continue the rulemaking but change aspects of the rule to reflect these new issues. If the changes are major, the*

[6]For purposes of this section, the terms *regulation* and *rule* are interchangeable.

[7]Created by the Federal Register Act of 1935 and provides the means through which the public is notified of proposed rules, executive orders, and other official documents that the president or Congress require to be published.

agency may publish a supplemental proposed rule. If the changes are minor, or a logical outgrowth of the issues and solutions discussed in the proposed rules, the agency may proceed with a final rule.

After the agency decides to issue a Final Rule (with "significant" regulations again subject to OIRA review), that rule must be published in the *Federal Register* no less than 30 days before it is to take effect. Even then, it is subject to legal challenges in the courts and potential Congressional disapproval. The Final Rule notice also specifies how the new language is to be integrated into the Code of Federal Regulations (CFR), which contains all of the generally applicable rules of the federal government (Carey, 2013, pp. 1–6; Office of the Federal Register, 2011).

As can be seen even in this brief description, the federal rulemaking process is highly complex and, similar to the legislative process, is subject to potential delays at virtually every step. In the field of transportation security, before 9/11, FAA security officials regarded rulemaking as the "bane" of effective security, and currently, the entire regulatory process is embroiled in the same type of partisan and ideological discord that has complicated legislative policymaking. A September 2013 article in The Hill's Regulation Blog reported:

Republicans and industry groups, who have bemoaned what they view as overly aggressive federal agencies, want more restrictions on the rule-making process and a greater reliance on economic analysis in decisions regarding new regulations. Democrats, unions and public interest groups, meanwhile, say agencies are already hamstrung by existing restrictions on their authority.

Johnstone, 2006, p. 27; Goad and Hattem, 2013

Maritime Security Regulations

Most federal maritime security regulations are contained in Title 33, Chapter I, Subchapter H of the Code of Federal Regulations (Table 5.4) and are designed to implement MTSA, subsequent amendments to that statute and relevant executive orders. One of the core purposes of Subchapter H was to bring U.S. security standards into compliance with the 2002 Safety of Life At Sea (SOLAS) Amendments and the International Ship and Port Facility Security Code. However, the U.S. regulations go beyond the international guidelines in being more detailed, applying to a much greater number of vessels and port facilities, and containing a list of control and compliance measures to be used by noncompliant facilities. Furthermore, there is no international equivalent of Part 103 establishing an Area Maritime Security system to coordinate port security at the local or regional level through the operation of special committees and plans (Bennett, 2008, pp. 174–176).

Land Security Regulations

Absent a counterpart to the foundational MTSA and ATSA, the promulgation of federal rules for the land mode has been more limited and uncoordinated. Although aviation security regulations are almost entirely contained within Chapter XII of Title 49 (the TSA title) of the CFR, such rules that do apply to land transportation security are distributed

Table 5.4 Code of Federal Regulations: Maritime Security Provisions

Part	Description
Title 33, Chapter I (Coast Guard), Subchapter H—Maritime Security	
101	*General*
101, Subpart A	General (purpose, definitions, etc.)
101, Subpart B	Maritime Security (MARSEC) Levels
101, Subpart C	Communication (Port—Facility—Vessel)
101, Subpart D	Control Measures for Security
101, Subpart E	Other Provisions (including Transportation Worker Identification Credential)
103	*Area Maritime Security*
103, Subpart A	General (applicability, definitions)
103, Subpart B	Federal Maritime Security Coordinator Designation and Authorities
103, Subpart C	Area Maritime Security (AMS) Committee
103, Subpart D	AMS Assessment
103, Subpart E	AMS Plan
104	*Vessels*
104, Subpart A	General (definitions, applicability, etc.)
104, Subpart B	Vessel Security Requirements
104, Subpart C	Vessel Security Assessment (VSA)
104, Subpart D	Vessel Security Plan (VSP)
105	*Facilities*
105, Subpart A	General (definitions, applicability, etc.)
105, Subpart B	Facility Security Requirements
105, Subpart C	Facility Security Assessment (FSA)
105, Subpart D	Facility Security Plan (FSP)
106	*Outer Continental Shelf (OCS) Facilities*
106, Subpart A	General (definitions, applicability, etc.)
106, Subpart B	OCS Facility Security Requirements
106, Subpart C	OCS Facility Security Assessment (FSA)
106, Subpart D	OCS Facility Security Plan (FSP)
107	*National Vessel and Security Control Measures and Limited Access Areas*
107, Subpart A	[Reserved]
107, Subpart B	Unauthorized Entry into Cuban Territorial Waters

(*Source*: Electronic Code of Federal Regulations. September 5, 2013. < http://www.ecfr.gov/ > (accessed 10.23.14.)

throughout Title 49, not only in Chapter XII, but also in the chapters covering the individual modal administrations responsible for the various land modes:

- Chapter I—Pipeline and Hazardous Materials Safety Administration
- Chapter II—Federal Railroad Administration
- Chapter III—Federal Motor Carrier Safety Administration
- Chapter IV—Coast Guard (covering its role in the testing and approval of containers)
- Chapter VI—Federal Transit Administration
- Chapter VII—Amtrak

TSA's land transportation role is reflected in Title 49, Subchapter D on Maritime and Land Transportation Security (Table 5.5). (The maritime role is largely confined to the

Table 5.5 Code of Federal Regulations: Transportation Security Administration Security Provisions

Part	Description
Title 49, Subtitle B, Chapter XII (Transportation Security Administration)	
Subchapter A	*Administrative and Procedural Rules*
1500	Applicability, Terms and Abbreviations
1502	Organization, Functions and Procedures
1503	Investigative and Enforcement Procedures
1507	Privacy Act—Exemptions
1510	Passenger Civil Aviation Security Service Fees
1511	Aviation Security Infrastructure Fee
1515	Appeal and Waiver Procedures for Security Threat Assessments for Individuals
Subchapter B	*Security Rules for All Modes of Transportation*
1520	Protection of Sensitive Security Information
Subchapter C	*Civil Aviation Security*
1540	Civil Aviation Security: General Rules
1542	Airport Security
1544	Aircraft Operator Security: Air Carriers and Commercial Operators
1546	Foreign Air Carrier Security
1548	Indirect Air Carrier Security (cargo security)
1549	Certified Cargo Screening Program
1550	Aircraft Security Under General Operating and Flight Rules (general aviation security)
1552	Flight Schools
1560	Secure Flight Program
1562	Operations in the Washington, DC Metropolitan Area
Subchapter D	*Maritime and Land Transportation Security*
1570	General Rules
1572	Credentialing and Security Threat Assessments
1572, Subpart A	Procedures and General Standards
1572, Subpart B	Standards for Security Threat Assessments
1572, Subpart C	Transportation of Hazardous Materials from Canada or Mexico To and Within the United States by Land Modes
1572, Subpart D	[Reserved]
1572, Subpart E	Fees for Security Threat Assessments for Hazmat Drivers
1572, Subpart F	Fees for Security Threat Assessments for Transportation Worker Identification Credential (Transportation Worker Identification Credential)
1580	Rail Transportation Security
1580, Subpart A	General (scope, terms, inspection authority)
1580, Subpart B	Freight Rail Including Freight Railroad Carriers, Rail Hazardous Materials Shippers, Rail Hazardous Materials Receivers, and Private Cars
1580, Subpart C	Passenger Rail Including Passenger Railroad Carriers, Rail Transit Systems, Tourist, Scenic, Historic and Excursion Operators, and Private Cars

(*Source*: Electronic Code of Federal Regulations. September 5, 2013. <http://www.ecfr.gov> (accessed 10.23.14.))

Transportation Worker Identification Credential, which applies to all modes.) These regulations center on passenger and freight rail and hazardous materials transportation (Electronic Code of Federal Regulations, 2013).

Aviation Security Regulations

The oldest and most elaborate of U.S. transportation security rules are those governing aviation. These were originally under the jurisdiction of the FAA but were transferred to TSA when that agency was created by ATSA. The regulations are found primarily in Chapter XII, Subchapter C of Title 49 CFR (Civil Aviation Security), but the passenger aviation security fees are located in Subchapter A. Many of these were holdovers from the FAA security program, but ATSA and subsequent legislation and presidential directives have added a number of additional requirements, including for air cargo, general aviation, TWIC, and the Secure Flight program for prescreening passenger names against watchlists. Perhaps the biggest change from the old system, however, was TSA's assumption of direct responsibility for the screening of commercial aviation passengers and baggage. This function used to be included as part of the air carrier security program in which FAA set standards for the airlines to carry out (Price and Forrest, 2009, pp. 121–125).

AIR CARGO SCREENING RULEMAKING

How a Regulation Was Made

The Implementing the Recommendations of the 9/11 Commission Act (9/11 Act) of 2007 contained a provision directing DHS "to establish a system to screen 100 percent of cargo transported on passenger aircraft operated by an air carrier or foreign air carrier," by August 3, 2010, with the system required to provide a level of security "commensurate with the level of security for the screening of passenger checked baggage." The legislation also explicitly authorized TSA to issue an interim final rule to implement these requirements, with a final rule issued not later than 1 year after the effective date of the interim rule. This allowed some of the preliminary stages of rulemaking to be bypassed.

TSA had been administering a risk-based system for securing cargo transported on passenger aircraft (which required air carriers to ensure that all cargo presenting an "elevated risk" was screened), but after the August 2007 enactment of the 9/11 Act, "TSA recognized that it needed to develop a program that could achieve the 9/11 Act's requirement for 100 percent screening while still allowing for the flow of commerce." Throughout the remainder of 2007 and the beginning of 2008, TSA examined similar programs in the United States (primarily Customs and Border Protection's container security programs) and internationally (especially the United Kingdom's Known Consignor program). As a result of these investigations, in February 2009, TSA launched a pilot program called the Certified Cargo Screening Program (CCSP), which began at the major domestic airports, accounting for more than 65% of all air cargo on passenger aircraft.[8] In developing the CCSP program, TSA consulted with 120 shippers and related entities.

On September 16, 2009, TSA published its interim final rule formally establishing the CCSP in the *Federal Register*, setting November 16, 2009, as the effective date. Although noting the interim rule had been adopted without prior notice or public comment, TSA solicited written comments on the regulation, also with a November 16, 2009, deadline. The September 16 publication summarized the interim rule as follows:

> *This rule establishes a program under which TSA will certify cargo screening facilities located in the U.S. that volunteer to screen cargo prior to tendering it to aircraft operators for carriage on passenger aircraft. This rule requires affected passenger aircraft operators to ensure that either an aircraft operator or certified cargo screening facility (CCSF) that does so in accordance with TSA standards, or TSA itself, screens all cargo loaded on passenger aircraft.... CCSF personnel must successfully undergo a TSA-conducted security threat assessment and pay a fee for that assessment.*

The 9/11 Act had mandated that a final rule be issued within 1 year of the effective date of the interim final rule, but TSA was unable to meet that deadline because of revisions that had to be made in the interim rule. Before issuing the final rule, TSA received comments from approximately 40 trade association and aircraft operators, as well as a few individuals. Two sets of comments produced changes in the interim rule. First, several commentators objected to a provision that required applicants for a CCSF to undergo "validation" of their fitness by a TSA-approved third party. TSA agreed to drop this requirement in the final rule, indicating that because there had been far fewer CCSF applicants than anticipated, TSA itself could handle this role. Second, a number of comments objected to the interim regulation's mandate that any air cargo screening facility not located at an airport, including those operated by aircraft operators, had to undergo the CCSF certification process. Thus, aircraft operators would have to comply with two separate security programs (for aircraft security operators and for the cargo screening program). The final rule removed this requirement as well because "the security programs for aircraft operators and foreign air carriers have been and will continue to be amended to ensure that the same level of security involving screened cargo are equivalent to that for CCSFs."

On August 5, 2011, OMB's Office of Information and Regulatory Affairs approved TSA's final rule on air cargo screening, and the rule was published in the August 18, 2011, *Federal Register*, with an effective date of September 19, 2011. In addition to the two changes in the interim final rule noted, the final rule also proposed a fee schedule for the TSA-conducted background check on CCSF personnel and invited public comments on that proposal. The final fee schedule was announced in the May 23, 2012, *Federal Register*, with an effective date of June 22, 2012.

It should be noted that, although the 9/11 Act cargo screening mandates applied to cargo loaded both within and outside the United States, the rulemaking process for air cargo screening did not apply to the latter. TSA's efforts on cargo loaded outside the United States has involved a "two-pronged approach" of working through International Civil Aviation Organization standards and using risk assessments to target the highest risk cargo for screening (Federal Register, 2009, pp. 47672–47675; 2011, pp. 51848–51853; 2012, pp. 30542–30543).

---

[8] The locations included airports in San Francisco, Chicago, Philadelphia, Seattle, Los Angeles, Dallas-Fort Worth, Miami, Atlanta, New York City, and Newark.

■ ■ Critical Thinking ■

What methods were used to shorten the policymaking process in the cases of the Implementing Recommendations of the 9/11 Commission Act and the rulemaking on air cargo screening? In each instance, were the methods effective and appropriate? Why or why not?

Conclusion

As the scope, intensity, and funding of transportation security programs have expanded over the early years of the 21st century, the role of policymaking that has mandated, guided, and provided resources for that expansion has also increased. Yet the instruments of policymaking for transportation security (and indeed other forms of homeland security) are still not well established, and in the United States—with its system of separated legislative and executive branch powers—have become increasingly entwined in the partisan gridlock that has characterized the second decade of this century.

There have been advances in U.S. transportation security policy since 2004, with the SAFE Port Act of 2006 building on and improving the maritime security program established by MTSA and the 2007 Implementing Recommendations of the 9/11 Commission Act establishing stricter requirements for the screening of maritime and air cargo and creating a framework for land transportation security programs. Furthermore, the body of transportation security regulations, which provide more specific guidance for policy implementation, has continued to grow.

However, authorizing legislation to provide better direction for transportation security policies has been sporadic, with external factors (now including political considerations as well as the traditional response to major incidents) serving as the prime determinant of action. The means for accounting for transportation security programs and policies within the budgetary and appropriations processes remain underdeveloped, and in the latter case, partisan divisions have seriously complicated the annual funding cycle, with delays and last minute across-the-board cuts becoming the new norm. Attempts have been made to streamline the regulatory process with respect to homeland security, but it remains to be seen how successful these efforts will be in making that system less cumbersome and time consuming.

Discussion Questions

1. Briefly describe the policymaking process in the United States, including the primary instruments used in adopting policy goals and means.
2. What were the main provisions of the SAFE Port Act and the Implementing Recommendations of the 9/11 Commission Act?
3. What are the two key documents pertaining to homeland security that are issued with the president's annual budget, and what information do they provide?
4. Describe the major trends in funding for transportation security since 2004.
5. What are the principal differences between the policymaking systems in the United States and in parliamentary systems?

References

AllGov.com. July, 2013. Obama administration hiding details of presidential policy directives. <http://www.allgov.com/news/controversies/obama-administration-hiding-details-of-presidential-policy-directives> (accessed 10.23.14.)

Bennett, J.C., 2008. In Bragdon, C.R. (Ed.), Transportation Security. Butterworth-Heinemann/Elsevier Burlington, MA, pp. 149–181.

BBC News. 2010. Election 2010 results. <http://news.bbc.co.uk/2/shared/election2010/results/> (accessed 10.23.14.)

Burrows, V.K., March, 2010. Executive orders: Issuance and revocation. Congressional Research Service, Washington, DC.

Carey, M., June, 2013. The federal rulemaking process: An overview. Congressional Research Service, Washington, DC.

Davis, C., Mansfield, J., November, 2012. Presidential appointee positions requiring Senate confirmation and committees handling nominations. Congressional Research Service, Washington, DC.

Electronic Code of Federal Regulations. September, 2013. <http://www.ecfr.gov/> (accessed 10.23.14.)

Federal Register. September, 2009. Transportation Security Administration: Air cargo screening. Washington, DC.

Federal Register. August, 2011. Transportation Security Administration: Air cargo screening. Washington, DC.

Federal Register. May, 2012. Transportation Security Administration: Air cargo screening. Washington, DC.

Federation of American Scientists. n.d. Presidential policy directives [PPDs] Barack Obama Administration. <https://www.fas.org/irp/offdocs/ppd/> (accessed 10.23.14.)

Goad, B., Hattem, J., September, 2013. Regulation nation: Obama rule-making seen as deeply flawed. [Web blog post]. <http://thehill.com/blogs/regwatch/other/319715-regulation-nation-proposals-abound-but> (accessed 10.23.14.)

GAO. October, 2007. Maritime security: The SAFE Port Act: Status and implementation one year later. Washington, DC.

GAO. March, 2013. Characteristics of presidential appointments that do not require Senate confirmation. Washington, DC.

Heniff, Jr., B., November, 2012a. Overview of the authorization-appropriations process. Congressional Research Service, Washington, DC.

Heniff, Jr., B., December, 2012b. Introduction to the federal budget process. Congressional Research Service, Washington, DC.

Heniff, Jr., B., February, 2014. Congressional budget resolutions: Historical information. Congressional Research Service, Washington, DC.

Johnstone, R.W., 2006. 9/11 and the Future of Transportation Security. Praeger, Westport, CT.

Kraft, M.E., Furlong, S.R., 2007. Public policy: Politics, analysis, and alternatives, 2nd ed. CQ Press, Washington, DC.

Library of Congress. 2007. Bill summary & status 110th Congress (2007-2008), H.R. 1: All congressional actions. <http://thomas.loc.gov/cgi-bin/bdquery/z?d110:HR00001:@@@X> (accessed 10.23.14.)

Library of Congress. 2011. Bill summary & status 112th Congress (2011-20012), S. 365: all information. <http://thomas.loc.gov/cgi-bin/bdquery/z?d112:SN00365:@@@L&summ2> (accessed 10.23.14.)

Library of Congress. 2013. Status of appropriations legislation for fiscal year 2013. <http://thomas.loc.gov/home/approp/app13.html> (accessed 10.23.14.)

Mapleleafweb.com. n.d. Parliamentary government in Canada: Basic organization and practices. <http://www.mapleleafweb.com/print/337> (accessed 10.23.14.)

McNicholas, M., 2008. Maritime Security. Butterworth-Heinemann/Elsevier, Burlington, MA.

NBCNews.com. July, 2007. Congress oks sweeping domestic security bill. <http://cpf.cleanprint.net/cpf/cpf?action=print&type=filePrint&key=msnbc&url=http%3A%> (accessed 10.23.14.)

Office of Management and Budget. March, 2013a. OMB report to the Congress on the Joint Committee sequestration for fiscal year 2013. Washington, DC.

Office of Management and Budget. April, 2013b. Budget of the United States Government, fiscal year 2014: Analytical perspectives. Washington, DC.

Office of Management and Budget. April, 2013c. Budget of the United States Government, fiscal year 2014: Appendix. Washington, DC.

Office of the Federal Register. 2011. A guide to the rulemaking process. <https://www.federalregister.gov/uploads/2011/01/the_rulemaking_process.pdf> (accessed 10.23.14.)

Painter, W.L., October, 2012. Department of Homeland Security: FY 2013 appropriations. Congressional Research Service, Washington, DC.

Painter, W.L., April, 2013. Department of Homeland Security: Appropriations: A summary of Congressional actions for FY 2013. Congressional Research Service, Washington, DC.

Painter, W.L., April, 2014. Department of Homeland Security: FY 2014 appropriations. Congressional Research Service, Washington, DC.

Price, J.C., Forrest, J.S., 2009. Practical Aviation Security. Butterworth-Heinemann/Elsevier, Burlington, MA.

Relyea, H.C., November, 2008. Presidential directives: Background and overview. Congressional Research Service, Washington, DC.

Rybicki, E., January, 2013. Senate consideration of presidential nominations: Committee and floor procedure. Congressional Research Service, Washington, DC.

SourceWatch. December, 2008. Congressional efforts to implement recommendations of the 9/11 Commission. <http://www.sourcewatch.org/index.php?title=Congressional_efforts_to_implement_recommendations_of_the_9/11_commission> (accessed 10.23.14.)

Tollestrup, J., February, 2012. The congressional appropriations process: An introduction. Congressional Research Service, Washington, DC.

U.K. Parliament. n.d. Parliament and government. <http://www.parliament.uk/about/how/role/parliament-government> (accessed 10.23.14.)

U.S. Department of Homeland Security. April, 2013. Budget-in-brief: fiscal year 2014. Washington, DC.

U.S. House of Representatives, Committee on Homeland Security. May, 2005. Department of Homeland Security Authorization Act for fiscal year 2006 (H. Report 109-71, Part 1). Washington, DC.

U.S. House of Representatives. July, 2007. Implementing Recommendations of the 9/11 Commission Act, conference report to accompany H.R. 1 (H. Report 110-259). Washington, DC.

U.S. House of Representatives, Committee on Homeland Security. January, 2008. Compilation of Homeland Security Presidential Directives (HSPD). Washington, DC.

White House. June, 2008. National Security Presidential Directive/NSPD-59, Homeland Security Presidential Directive/HSPD-24: Biometrics for identification and screening to enhance national security. <http://www.fas.org/irp/offdocs/nspd/nspd-59.html> (accessed 10.23.14.)

White House. February, 2013. Countering improvised explosive devices. Washington, DC.

6

Implementing Maritime Security

CHAPTER OBJECTIVES:

In this chapter, you will learn about plans and programs for the implementation of maritime security internationally and in the United States, including:

- Port and vessel security
- Supply chain security (including container security)
- Maritime domain awareness and intelligence

Introduction

Maritime security in the 21st century rests on a combination of international standards and the policies and implementation of national governments. It can be subdivided into three sometimes overlapping constituent parts: the security of the ports and vessels that comprise the maritime transportation system, the security of the international maritime supply chain, and the intelligence and domain awareness activities that support maritime security.

In the United States, a National Strategy for Maritime Security was released in September 2005 pursuant to the requirements of Homeland Security Presidential Directive-13 (HSPD-13) issued by President Bush in December 2004. The National Strategy sought to "align all Federal government maritime security programs and initiatives into a comprehensive and cohesive national effort involving appropriate Federal, State, local, and private sector entities." The plan listed four strategic objectives (prevent terrorist attacks and criminal or hostile acts, protect maritime-related population centers and critical infrastructures, minimize damage and expedite recovery, and safeguard the ocean and its resources) and outlined five "strategic actions" to achieve the objectives:

- Enhance international cooperation to ensure lawful and timely actions against maritime threats.
- Maximize domain awareness to support effective decision making.
- Embed security into commercial practices to reduce vulnerabilities and facilitate commerce.
- Deploy layered security to unify public and private security measures.
- Assure continuity of the marine transportation system to maintain vital commerce and defense readiness . . . in the aftermath of any terrorist attack or other similarly disruptive incidents that occur within the maritime domain (U.S. Department of Homeland Security, 2005a, pp. 7–24).

The National Strategy was supplemented by "eight supporting plans to address the specific threats and challenges of the maritime environment" that were issued in 2005 and 2006:

1. National Plan to Achieve Domain Awareness
2. Global Maritime Intelligence Integration Plan
3. Interim Maritime Operational Threat Response Plan
4. International Outreach and Coordination Strategy
5. Maritime Infrastructure Recovery Plan
6. Maritime Transportation System Security Plan
7. Maritime Commerce Security Plan
8. Domestic Outreach Plan (p. ii)

In a June 2008 report on the National Strategy and supporting plans, the Government Accountability Office (GAO) indicated the documents "were generally well-developed and, collectively, included desirable characteristics, such as (1) purpose, scope, and methodology; (2) problem definition and risk assessment; (3) organizational roles, responsibilities, and coordination; and (4) integration and implementation" (GAO, 2012b, p. 5).

Port and Vessel Security

Ports and the vessels that service them are responsible for transporting a large portion of all international trade, with the ports also being located at or near many of the world's population centers and major industrial facilities, such as power plants and oil refineries. In addition to their economic importance, their size and openness make them attractive potential targets for pirates, terrorists, and other criminals.

2002 SOLAS Amendments and the International Ship and Port Facility Security Code

The principal international instruments in the field of port and vessel security are the 2002 amendments to the International Convention for the Safety of Life at Sea (SOLAS), which created a new regulatory system for international maritime security, and the International Ship and Port Facility Security Code (ISPS), which provides further details for the implementation of that system.[1] According to Bennett (2008), the ISPS Code seeks to:

- Establish a framework for international cooperation between governments, their agencies, and the shipping and port industries in order to detect security threats and take preventive measures against security incidents affecting ships and port facilities in international trade.

[1]Adherence to the SOLAS Convention and the associated ISPS Code is nearly universal among nations of the world. As of October 17, 2014, 162 countries—representing 99% of global shipping—had signed onto the Convention as "'contracting states' bound by its provisions, including the ISPS Code (International Maritime Organization, n.d.a).

- Establish roles and responsibilities for the various players in the international maritime transportation system.
- Ensure early and efficient collection and exchange of security-related information.
- Provide a methodology for security assessments in order to have ship security plans and port facility security plans, including procedures to react to changing security levels.
- Ensure confidence that adequate and proportionate security measures are in place (p. 165).

Among the key components of the ISPS Code are the requirements that passenger ships and large cargo ships (of 500 gross tons or more) "engaged in international voyages" obtain from the government whose flag they fly approval of their security plans and issuance of an International Ship Security Certificate (ISSC) verifying their compliance with the ISPS Code. Port facilities serving such vessels must receive approval of their security plans from their host governments. The security plans for ships and port facilities are similar and, among other requirements, must cover measures for preventing the introduction of unauthorized weapons and dangerous substances or devices and unauthorized access to restricted areas; responding to security threats and governmental security instructions; organizing security duties and identifying relevant security officers; conducting security audits, training, and exercises; and providing for review and update of the security plan, reports of security incidents and inspection, and maintenance of security equipment. In addition, the ship security plans must provide guidance on the use of the Ship's Security Alert System (SSAS), and the port facility plans must also address the security of cargo and cargo-handling equipment at the facility, response to SSAS activation by a vessel at the port, and the facilitation of shore leave for ships' crews (pp. 166–170).

Although national governments are primarily responsible for ensuring compliance with the ISPS Code by their own ships and ports, the SOLAS Convention makes provision for compliance actions for foreign ships. In the case of a foreign vessel covered by the ISPS Code seeking to enter a port, the port's government is authorized to require validation of the ship's ISSC, a description of its current security level and the levels it operated under in its previous 10 port calls,[2] and other "practical security-related information." If the government has "clear grounds" for believing the vessel is noncompliant (e.g., an invalid ISSC, evidence of "serious deficiencies" in security equipment or procedures, or reliable reports of noncompliance), it may require rectification of the noncompliance, inspect the ship if in its territorial waters, require the vessel to proceed to a specified location in its waters for such an inspection, or deny entry into port (pp. 170–172).

[2]The ISPS Code identifies three security levels that are to be applied by national governments: Security level 1: normal; the level at which the ship or port facility normally operates, for which minimum appropriate protective security measures shall be maintained at all times Security level 2: heightened; the level applying as long as there is a heightened risk of a security incident, for which appropriate additional protective security measures shall be maintained for a period of time Security level 3: exceptional; the level applying for the period of time when there is the probable or imminent risk of a security incident, for which further specific protective security measures shall be maintained for a limited period of time (International Maritime Organization, n.d.b).

The U.S. Coast Guard is responsible for monitoring and enforcing ISPS compliance for foreign ships entering U.S. waters. In 2005 (the first full year of ISPS enforcement), it conducted 9117 ISPS inspections, which resulted in the detection of 115 security deficiencies that required 51 major control actions (detention, denial of entry or expulsion). The rate of major control actions has declined steadily since then, with the 2012 figures indicating that of the 8627 security examinations that identified 207 security deficiencies, only eight resulted in a major control action. Over the past 4 years, access control has been the leading source of security deficiencies followed by restricted areas, ship security officers, and ship security plans, and bulk carriers have been the vessel type most often subjected to major control actions followed by general dry cargo ships and container ships (U.S. Coast Guard, 2013a, pp. 4, 17–20).

International Port Security Program

The Maritime Transportation Security Act of 2002 (MTSA) directed the U.S. Coast Guard to assess the security of foreign ports and make recommendations for improving their security measures. In response, in 2004 the Coast Guard established the International Port Security Program under which the Coast Guard and its counterpart agency in other countries visit the host country's ports to evaluate the implementation of security measures, especially those required by the ISPS Code. By October 2007, the Coast Guard had visited ports in more than 100 countries and found that most of them had substantially implemented the ISPS Code.[3] However, a 2010 GAO report indicated that the program faced a number of challenges, including "reluctance by some countries to allow the Coast Guard to visit their ports due to concerns over sovereignty," and that "other than sharing best practices or providing presentations on security practices, the program does not currently have the resources to directly assist countries, particularly those that are poor, with more in-depth training or security assistance." A 2012 update by the same agency noted the Coast Guard had made progress on both fronts. First, sovereignty concerns have been alleviated by the creation of reciprocal visits in which the Coast Guard hosts foreign delegations in observing ISPS Code implementation in the United States, and, second, it has worked with other U.S. agencies, as well as international organizations, in obtaining some funding for training and other forms of assistance to lower income countries that need to strengthen their port security efforts (GAO, 2010, pp. 10–11; 2012b, p. 49).

To aid in the assessment of risks and allocation of assistance, the Coast Guard began developing a risk model for its IPS program in 2005. The results from the model are used to ensure that foreign ports in high-risk countries are visited more frequently than other ports and to target foreign assistance based on country threat information, results from port visits, a determination of which countries would most benefit from port security assistance, and an assessment of a country's ability to best use such assistance (GAO, 2013c, pp. 25–26).

[3]As of June 2013, the Coast Guard reported visits to 151 foreign port facilities (GAO, 2013c, p. 12).

A June 2014 "Port Security Advisory" issued by the Coast Guard listed 16 countries (Cambodia, Cameroon, Comoros, Cote d'Ivoire, Cuba, Equatorial Guinea, Guinea-Bissau, Iran, Liberia, Madagascar, Nigeria, Sao Tome and Principe, Syria, Timor-Leste, Venezuela, and Yemen), all or some of whose ports were judged to be "not maintaining effective anti-terrorism measures." The advisory directed that vessels travelling to the listed countries shortly before arrival in the United States take the following actions while in those countries (with compliance subject to Coast Guard examination upon the vessel's entry into U.S. waters):

- Implement measures per the ship's security plan equivalent to Security Level 2.
- Ensure that each access point to the ship is guarded and that the guards have total visibility of the exterior of the vessel.
- Attempt to execute a Declaration of Security.[4]
- Log all security actions in the ship's security records.
- Report actions taken to the cognizant U.S. Coast Guard Captain of the Port before arrival in the United States (U.S. Coast Guard, 2014).

Area Maritime Security Plans

MTSA developed the concept of area maritime security, which is not addressed in the ISPS Code and is organized among relevant stakeholders in regionally meaningful subdivisions of the nation's ports. One of its key components is the Area Maritime Security Plan (AMSP), a document that the Coast Guard is to produce for each identified area and update every 5 years. The plans are to describe the area and infrastructure covered and how the plans are to be integrated with each other. The SAFE Port Act added the requirement that the AMSP should address the issues of trade resumption after a transportation security incident and salvage response plans. The AMSPs are the principal means by which Coast Guard procedures concerning prevention, protection, and security response for U.S. ports are identified and coordinated (GAO, 2012a, p. 4; 2012b, p. 30).

In implementing the statutory requirements and DHS guidelines, the Coast Guard developed AMSPs for each of its 43 "Captains of Port Zones," covering area ports in regions along both coasts as well as major inland waterways (e.g., the Great Lakes and Mississippi River). The plans address the following elements, as applicable:

- Details of operational and physical security measures in place in the port at Maritime Security (MARSEC) levels 1, 2, and 3[5]
- Details of the security incident command-and-response structure
- Details for regular audit and amendment of the plan

[4]Under the ISPS Code, a Declaration of Security (DoS) is an agreement entered into between ships and port facilities or their host government that defines their respective security responsibilities. "Contracting governments and ships will determine when a DoS is required, often based on the information provided by the port facility risk assessment" (McNicholas, 2008, pp. 102–103).

[5]MARSEC levels are the U.S. equivalent to the three security levels established by the ISPS Code.

- Measures to prevent the introduction of dangerous substances and unauthorized access into designated restricted areas within the port
- Procedures and expected timeframes for responding to security threats or breaches of security, including provisions for maintaining infrastructure and operations in the port
- Procedures for responding to any security instructions the Coast Guard announces at MARSEC level 3
- Procedures for evacuation within the port in case of security threats or breaches of security
- Identification of and methods to communicate with designated security officers, public safety officers, emergency response personnel, and crisis management representatives
- Security measures to ensure effective security of infrastructure, special events, vessels, passengers, cargo, and cargo-handling equipment at facilities within the port not otherwise covered by a Vessel or Facility Security Plan
- Procedures to be taken when a vessel is at a higher security level than the facility or port it is visiting
- Procedures for responding if a vessel security alert system on board a vessel within or near the port has been activated
- Procedures for communicating appropriate security and threat information to the public and for receiving and handling reports from the public and maritime industry regarding suspicious activity
- The jurisdiction of federal, state, Indian tribal, and local government agencies, and law enforcement entities over area security-related matters
- Security resources available for incident response and their capabilities
- Procedures for responding to a transportation security incident and to facilitate the recovery of the Marine Transportation System after such an incident
- Identification of any facility otherwise subject to port facility security requirements that has been designated by the Captain of the Port as a public access facility within the area, the security measures that must be implemented there at each MARSEC level, and who is responsible for implementing those measures (Figure 6.1) (GAO, 2012a, pp. 4–5; U.S. Coast Guard, n.d.a).

The GAO reported in 2012 that the Coast Guard had made progress in developing AMSPs "that were focused on preventing terrorism and included discussion regarding natural disasters with detailed plans for recovery after an incident" (GAO, 2012b, p. 30).

Port Facility Security Plans

The MTSA expanded the reach of the port facility security plans required by the ISPS Code well beyond the latter's applicability only to facilities serving ISPS-covered vessels. Thus, in the United States, in addition to the ISPS-covered facilities, security plans are mandatory for port facilities that receive vessels certified to carry more than 150 passengers, foreign

FIGURE 6.1 Maritime Security Level (MARSEC) notice. *(Courtesy of the U.S. Coast Guard, http://www.uscg.mil/hq/cg5/cg544/img7.jpg.)*

and U.S. cargo vessels of greater than 100 tons, barges carrying bulk cargoes, and facilities that are subject to regulation for dangerous or hazardous cargoes (Bennett, 2008, p. 175).

Maritime facility security plans in the U.S. must include details on:

1. Security administration and organization of the facility
2. Personnel training
3. Drills and exercises
4. Records and documentation.
5. Response to change in MARSEC level
6. Procedures for interfacing with vessels
7. Declaration of Security
8. Communications
9. Security systems and equipment maintenance
10. Security measures for access control, including designated public access areas
11. Security measures for restricted areas
12. Security measures for handling cargo
13. Security measures for delivery of vessel stores and bunkers
14. Security measures for monitoring
15. Security incident procedures
16. Audits and security plan amendments
17. Facility Security Assessment Report
18. Facility Vulnerability and Security Measures Summary (33 C.F.R. 105.405)

The MTSA required the Department of Homeland Security (DHS) to approve and oversee the port facility security plans, and DHS delegated that responsibility to the Coast Guard. In 2006, the Coast Guard estimated that approximately 3200 facilities required annual inspection, but records indicate that only 2126 such inspections were conducted (with officials reporting some of the inspections may not have been recorded and others were delayed).[6] In addition, another 4500 spot checks were performed that year at about 1200 facilities. Deficiencies were found in roughly one-third of all inspections, with security measures for access control (18% of all deficiencies), facility record-keeping requirements (17%), and security measures for restricted areas (14%) the leading problem areas. In more than 80% of these cases, the deficiencies were subsequently corrected by the facility operators, with no additional Coast Guard action required. (The SAFE Port Act of 2006 directed the Coast Guard to conduct at least two inspections–one of which must be unannounced—of each covered maritime facility annually to verify compliance with the terms of the facility's security plan.) As of 2012, the GAO found "the Coast Guard has made progress by generally requiring maritime facilities to develop security plans and conducting required annual inspections," with the inspections leading to the identification and correction of deficiencies (GAO, 2008, p. 4; 2012b, p. 32).

Port Security Grant Program

A program to provide federal financial assistance for local port security costs was established by a January 2002 Congressional appropriations act that allocated $93.3 million for grant awards to "critical national seaports." The MTSA codified this effort as the Port Security Grant Program (PSGP) in November 2002. Since that time, it has been subject to a number of changes in administration. Initially, TSA was responsible for managing the grant awards in partnership with the Coast Guard and the Department of Commerce's Maritime Administration. In March 2004, DHS created the Office of State and Local Government Coordination and Preparedness, which administered the PSGP in fiscal year (FY) 2005. Another DHS reorganization (in October 2005) established the Preparedness Directorate, within which was the Office of Grants and Training that was responsible for running the PSGP in FYs 2006 and 2007. Finally, the Post-Katrina Emergency Management Reform Act of 2007 transferred many of the Preparedness Directorate's functions, including the PSGP, to Federal Emergency Management Agency (FEMA), which has administered the program since then. At present, FEMA "is responsible for designing and operating the administrative mechanisms needed to implement and manage the grant program [and] the Coast Guard . . . provides subject matter expertise [about] the maritime industry and participates in project award decisions." All told, PSGP has received more than $2.5 billion in funding from FY 2002 through FY 2013 (GAO, 2011, p. 7).

The PSGP provides funding to eligible entities in the U.S. port areas deemed most at risk. Before FY 2007, all eligible entities competed for the available funds, but in response to concerns that the awards were not being adequately apportioned according to risk assessment, in that year DHS began classifying port areas into four groups, based on risk, with the port areas in the highest risk categories (groups I and II) competing for awards from the larger

[6]According to later Coast Guard figures, as of January 2011, 2509 port facilities were subject to MTSA regulations (GAO, 2012b, p. 34).

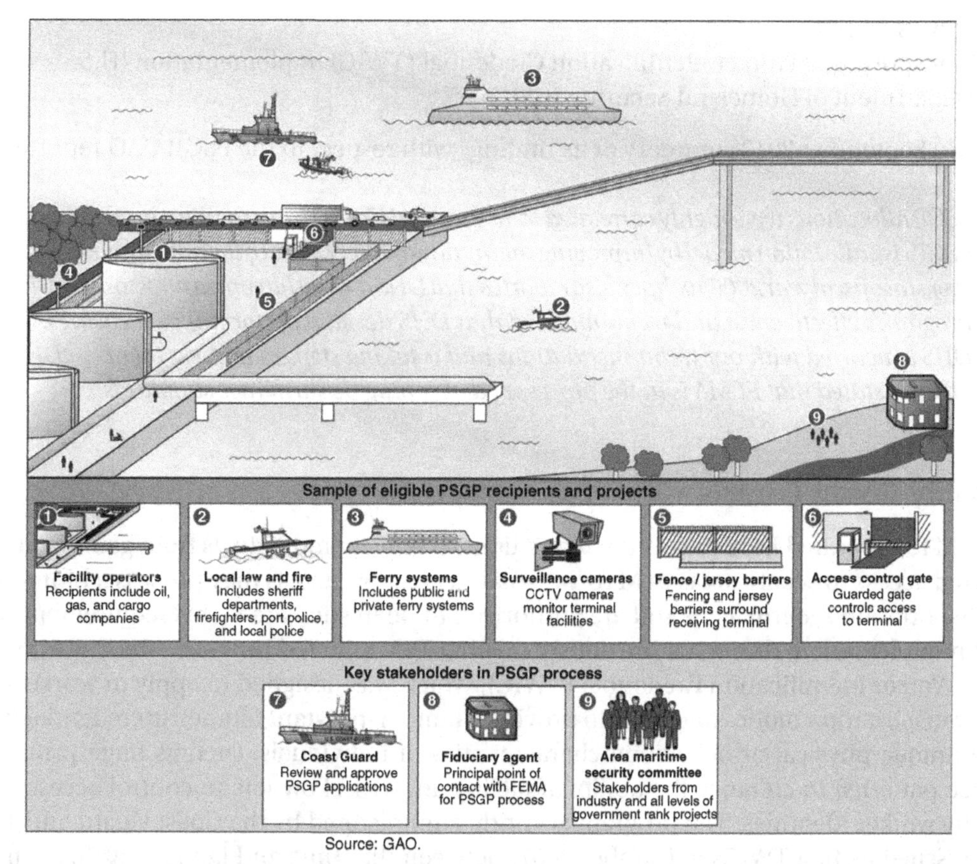

FIGURE 6.2 Sample of eligible Port Security Grant Program (PSGP) recipients and projects. CCTV, closed-circuit television; FEMA, Federal Emergency Management Agency. *(Source: GAO. July, 2010. Maritime security: DHS progress and challenges in key areas of port security. GAO, Washington, DC, p. 6.)*

allocation set aside for them and the remaining port areas (group III and All Other Port Areas) competing for funding from the smaller allocation. In FY 2013, the four categories were consolidated into two: group I (the eight highest risk port areas[7] that are to compete for 60% of available funding) and group II (the 82 additional eligible port areas that are to compete for the remaining 40% of funds) (Figure 6.2) (GAO, 2011, pp. 12–14; FEMA, 2013, Appendix A).

PSGP funding is focused on supporting:

- Enhanced domain awareness
- Training and exercises
- Expansion of port recovery and resiliency capabilities
- Enhanced capabilities to prevent, detect, respond to, and recover from attacks involving improvised explosive devices and other nonconventional weapons

[7]Los Angeles–Long Beach, San Francisco Bay, San Diego, New Orleans, Delaware Bay, New York–New Jersey, Houston–Galveston, and Puget Sound (FEMA, 2013, pp. 24–25).

- Cybersecurity
- Transportation Worker Identification Credential (TWIC) implementation (U.S. Department of Homeland Security, 2014, p. 2).

In a September 2012 summary of its findings with respect to the PSGP, GAO reported:

PSGP allocations were highly correlated to risk and DHS has taken steps to strengthen the PSGP risk allocation model by improving the quality and precision of the data inputs. However, since fiscal year 2006, we have also reported that DHS did not have measures to assess the program's effectiveness and recommended that DHS develop performance measures. ... DHS concurred with our recommendations and is taking steps to address them. ... DHS officials stated that FEMA is in the process of developing performance standards.

GAO, 2012b, p. 33

Facility Access Control

The MTSA required DHS to develop regulations to prevent individuals from gaining unescorted access to secure areas of MTSA-regulated facilities unless they possess a biometric transportation security card and are authorized to be in such area. This requirement has been implemented through a previously existing TSA-initiated program, the Transportation Worker Identification Credential (TWIC), which was designed to apply to workers in all transportation modes and aims to provide a tamper-resistant biometric credential (using unique physical or behavioral characteristics of individuals, such as fingerprints or voice patterns) to enhance the ability of facility and vessel owners to control access and verify worker identities. The program is jointly administered by the Coast Guard and TSA and issued its first TWICs in October 2007. Between that time and January 2013, 2.1 million maritime workers in U.S. ports (including longshoremen, truck drivers, railroad workers, mechanics, merchant mariners, and others requiring access to secure areas of MTSA-regulated ports and vessels) had obtained the card. The cards must be renewed once 5 five years, so many credentialed workers are facing first-time renewals (Peterman et al., 2013, pp. 11–12; GAO, 2012b, p. 34).

Before issuing the TWIC card, TSA must conduct a security threat assessment of each applicant, including a background check covering the applicant's criminal history, immigration status, and possible links to terrorist organizations. The applicant is charged a fee (currently $132.50) to cover the cost of program administration. Port facility and vessel operators are required to visually inspect each worker's TWIC before granting him or her unescorted access to secure areas (Peterman et al., 2013, pp. 11–12; GAO, 2013a, pp. 1–2).

The SAFE Port Act mandated that port terminal operators deploy electronic card readers at the gates of their facilities to scan a worker's TWIC each time he or she enters the port area and set a deadline of April 13, 2009, for the issuance of a final rule to implement this requirement. Without the electronic readers, the TWIC cards are currently subjected to visual inspection by security personnel, and the biometric data they contain are not being used to positively identify the worker. The Coast Guard published a Notice of Proposed

Rulemaking (NPRM) in the March 22, 2013 Federal Register outlining its plan for implementation of the electronic reader requirement. Under the Coast Guard proposal, this requirement would be limited to higher-risk vessels and facilities whereas those deemed to be at lower risk would be allowed to continue to utilize visual inspection of the TWIC. The final rule was expected to be issued in January 2015 (Peterman et al., 2013, p. 12; Federal Register, 2014).

The GAO reported that TSA and the Coast Guard "have made progress in enrolling workers and activating TWICs," with almost all covered workers issued the credential by the middle of 2012, but face challenges in implementing the program, including "enrolling and issuing TWICs to a larger population than was originally anticipated, ensuring that TWIC access control technologies perform effectively in the harsh maritime environment, and balancing security requirements with the need to facilitate the flow of legitimate maritime commerce." In June 2013, the GAO issued a report on a TWIC electronic reader pilot project conducted by TSA from August 2008 through May 2011 that tested several different reader technologies as well as the credential authentication and validation process. The report "identified several challenges related to pilot planning, data collection, and reporting, which affected the completeness, accuracy and reliability of the results" (GAO, 2013a, pp. 4–9).

Vessel Security Plans

As was the case with port facilities, the MTSA greatly enlarged the number of vessels subject to security regulation. One estimate indicated that more than 90% of the vessels subject to MTSA regulation are not covered under SOLAS and the ISPS Code. (According to Coast Guard figures, 12,908 vessels were subject to MTSA regulation as of January 2011.) In addition to those subject to the ISPS Code, MTSA requirements apply to foreign commercial vessels greater than 100 gross tons, almost all U.S. flag commercial vessels, passenger vessels certified to carry 12 or more passengers, and certain tugs and barges (primarily those carrying dangerous cargoes) (Figure 6.3) (Bennett, 2008, pp. 174–175; GAO, 2012b, p. 34; McNicholas, 2008, pp. 117–118).

The vessel security plans required under MTSA are virtually identical to those mandated for maritime facilities (33 C.F.R. 104.405).

The Coast Guard is responsible for determining which vessels must develop security plans and for reviewing, approving, and conducting compliance inspections of those plans. To assist vessel owners and operators in understanding and complying with these requirements, the Coast Guard has (1) issued regularly updated guidance and created a "help desk" to provide owners and operators, as well as other stakeholders, with a single point of contact for assistance; (2) hired outside contractors to provide additional expertise in reviewing the security plans; and (3) conducted regular compliance inspections. Furthermore, Coast Guard enforcement has adapted to changing circumstances. For example, in 2010, it required that the security plans for U.S.-flagged vessels transiting through areas with a high risk of piracy (e.g., the Horn of Africa) include a piracy annex, with compliance to be monitored by the Coast Guard when the vessels dock at certain ports (GAO, 2012b, pp. 9, 35).

FIGURE 6.3 Container ship *Edith Maersk*. *(Source: Wikimedia Commons/Heb, http://upload.wikimedia.org/wikipedia/commons/0/00/Edith_Maersk_Suez.jpg.)*

Vessel Crew Screenings

The Coast Guard and CBP obtain and evaluate advance information on commercial vessels and their crew before their arrival at U.S. ports and perform risk assessments based on this information. Key factors used in assessing the risk posed by a given vessel include previous instances of invalid or incorrect crew manifests, history of its seafarers unlawfully entering the United States, and whether the vessel is making its first visit to a U.S. seaport within the past year. In addition, based on intelligence reports and other relevant information, the Coast Guard is authorized to conduct an armed security boarding of commercial vessels arriving in U.S. waters to examine crew passports and visas and determine if the submitted crew lists are accurate (GAO, 2012b, pp. 9–10).

A particular security concern is access controls for foreign seafarers, who constitute an "overwhelming majority" of the approximately 5 million maritime crew entries into the United States each year on commercial cargo and passenger vessels. Roughly 80% of these entries are from passenger vessels, such as cruise ships, and most of the seafarers come from a limited number of nations, headed by the Philippines, India, and Russia. Because the U.S. has no control over the credentialing practices of other nations, the federal government has developed a multi-agency approach to address potential risks posed by foreign seafarers. The State Department reviews seafarer applications for U.S. visas, including evaluating applications, interviewing applicants, checking applicant information against federal databases, and reviewing supporting documentation to determine whether the applicant poses a potential security risk. The Coast Guard and CBP conduct advance screening inspections of seafarers entering the U.S., but the GAO reported discrepancies between the data collected by the two agencies on illegal seafarer entry at domestic ports, a problem that DHS is working to correct (GAO, 2012b, pp. 9–10, 37).

Small Vessel Security

Even before the November 2008 sea-based attack on Mumbai, the 2000 assault on the *USS Cole* had demonstrated the potential risk posed by terrorists using small boats. In April 2008, DHS issued its *Small Vessel Security Strategy*, which was designed "to reduce potential security risks from small vessels [less than 300 gross tons] through the adoption and implementation of a coherent system of regimes, awareness, and security operations that strike the proper balance between fundamental freedoms, adequate security, and continued economic stability. Additionally, the strategy is intended to muster the help of the small vessel community in reducing risks in the maritime domain" (U.S. Department of Homeland Security, 2008, p. 1).

The *Security Strategy* outlined the "scenarios of gravest concern in using small vessels in terrorist-related attacks:" (1) use in the United States of waterborne improvised explosive devices, (2) conveyance of smuggled weapons (including weapons of mass destruction) into the United States, (3) conveyance of terrorists into the United States, and (4) use as a waterborne platform for conducting a stand-off attack (such as using portable rocket launchers). To combat these and other threats, the *Strategy* set the following security goals:

1. Develop and leverage a strong partnership with the small vessel community and public and private sectors in order to enhance maritime domain awareness.
2. Enhance maritime security and safety based on a coherent plan with a layered, innovative approach (including improved detection and tracking, enhanced radiologic and nuclear detection capabilities, and improved data collection and analysis).
3. Leverage technology to enhance the ability to detect, determine intent, and when necessary, interdict small vessels.
4. Enhance coordination, cooperation, and communications among federal, state, local, and tribal partners and the private sector as well as international partners (U.S. Department of Homeland Security, 2008, pp. 11, 16–21).

DHS published the *Small Vessel Security Implementation Plan Report to the Public* in January 2011. The *Implementation Plan* "is a roadmap for realizing the goals and objectives of the DHS National Small Vessel Security Strategy [and] identifies possible and proven means of managing and controlling risks posed by the potential threat and possibly dire consequences of small vessel exploitation by terrorists." The plan contains no mandates or specific requirements for action but rather is intended to provide guidance to the small vessel community, as well as governmental authorities "about how programs may be developed and coordinated to achieve the goals and objectives outlined in the strategy." Because of its sensitive nature, distribution of the plan itself is restricted to stakeholders with clearances to receive sensitive security information (including members of Area Maritime Security Committees), but the *Report to the Public* about the plan provides examples of activities to help achieve the *Security Strategy's* goals (U.S. Department of Homeland Security, 2011, pp. 1–12).

In addition to the strategy, DHS (including the Coast Guard and CBP) has undertaken "community outreach, the establishment of security zones in U.S. ports and waterways, escorts of vessels that could be targeted for attack and port-level vessel tracking with radars and cameras since other vessel tracking systems—such as the Automatic Identification System[8]—are only required on larger vessels." On this latter point, the GAO reported "the expansion of vessel tracking to all small vessels may be of limited utility because of, among other things, the large number of small vessels, the difficulty identifying threatening actions, and the challenges associated with getting resources on scene in time to prevent an attack once it has been identified" (GAO, 2012b, p. 36).

■ ■ Critical Thinking ■

It can be argued that the Homeland Security Act's removal of responsibility for maritime security from TSA, which had originally been assigned the lead role for securing all modes of transportation by the Aviation and Transportation Security Act, significantly diminished the prospects for the development of a well-coordinated and prioritized transportation security policy in the United States. What are some of the advantages and disadvantages of having the Coast Guard take the lead in maritime security?

Supply Chain Security

The international supply chain refers to "the worldwide network of transportation, postal, and shipping assets, and infrastructures by which goods are moved from the point of manufacture until they reach an end consumer, as well as supporting communications infrastructure and systems" (White House, 2012, p. 1).

Although the concept of global supply chain security applies to all transportation modes, the prominence of the maritime sector in the international dimension of the movement of goods has meant, in practice, that much of the attention to this subject falls within that sector.

SAFE Framework of Standards to Secure and Facilitate Global Trade

The 2002 World Customs Organization (WCO) SAFE Framework of Standards to Secure and Facilitate Global Trade was modified in 2007, 2010, and 2012. The SAFE Framework is designed "to secure the movement of global trade in a way that does not impede but, on the contrary, facilitates the movement of that trade" and operates through its member national customs administrations (including CBP in the United States) that account for 99% of global trade. The utility of this approach is explained in the Foreword to the 2012 Framework:

[8]Under a 2000 amendment to the SOLAS Convention, all ships of 300 gross tons or greater engaged in international voyages, cargo ships of 500 gross tons or greater not engaged in international voyages, and all passenger ships, are required to carry onboard an Automatic Identification System that shall "provide information—including the ship's identity, type, position, course, speed, navigational status and other safety-related information—automatically to appropriately equipped shore stations, other ships, and aircraft" (International Maritime Organization, n.d.c).

MARITIME SECURITY IN THE UNITED KINGDOM

The European Dimension

As in the United States (and most other countries), a large part of the United Kingdom's maritime security program involves implementation of the international SOLAS Convention and ISPS Code for vessel and port security and the SAFE Framework for supply chain security, with the U.K. Department for Transport responsible for the former and Her Majesty's Revenue and Customs carrying out the latter.

An important difference between the two nations, however, is the U.K.'s membership in the European Union (EU),[9] whose regulations generally transpose the international transportation security agreements into the national laws or regulations of its member states. A 2012 EU Commission staff working paper discussed the European role across all transportation modes:

> *Whilst transport security policy should be developed and implemented at the national and local level...a large proportion of transport operations occur among Member states and it is clear that there is an added value to certain actions being taken at the EU level. Good EU-wide baseline levels of security are relevant to all Member states, especially with the free movement of persons and cargo [across Member state national borders]. The risk of criminality has, potentially, a cross-border dimension, therefore common approaches to ensure a good baseline of transport security throughout the EU is desirable.*
>
> *European Commission, 2012, p. 3*

The ISPS Code served as the basis for the EU regulation on enhancing ship and port facility security (Regulation 725/2004) that "emphasizes the application of security controls to passengers, staff, vehicles and cargo entering ports or port facilities or boarding a vessel." Among other things, the rule requires ships and port facilities to have security plans that accommodate multiple levels of threat and provides a list of recommended best management practices for use by ships in curbing piracy. The applicability of the ISPS Code is expanded by the EU regulation to cover domestic passenger ships that travel more than 20 miles "from a place of refuge," domestic ships required to comply by an EU member state's risk assessment (in the United Kingdom, this includes ships certified to carry more than 250 passengers as well as tankers), and port facilities serving any of these vessels. Furthermore, the EU is given the authority to conduct security inspections of registered ships of member states while in member state ports (European Commission, 2012, pp. 9–10; Government of the United Kingdom, 2013).

The EU's Safety and Security Amendment to the Customs Code (Regulation 648/2005) and its Implementing Provisions (Regulation 1875/2006) help to carry out the WCO SAFE Framework objectives by requiring advance electronic submission of customs information before arrivals to and departures from the EU, establishing a common risk management framework, and creating the Authorised Economic Operator program that confers expedited customs processing for companies meeting certain security conditions (SITPRO, 2008).

---

[9] The EU is an association of sovereign states currently composed of 28 European members. Begun as an economic union, it has been evolving toward a larger role in other policy areas (European Union, n.d.).

> *Customs administrations have important powers that exist nowhere else in government—the authority to inspect cargo and goods shipped into, through and out of a country. Customs also have the authority to refuse entry or exit and the authority to expedite entry. Customs administrations require information about goods being imported, and often require information about goods exported. They can, with appropriate legislation, require that information to be provided in advance and electronically.*
>
> *World Customs Organization, 2012, preface, p. 2*

The SAFE Framework seeks to:

- Establish standards that provide supply chain security and facilitation at a global level to promote certainty with predictability.
- Enable integrated and harmonized supply chain management for all modes of transport.
- Enhance the role, functions and capabilities of Customs to meet the challenges and opportunities of the 21st century.
- Strengthen cooperation between Customs administrations to improve their capability to detect high-risk consignments.
- Strengthen Customs–business cooperation.
- Promote the seamless movement of goods through secure international trade supply chains.

To accomplish these objectives, the framework relies on networking between national customs administrations and partnerships between Customs authorities and businesses. The framework itself contains four "core elements:" (1) coordination and harmonization of national advance electronic cargo information requirements for inbound, outbound, and transit shipments; (2) commitment of participating nations to use "a consistent risk management approach to address security threats;" (3) inspections of high-risk outbound cargo or transport conveyances (preferably using nonintrusive detection equipment) by "sending" Customs administrations "at the reasonable request of the receiving nation, based upon a comparable risk targeting methodology;" and (4) suggestion of specific benefits that should be provided to businesses meeting minimal supply chain security standards and using best industry practices (World Customs Organization, 2012, p. 3).

National Strategy for Global Supply Chain Security

The Safe Port Act of 2006 directed DHS to develop and implement a strategy for improving the security of the global supply chain. In 2007, DHS released an interim report that contained a series of strategic objectives and programs for accomplishing that purpose. The *National Strategy for Global Supply Chain Security* was released in January 2012, and it outlined two key objectives.

> *Goal 1: Promote the efficient and secure movement of goods. The first goal of the strategy is to promote the timely, efficient flow of legitimate commerce while protecting*

> *and securing the supply chain from exploitation and reducing its vulnerability to disruption. To achieve this goal we will enhance the integrity of goods as they move through the global supply chain. We will also understand and resolve threats early in the process, and strengthen the security of physical infrastructures, conveyances and information assets, while seeking to maximize trade through modernizing supply chain infrastructures and processes.*
>
> *Goal 2: Foster a resilient supply chain. The second goal of the strategy is to foster a global supply chain system that is prepared for, and can withstand, evolving threats and hazards and can recover rapidly from disruptions. To achieve this we will prioritize efforts to mitigate systemic vulnerabilities and refine plans to reconstitute the flow of commerce after disruptions.*
>
> *White House, 2012, p. 1*

One year after the strategy was released, the White House issued an update on its implementation. Key accomplishments during 2012 included:

- Refining the government's understanding of supply chain threats and risks through assessments of the system as an interconnected network
- Establishing technology development priorities (e.g., tracking and intrusion detection capabilities for containers in transit)
- Increasing the number of radiation detection systems provided to foreign governments
- Creating incentives to encourage industry stakeholders to build resilience into their supply chains
- Creating common resilience measures and standards for worldwide use and implementation
- Promoting the development and utilization of supply chain standards for radiation detection through engagement with relevant stakeholders
- Completing mutual recognition arrangements with the European Union for air, land, and sea cargo security programs
- Establishing a Cross Sector Supply Chain Working Group of private sector representatives from various domestic critical infrastructure sectors that is to develop a "Global Supply Chain Findings and Recommendations Report" (White House, 2013, pp. 3–4).

Cargo Targeting and Screening

Customs and Border Protection is responsible for reducing U.S. supply chain vulnerabilities. Among the means used to achieve this objective is the analysis of information to identify shipments that may pose a high-risk of transporting weapons of mass destruction (WMD) or other contraband to the United States. CBP uses its Automated Targeting System (ATS) to assist in assessing the risk posed by incoming cargo and to help determine which shipments should be subjected to a physical examination (by nonintrusive inspection technology, such as imaging equipment, or direct physical inspection). In this way, the agency seeks to allocate its resources toward addressing the greatest threats while minimizing disruptions to the free flow of commerce (GAO, 2012c, pp. 1, 4).

The ATS "is an intranet-based enforcement and decision support system that compares traveler, cargo, and conveyance information against intelligence and other enforcement data. Among other things, ATS uses a set of rules that assess different factors in data provided by supply chain parties, such as importers, to determine risk level of a shipment." Much of the information used by ATS comes from advance cargo information obtained by CBP under two regulations:

- The "24-hour rule," begun in February 2003, which requires vessel carriers to electronically submit complete and accurate manifest information to CBP 24 hours before the loading of cargo onto U.S.-bound vessels at a foreign port
- The Importer Security Filing and Additional Carrier Requirements, known as the "10+2 rule," started in 2009 as a result of a mandate in the SAFE Port Act, which directs importers to provide CBP with 10 shipping data elements (including country of origin) and vessel carriers to supply it with container status messages and the vessel's stow plan in advance of the arrival of the shipment at a U.S. port (GAO, 2012b, p. 13; 2012c, p. 8)

Customs and Border Protection considers ATS as the "cornerstone" of its efforts to effectively target its resources in implementing its security responsibilities, including in such programs as the Container Security Initiative (CSI) and Customs-Trade Partnership Against Terrorism (C-TPAT). The GAO has reported that progress has been made over the years in improving ATS through refinements in its targeting methods, enhanced training, and improved information inputs. However, the GAO has indicated that CBP needs to better use performance assessments of the system to achieve further improvements (Figure 6.4) (GAO, 2012b, 42; 2012c, pp. 25–26).

All U.S.-bound cargo shipments are subjected to analysis by ATS. The ATS risk score is used to help classify a particular shipment as low, medium, or high risk, with all relevant information on medium- and high-risk shipments reviewed by CBP targeting officials stationed at ports and only high-risk shipments subjected to examination by either nonintrusive inspection technology (using x-rays or gamma rays to create an image of the contents inside a container or other conveyance) or physical inspection. Thus, "because CBP does not [examine] 100 percent of U.S.-bound containers, the effectiveness of CBP's security strategy depends on CBP's ability to use ATS, among other tools, to effectively target shipments in the supply chain that pose the greatest security risks" (GAO, 2012c, Highlights, p. 4).

One of the most controversial issues in U.S. maritime security policy in recent years has been the provision in the Implementing the Recommendations of the 9/11 Commission Act of 2007 requiring that, by July 1, 2012, all U.S.-bound, ship-borne containers leaving a foreign port must be scanned by nonintrusive imaging and radiation detection equipment at that port before being loaded onto U.S.-bound vessels. The Act allowed DHS to extend the deadline, on a port-by-port basis, by 2 years, provided it could demonstrate that earlier implementation is not feasible (Peterman et al., 2013, p. 11).

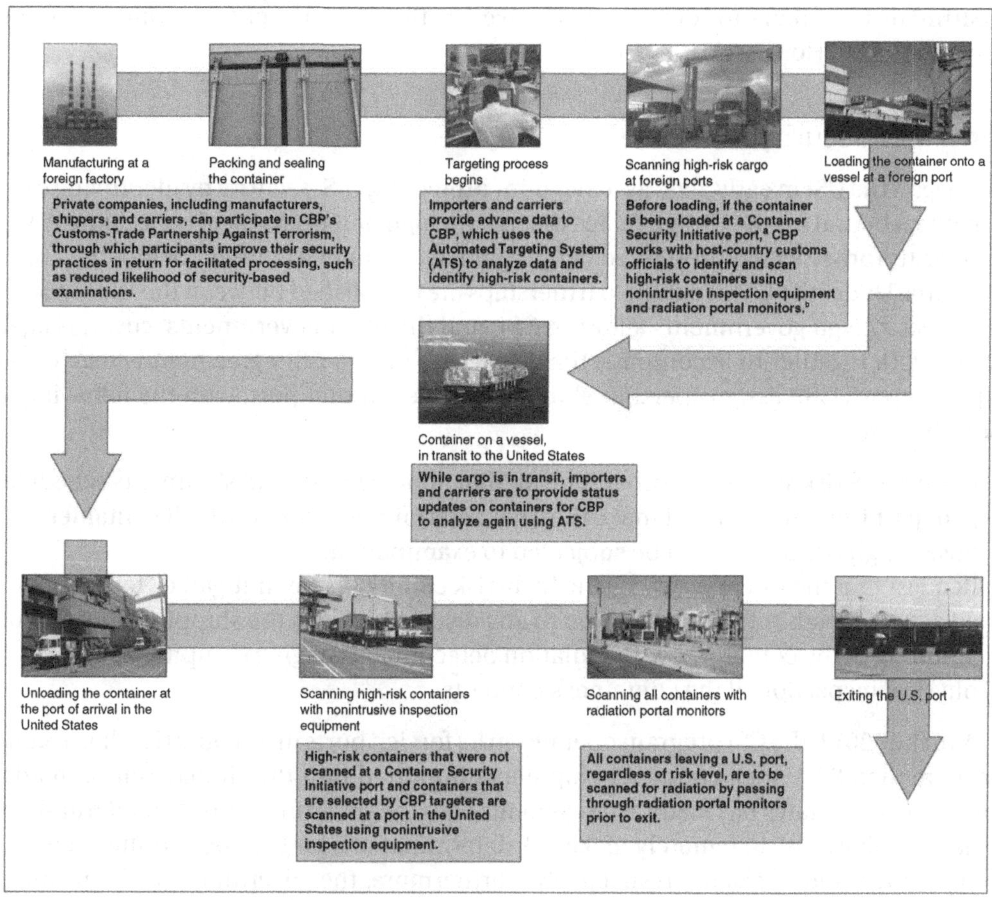

FIGURE 6.4 Key steps for targeting high-risk shipments throughout the maritime supply chain. *(Source: GAO. October, 2012c. Supply chain security: CBP needs to conduct regular assessments of its cargo targeting system. GAO, Washington, DC, p. 10.)*

The SAFE Port Act of 2006 previously authorized pilot projects to test the feasibility of scanning 100% of U.S.-bound cargo containers in three foreign ports. CBP undertook such projects but faced a large number of challenges (including operational, technical, logistical, financial, and diplomatic obstacles) in implementing the requirement, achieving full success in only one foreign port (Port Qasim, Pakistan) where it had been attempted. In addition, major U.S. trading partners, including the European Union and Japan, voiced objections to the requirement. In response, DHS announced in May 2012 it was extending the deadline until July 2014 (GAO, 2013c, pp. 14–16; Peterman et al., 2013, p. 11).

In January 2013, the White House indicated "DHS and other relevant Federal departments and agencies have determined . . . that significant challenges will likely preclude full implementation of the provision in the specific manner prescribed" and that the department is working to identify alternatives to the "100 percent maritime scanning" mandate (White House, 2013, p. 14).

Container Security Initiative

CBP began the CSI in early 2002 as a means of protecting U.S. security by identifying cargo containers bound for the United States that pose a high potential risk of concealing WMDs or other terrorist-related prohibited cargo and having them examined before they reach U.S. ports. Under the CSI program, partnerships are established between the United States and select foreign governments allowing CBP and the other governments' customs agencies to work together to accomplish that objective. After a CSI agreement is reached for a specific port, CBP assigns personnel to be stationed at that port, with the following responsibilities:

- Review ATS risk scores and other information about U.S.-bound shipments scheduled to depart from the port and make the final determination about which containers pose a high risk and should be subjected to examination.
- Request that the threat posed by the high-risk containers be "mitigated" by the host customs officials through resolution of any discrepancies in the shipping information, scanning of the containers with radiation detection or imaging equipment, or physical inspection of the container's contents.

As of July 2013, the CSI program operates in 58 foreign ports in 32 countries that account for more than 80% of all container shipments imported into the United States. In addition, the United States has made arrangements with the governments of Australia and New Zealand to allow CBP to remotely make risk determinations for U.S.-bound shipments out of Melbourne and Auckland, respectively. Furthermore, the government of China allows CBP to remotely monitor U.S.-bound shipments from Shekou, China (and physically visit that port for any examinations) from the CBP station in Shenzhen, China. Thus, there are a total of 61 foreign ports that participate in the CSI (Table 6.1) (GAO, 2013c, pp. 9–11).

The GAO reported in September 2012, "Our work on CSI showed that the program has matured and improved, meeting its strategic goals by increasing both the number of CSI locations and the proportion of U.S.-bound containers passing through CSI ports. In addition, relationships with host governments have improved over time, leading to increased information sharing between governments and a bolstering of host government customs and port security practices" (GAO, 2012b, p. 44).

One of the largest trading nations not participating in CSI thus far is India. A 2007 article by the Indian Institute for Defence Studies and Analyses pinpointed the considerations that continue to inform the government of India's decision making on this issue while also highlighting more general concerns about global maritime security and national interests.

> *Initially, the CSI was a conundrum for many countries that had the USA as a major export destination. India was one such country. On the one hand, there were security*

Table 6.1 Ports Participating in the Container Security Initiative (as of July 2013)

Port	Began CSI Operations	# of U.S.-Bound Shipments (FY 2012)*	Port	Began CSI Operations	# of U.S.-Bound Shipments (FY 2012)*
Vancouver, Canada	02/20/2002	75,226	Zeebrugge, Belgium	10/29/2004	25
Halifax, Canada	03/25/2002	11,731	Livorno, Italy	12/16/2004	77,299
Montreal, Canada	03/25/2002	257	Marseilles, France	01/07/2005	16,378
Rotterdam, Netherlands	09/02/2002	177,448	Dubai, United Arab Emirates	03/26/2005	13,350
Le Havre, France	12/02/2002	130,577	Shanghai, China	04/12/2005	1,900,294
Bremerhaven, Germany	02/02/2003	379,662	Shenzhen, China	06/24/2005	1,475,210
Hamburg, Germany	02/09/2003	184,163	Kaohsiung, Taiwan	07/25/2005	630,732
Antwerp, Belgium	02/23/2003	268,479	Santos, Brazil	09/21/2005	50,816
Singapore, Singapore	03/10/2003	428,730	Colombo, Sri Lanka	09/29/2005	127,432
Yokohama, Japan	03/24/2003	42,953	Buenos Aires, Argentina	11/17/2005	20,791
Hong Kong, China	05/05/2003	938,821	Lisbon, Portugal	12/14/2005	36,903
Gothenburg, Sweden	05/23/2003	14,007	Port Salalah, Oman	03/08/2006	97,450
Felixstowe, United Kingdom	05/24/2003	54,926	Puerto Cortes, Honduras	03/25/2006	67,996
Genoa, Italy	06/16/2003	151,464	Auckland, New Zealand†	04/01/2006	47,244
La Spezia, Italy	06/23/2003	139,382	Chi-lung, Taiwan	09/25/2006	97,476
Busan, South Korea	08/04/2003	867,627	Valencia, Spain	09/25/2006	106,118
Durban, South Africa	12/01/2003	11,807	Caucedo, Dominican Republic	09/26/2006	24,843
Port Kelang, Malaysia	03/08/2004	7393	Barcelona, Spain	09/27/2006	41,763
Tokyo, Japan	05/21/2004	139,659	Kingston, Jamaica	09/28/2006	75,607
Piraeus, Greece	07/27/2004	9746	Freeport, Bahamas	09/29/2006	66,912
Algeciras, Spain	07/30/2004	33,733	Qasim, Pakistan	04/30/2007	46,486
Kobe, Japan	08/06/2004	77,790	Shekou, China†	08/01/2007	60,019
Nagoya, Japan	08/06/2004	74,402	Chiwan, China	08/01/2007	138,069
Laem Chabang, Thailand	08/13/2004	95,551	Balboa, Panama	08/27/2007	76,380
Tanjung Pelepas, Malaysia	08/16/2004	84,337	Cartagena, Colombia	09/13/2007	52,682
Naples, Italy	09/30/2004	19,024	Ashdod, Israel	09/17/2007	543
Liverpool, United Kingdom	10/19/2004	35,273	Haifa, Israel	09/25/2007	36,490
Thamesport, United Kingdom	10/19/2004	27,818	Colon, Panama	09/28/2007	50,481
Southampton, United Kingdom	10/19/2004	50,357	Manzanillo, Panama	09/28/2007	77,030
Tilbury, United Kingdom.	10/19/2004	2,382	Melbourne, Australia†	11/01/2011	37,730
Gioai Tauro, Italy	10/29/2004	12,381			

FY, fiscal year.

*U.S.-bound shipments = maritime container shipments.

†No formal Container Security Initiative (CSI) agreement but allows Customs and Border Protection to remotely monitor U.S.-bound shipments and recommend mitigation of high-risk shipments

(*Source*: GAO. September, 2013c. Supply chain security: DHS could improve cargo screening by periodically assessing risks from foreign ports . GAO, Washington, DC, pp. 39–41.)

and sovereignty concerns attendant to the stationing of U.S. officials in their ports (notwithstanding the fact that CSI is a reciprocal arrangement). Besides, compliance with CSI standards entailed enormous financial investment for advanced technology and port operations, besides time delays due to container checks. On the other hand, if their ports were not CSI-compliant, their exports would have to be re-routed through trans-shipment ports that were CSI-compliant, which would have led to delays and possibly even disruptions due to congestion in those few ports. This would have resulted in increased costs and ensuing losses, including in terms of competitiveness... India has been contemplating joining the CSI since 2003, and many rounds of Indo-US discussions have been held in this regard. However, the primary impediment so far has been the concerns expressed by intelligence and customs agencies about stationing U.S. officials in Indian ports, including their possible intrusion into local port jurisdiction, enforcement and strategic imports.

Khurana, 2007

Customs-Trade Partnership Against Terrorism

CBP's C-TPAT, which also became operational in early 2002, is a voluntary program through which the agency works with private companies involved in international trade to review and improve the security of their supply chains while facilitating the flow of legitimate commerce. The SAFE Port Act codified the C-TPAT program and added new elements to it. In its current form, companies applying to join C-TPAT must conduct a self-assessment of their security procedures using C-TPAT guidelines. The assessment is reviewed by CBP and approved only if it meets the agency's minimum security criteria, including compliance with MTSA and the ISPS Code, as applicable, and procedures for:

- Container security (including inspection, seals, and storage)
- Physical access controls
- Personnel security (including background checks for certain personnel)
- Procedural security for the transportation, handling, and storage of cargo
- Security training and awareness
- Physical security (e.g., fencing or locking devices)
- Information technology security
- Security assessment, response and improvement (McNicholas, 2008, pp. 118–129)

Companies accepted into C-TPAT must agree to provide CBP with further information on their specific security measures and to allow it to periodically verify that these measures meet or exceed the minimum security requirements and are actually in place and effective. As an incentive for the company's participation, CBP offers certain benefits, such as a reduction in the number of inspections (based on lower risk scores) or shorter wait times for the processing of their cargo shipments. As of February 2012, CBP had awarded initial C-TPAT certification to more than 10,000 companies. Although noting that the program allows CBP to develop partnerships within the international trade industry and obtain information from them that would not otherwise be available, the GAO has pointed out the challenges posed "given the international nature of the industry and resulting limits on CBP's jurisdiction and activities" (GAO, 2012b, pp. 14–15, 47).

To broaden the scope of the C-TPAT program, CBP has developed Mutual Recognition Arrangements (MRAs), which are bilateral understandings between CBP and other national customs administrations indicating "that the security requirements or standards of the foreign industry partnership program, as well as its verification procedures, are the same or similar with those of the C-TPAT program. ... The goal of Mutual Recognition is to link the various international industry partnership programs so that together they create a unified and sustainable security posture that can assist in securing and facilitation global cargo trade."

To be considered for an MRA with CBP, a foreign customs administration must have a "full-fledged operational program in place," a strong validation process, strong security procedures, and an existing customs mutual assistance agreement with the United States. If these conditions are met, CBP and the other customs administration compare the two programs' security requirements (with CBP receiving detailed information about the other program, including its eligibility criteria and auditing procedures) and conduct joint validation visits and meetings "to determine if the programs align in basic practice." If both sides are satisfied as to the comparability of C-TPAT and its foreign counterpart, an MRA is signed followed by the development of operational procedures primarily involving information sharing. (In addition to the arrangements with the United States, similar bilateral agreements have been reached between other national customs administrations.)

Among the benefits expected to be realized from the MRA are greater efficiency, expanded information exchanges, less redundancy and duplication of effort, common standards for the facilitation of trade, and enhanced transparency in international commerce between and among customs authorities and the private sector. As of October 2014, CBP has signed nine MRAs:

- New Zealand Customs Service's Secure Export Scheme Program (June 2007)
- Canada Border Services Agency's Partners in Protection Program (June 2008)
- Jordan Customs Department's Golden List Program (June 2008)
- Japan Customs and Tariff Bureau's Authorized Economic Operator Program (June 2009)
- Korea Customs Service's Authorized Economic Operator Program (June 2010)
- European Union's Taxation and Customs Union Directorate Authorized Economic Operator Program (May 2012)
- Taiwan's Directorate General of Customs Authorized Economic Operator Program (November 2012)
- Israel Tax Authority's Authorized Economic Operator Program (June 2014)
- Mexico Tax Administration Service's New Certified Companies Scheme (October 2014) (U.S. Customs and Border Protection, n.d.a; 2014).

The success of the MRAs has led CBP and other national customs authorities to view them as a possible model for the further expansion of the CSI program (GAO, 2012b, p. 48).

Cargo Scanning Technologies

DHS has concentrated its maritime security technology efforts on developing and deploying equipment to scan cargo containers nonintrusively for nuclear materials and certain other dangerous contents. Such technology in current use includes handheld probes that detect possible human occupancy within a container by measuring carbon dioxide levels, truck-mounted nonintrusive gamma ray imaging systems that produce radiographic images of container contents, x-ray imaging portals that produce high-resolution images of container contents, and radiation portal monitors that detect radiation sources. The latter have been a subject of particular interest, given concerns about the possible use of containers to smuggle nuclear materials into ports. In coordination with DHS's Domestic Nuclear Detection Office (DNDO), CBP has deployed more than 1400 radiation portal monitors at U.S. ports of entry, where they are positioned in primary inspection lanes through which virtually all shipping containers must pass (McNicholas, 2008, pp. 361–365; GAO, 2012b, pp. 13–14).

In an effort to develop an improved radiation portal monitoring system, DNDO began the Advanced Spectrographic Portal (ASP) program in 2005. The new portals were intended to better detect key nuclear threat materials while increasing the flow of commerce by reducing the need for secondary inspections. Between 2005 and 2011, GAO reported a number of problems, including testing inadequacies, limited capability to detect certain nuclear materials, excessive false alarms, and a significantly higher lifecycle cost ($822,000) per portal compared with the existing units ($308,000). In 2012, GAO reported, "Once ASP testing became more rigorous, these machines did not perform well enough to warrant deployment." DHS subsequently cancelled the program in July 2012 (GAO, 2012b, p. 43).

The detection and interdiction of nuclear or other radiologic materials smuggled through foreign ports is also the objective of the Megaports Initiative created by the U.S. Department of Energy's National Nuclear Security Administration (NNSA) in 2003. This program funds, or helps fund, the installation of radiation detection equipment at select foreign seaports, provides training to foreign personnel in using the equipment to scan shipping containers for radioactive materials, and includes a sustainability component to assist countries in operating and maintaining the equipment over time. The initiative is designed to work in coordination with the CSI program by providing equipment to be used by host country personnel in scanning containers identified as high risk by CSI officials. Originally, NNSA had intended to provide radiation detection capability at 100 foreign ports, including all 58 CSI ports. As of August 2012, equipment installation had been completed at 42 ports (including 29 CSI ports), with three more scheduled for completion by the end of FY 2012. However, budget limitations have called into question NNSA's ability to equip additional ports and have led to a shift in focus toward sustaining existing deployments (GAO, 2012d, Highlights, pp. 5–6, 9–10).

■ ■ Critical Thinking Question ■

Compare and contrast the CBP/DNDO and DOE programs for radiation detection at ports. What are some of the advantages and disadvantages of each?

DIGGING DEEPER
CARGO CONTAINERIZATION

"One of the Most Significant Events in Ocean Shipping Commerce"

The era of containerized shipping began in 1956 when trucking owner Malcolm McLean had 58 containers loaded onto a converted tanker bound from Newark, New Jersey, to Houston, Texas. The potential advantages of this new form of transportation were not immediately apparent (with resistance from the major traditional shipping lines and longshoremen unions and a lack of standardization in container size), but by the 1980s, container shipping had become the dominant means of nonbulk (i.e., excluding oil, grain, and other bulk shipments) maritime cargo transportation (Table 6.2) (McNicholas, 2008, pp. 34–35).

McNicholas (2008) explains the causes and results of the transformation:

> *What was new about McLean's innovation was the idea of using large containers that were never opened in transit between shipper and consignee and that were transferrable on an intermodal basis between trucks, ships, and railroad cars. The reduction in loading time (about 1/20 of the time used for the same quantity of [non-containerized] cargo) and the reduction in overall port/shipping costs (about 1/50 of the cost of loading and shipping the same quantity of [non-containerized] cargo) had a huge financial impact on the cost of the operation…Containerization has truly revolutionized cargo shipping.… Today, between 90 and 95% of all nonbulk goods are shipped in containers.*
>
> *McNicholas, 2008, pp. 29, 35*

However, the continued growth in container shipping has also been viewed as a potential security threat, with its characteristics providing significant opportunities for concealing weapons and terrorists.

- Container shipping is highly complex, with the vessels typically carrying container cargo that originated from hundreds of companies and individuals, stored in and transported from numerous inland warehouses, and secured by a "rudimentary" locking system. Each of these links in the supply chain is potentially susceptible to exploitation by terrorists (and other criminals).
- A particular vulnerability is "the ineffectiveness of point of origin inspections. Many [coastal] states fail to routinely vet dock workers, do not require that truck drivers present valid identification before entering an offloading facility, and frequently overlook the need to ensure that all cargo is accompanied by an accurate manifest. The absence of uniform and concerted dockside safeguards works to the advantage of the terrorist, both because it is virtually impossible to inspect containers once they are on the high seas and due to the fact that only a tiny fraction of boxed freight is actually checked on arrival at its destination" (Chalk, 2008, pp. 26–29).

In particular, government officials in the United States and elsewhere have been concerned about the possible use of containers to smuggle WMDs into their countries. Although DHS officials consider the likelihood of such an occurrence to be relatively low, the consequences would be "catastrophic." Furthermore, although there have been no known incidents of cargo containers being used to transport WMD, there have been cases around the world of terrorists and other criminals using containers to illicitly carry people, weapons, and other illegal substances. A 2012 DHS risk assessment concluded that attacks using containers could cause major disruptions in the maritime transportation system (GAO, 2013c, p. 1).

Based on such considerations, a large portion of maritime security efforts have been aimed at improving container security. As with most security measures, these programs have both direct (governmental) and indirect (business compliance) costs. A 2008 report prepared for the U.K. government opined:

> *The trading community faces a serious security threat—not from terrorists, but from increasing measures being put in place in the name of security that affect the international supply chain…Over the last few years we have seen an avalanche of such initiatives presented by governments, nationally, regionally and internationally and new requirements are being introduced all the time. This creates uncertainty for traders that have to comply with them, together with a compliance cost in the form of changes to their own control and data management systems to meet them—costs which ultimately get passed onto the consumer.*
>
> *SITPRO, 2008, p. 2*

This debate brings into focus the question of what the proper balance is in securing against low-probability, high-consequence events. In the case of container security, what do you think? Have the benefits (no known major terrorist incidents) been worth the costs? Why or why not?

Table 6.2 Top 20 World Container Ports, 2010

Port	Country	TEUs
Shanghai	China	29,069,000
Singapore	Singapore	28,431,000
Hong Kong	China	23,669,000
Shenzhen	China	22,510,000
Busan	South Korea	14,194,000
Ningbo	China	13,144,000
Guangzhou	China	12,487,000
Qingdao	China	12,012,000
Dubai Ports	United Arab Emirates	11,576,000
Rotterdam	Netherlands	11,146,000
Tianjin	China	10,080,000
Kaohsiung	Taiwan	9,121,000
Port Kelang	Malaysia	8,872,000
Antwerp	Belgium	8,468,000
Hamburg	Germany	7,896,000
Los Angeles	United States	7,832,000
Tanjung Pelepas	Malaysia	6,299,000
Long Beach	United States	6,263,000
Xiamen	China	5,824,000
New York/New Jersey	United States	5,292,000

TEUs = 20-foot equivalent unit = volume carried by a 20-foot long container.
(*Source*: U.S. Department of Transportation, Research and Innovative Technology Administration. 2013. Pocket guide to transportation 2013. Washington, DC, p. 40.)

Maritime Domain Awareness and Intelligence

The 2005 National Plan to Achieve Maritime Domain Awareness (MDA) defines its subject as "the effective understanding of anything associated with the maritime domain that could impact the security, safety, economy, or environment of the United States." The plan outlines the key objectives that are to guide the development of U.S. MDA capabilities:

- Persistently monitor vessels and other watercraft, cargo, vessel crews and passengers, and all identified areas of interest in the global maritime domain.
- Access and maintain data on vessels, facilities, and infrastructure.
- Collect, fuse, analyze, and disseminate information to decision makers to facilitate effective understanding.
- Access, develop, and maintain data on MDA-related mission performance (U.S. Department of Homeland Security, 2005b, pp. 1, 3).

The U.S. Coast Guard is the principal federal agency responsible for achieving maritime domain awareness in collaboration with the U.S. Navy, CBP, U.S. Immigration and Customs Enforcement, and other stakeholders in the maritime community. The primary means for providing the necessary information sharing, situational awareness, and collaborative planning is the national maritime Common Operational Picture (COP), which is a map-based, "near-real time, dynamically tailorable, network-centric virtual information grid shared by all U.S. Federal, state, and local agencies with maritime interests and responsibilities" (Randol, 2010, pp. 43–44; U.S. Department of Homeland Security, 2005b, p. ii).

A 2013 GAO analysis reported that the Coast Guard has made progress in developing the COP by "adding internal and external data sources that allow for better understanding of anything associated with the global maritime domain" and "increasing user access to this information." On the other hand, the agency continues to face challenges in deploying systems that effectively display and share COP data (GAO, 2013b, Highlights).

Maritime Security Risk Analysis Model

The Maritime Security Risk Analysis Model (MSRAM) was developed by the Coast Guard in 2005 and has been refined since then. It serves as the agency's major method for evaluating and managing maritime security risks:

> *MSRAM calculates the risk of a terrorist attack based on scenarios—a combination of target and attack modes—in terms of threats, vulnerabilities, and consequences at more than 28,000 maritime targets. The model focuses on individual facilities and cannot model system impacts or complex scenarios involving adaptive or intelligent adversaries.*

The Coast Guard's risk management efforts, including its use of risk management in allocating resources across all its mission areas, have been viewed as more developed compared with other DHS components (GAO, 2012b, pp. 10–11, 38).

Area Maritime Security Committees

In implementing MTSA requirements concerning regional port security, the Coast Guard has organized 43 Area Maritime Security Committees covering both coastal and inland U.S. ports. The committees were designed to enhance communications among federal, state, local, and private sector stakeholders within the port area and are responsible for identifying critical port infrastructure and operations, identifying risks, determining mitigation strategies and implementation methods, developing and describing the process for continual evaluation of overall port security, providing advice and assistance to the Captain of the Port in developing the Area Maritime Security Plan, and facilitating the communication of threats and changes in maritime security. Each committee must meet at least once a year and be composed of at least seven members (with at least seven of the total membership having 5 or more years of experience related to maritime or port security operations). The Coast Guard makes the appointments, with the membership drawn from (but not limited to) representatives of:

- Federal, territorial, or tribal governments
- State governments and political subdivisions
- Local public safety, crisis management, and emergency response agencies
- Law enforcement and security organizations
- Maritime industry, including labor
- Port stakeholders affected by security practices and policies (U.S. Coast Guard, n.d.b).

The AMSCs have advanced information sharing, and helped to improve the timeliness, completeness, and usefulness of such data, although the lack of federal security clearances for some committee members has sometimes hindered the dissemination of certain information (GAO, 2012b, pp. 11, 39).

■ ■ Critical Thinking ■

The concept of area maritime security is not part of the ISPS Code. What advantages does use of this security layer provide? What are the disadvantages, if any?

Vessel Tracking Systems

The goal of vessel tracking for security purposes is to obtain information about the identification and location of ships to assess the degree of risk they pose while minimizing disruption of the maritime transportation system. The MTSA established the United States' first vessel tracking requirements by directing that automatic identification systems be operated on certain vessels in U.S. waters and authorizing the Coast Guard to develop a long-range, automated tracking system that could track vessels at sea using existing onboard radio and data communications equipment. In 2006, the

SOLAS Convention was amended to mandate the establishment of an international system for the identification and tracking of vessels, including requirements that vessels of International Maritime Organization (IMO) member nations on international voyages transmit certain identification information, that member nations create and operate data centers to receive such information, and that an international information exchange network be created for the sharing of data on vessel identification and tracking (GAO, 2012b, pp.11n24, 41).

To carry out its responsibilities for long-range tracking under MTSA and the 2006 SOLAS amendments, the Coast Guard developed the Long Range Identification and Tracking (LRIT) system to collect and disseminate vessel position information. The LRIT system is "a satellite-based, real-time reporting mechanism that allows unique visibility to position reports of vessels that would otherwise be invisible and potentially a threat to the United States." The Coast Guard's National Data Center supports the LRIT system and "monitors IMO member state ships that are 300 gross tons or greater on international voyages and either bound for a U.S. port or travelling within 1000 nautical miles of the U.S. coast" (U.S. Coast Guard, n.d.c).

For tracking vessels in U.S. coastal waters, inland waterways and ports, the Coast Guard created the Nationwide Automatic Identification System (NAIS) in 2004 to collect information from vessels that are equipped with an automatic identification system providing data on the vessel's name, course, speed, and registration number, among other things. At present, the system consists of approximately 200 receiver sites along the U.S. coast (including Alaska and Hawaii) and inland waterways, plus the territory of Guam (U.S. Coast Guard, n.d.d).

The GAO has reported that the Coast Guard's tracking systems have worked well with respect to larger vessels adequately equipped with tracking technologies but have not been as effective in monitoring smaller vessels (GAO, 2012b, p. 41).

Interagency Operations Centers

The SAFE Port Act directed DHS to establish Interagency Operations Centers (IOCs) to enhance security information sharing in key ports. The 2010 Coast Guard Authorization Act further required that the IOCs should provide, where practicable, for the physical colocation of the Coast Guard with its major partners in a given port and include information management systems. In response, the Coast Guard is working to develop such centers at 35 U.S. ports (GAO, 2012b, p. 40).

The IOCs are designed to assist the Coast Guard and other port agencies to collaborate in the conduct of first response, law enforcement and homeland security operations; collaborate and jointly plan operations; share targeting, intelligence, and scheduling information; develop improved real-time awareness, threat evaluation, and resource deployment; and minimize the economic impact of any disruption of port operations (U.S. Coast Guard, n.d.e).

Information management at the IOCs is to be handled by the Coast Guard's Watch-Keeper program, which was initiated in 2005 and is considered by the agency to be the "heart" of the IOCs. It is currently deployed as a "technology demonstration" project at 22 of the planned IOCs (with full deployment to be completed by October 2014). Watch-Keeper is "a data fusion and information management system [that] provides a common tool that can be accessed by Coast Guard users and the full spectrum of port partners, including CBP, Immigration and Customs Enforcement, state agencies and local law enforcement." The system uses data from a number of sources, including the Nationwide Automatic Identification System and the Common Operational Picture (U.S. Coast Guard, 2013b).

The Coast Guard has faced a number of challenges in developing the IOCs, caused in part by insufficient funding and changing requirements for the centers. As a result, GAO reported that although the IOCs have "provided promise in improving maritime domain awareness and information sharing," the Coast Guard "has experienced coordination challenges that have limited implementation" of the program (GAO, 2012b, pp. 12, 40).

Coast Guard Intelligence Elements

The primary objective of the Coast Guard's various intelligence components is the pursuit of maritime domain awareness. At the agency level, Coast Guard intelligence is organized into two distinct parts. One is the National Intelligence Element, which conducts traditional intelligence activities in gathering and disseminating foreign intelligence and counterintelligence and includes the *Cryptologic Program* (responsible for signals intelligence) and the *Counterintelligence Service*. The second is the Law Enforcement Intelligence Program, which supports the Coast Guard's law enforcement and regulatory activities and includes the *Coast Guard Investigative Services* unit (Randol, 2010, pp. 43–46).

Operationally, Coast Guard intelligence organizations provide support at the national, area (Atlantic and Pacific), district (subdivisions of the two areas), and sector (subdivisions of districts) levels. The *Intelligence Coordination Center (ICC)* is the national-level entity responsible for:

- Managing, analyzing, and producing maritime intelligence for the Coast Guard with respect to law enforcement, military readiness, counterterrorism, force protection, marine environmental protection, and port and maritime security
- Maintaining a 24-hour Indications and Warning Center and a current intelligence watch that includes the COASTWATCH Branch[10]

Maritime Intelligence Fusion Centers (one for the Atlantic and one for the Pacific areas) provide Coast Guard operational commanders, the Department of Defense, the U.S.

[10]COASTWATCH reviews notices of arrival (NOAs) information required to be submitted 96 hours in advance of arrival in the United States by large (greater than 300 gross tons) incoming commercial vessels. It checks this information against federal databases to identify potential security and criminal threats (Randol, 2010, p. 47).

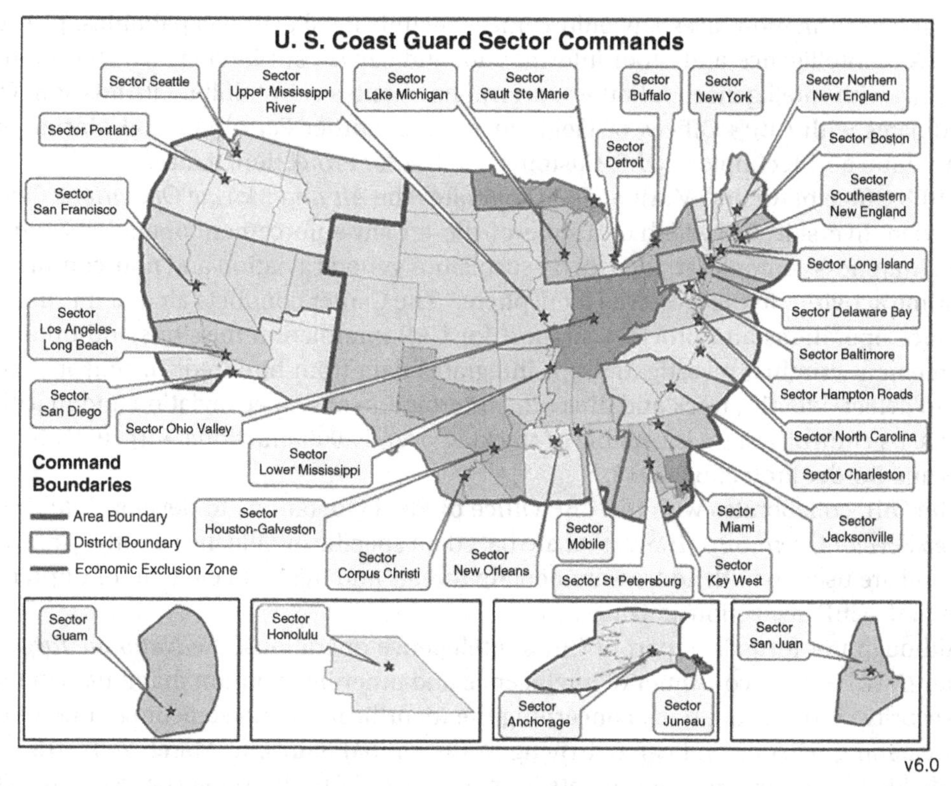

FIGURE 6.5 U.S. Coast Guard sector commands. *(Courtesy of U.S. Coast Guard, http://upload.wikimedia.org/wikipedia/commons/6/60/USCG_Sector_Map.jpg.)*

intelligence community,[11] and other law enforcement agencies with intelligence analysis of geopolitical issues, terrorism, vessel movements, transnational crimes (including piracy), port security, and marine biological resources (Figure 6.5).

Area and District Intelligence Staffs provide intelligence support for their respective commands. *Sector Intelligence Staffs* are the key intelligence support element at the port level and provide threat assessments, as well as obtain intelligence on foreign ports via debriefings of ship crews returning to the U.S. from overseas (Randol, 2010, pp. 46–48).

Customs and Border Protection Intelligence Elements

CBP's Office of Intelligence and Investigative Liaison (OIIL) "serves as a coordinating facilitator that integrates U.S. Customs and Border Protection's diverse intelligence capabilities into a single, cohesive intelligence enterprise." It collects and analyzes advance traveler

[11]Statutory members of the U.S. Intelligence Community include the CIA; FBI; Defense Intelligence Agency; National Reconnaissance Office; National Security Agency; the Departments of Energy, Justice, Homeland Security, State and Treasury; the intelligence components of the U.S. Army, Navy, Marines, and Air Force; and the Coast Guard (Randol, 2010, p. 4n21).

and cargo information, uses law enforcement technical collection capabilities, provides analysis of intelligence and other information, and carries out intelligence-sharing relationships with intelligence and other government agencies. In fulfilling its mission, OIIL coordinates with CBP's Offices of Field Operations, Border Patrol, Air and Marine, and Information and Technology (U.S. Customs and Border Protection, n.d.b).

CBP's Office of Air and Marine (OAM) operates the *Air and Marine Operations Center,* located in Riverside, CA, which is "a state-of-the-art law enforcement operations and domain awareness center [that] focuses on suspicious general aviation and non-commercial maritime activities in the Western Hemisphere." The Center conducts air and marine surveillance operations and provides support for CBP interdiction missions and other law enforcement criminal investigations. It integrates data from hundreds of sensors and is able to detect, identify, track and direct the interdiction of suspect aviation and maritime targets. Its control system is capable of tracking over 50,000 individual targets (U.S. Customs and Border Protection, 2013).

The OIIL collaborates with the CBP Office of Field Operations to develop *Intelligence Driven Special Operations (IDSO)* that are based on specific intelligence or other information and are used to both address immediate threats and inject an element of unpredictability into CBP inspection activities.

Although not formally a part of CBP's intelligence operations, the *National Targeting Center (NTC)* is a key consumer of intelligence and other information that it uses in making security recommendations concerning the identification and screening of potentially threatening persons and cargo arriving in the United States. In March 2007, the NTC was divided into separate components for passengers (NTC-Passenger) and cargo (NTC-Cargo). Both divisions help formulate targeting rules for the Automated Targeting System, and perform analysis of advance passenger and cargo information required to be submitted to CBP (Randol, 2010, pp. 20–28).

Conclusion

Programs now underway have improved the security of maritime transportation systems, their cargo, and passengers. Compared with the pre-9/11 world, there is much more widespread recognition and acceptance of international standards for vessel, port, and customs security (albeit with variable national-level implementation); better and more widely disseminated information on the maritime threat environment; and heightened security awareness.

Yet the number of terrorist attacks on the maritime mode has remained fairly constant over time, and the incidence of sea piracy remains high (equal to or greater than the combined number of terrorist attacks on all transportation modes in recent years). Bennett (2008) observed that the current security system has "hardened the target through standardized procedures for coordinating security measures and international cooperation on maritime security. It would be unrealistic to expect more than modest gains. The open nature of the maritime transportation system ensures its vulnerability" (p. 178).

Beyond its openness, maritime security faces certain unique challenges. First, unlike the case with respect to aviation and land transportation, cargo—not passengers—has been the focus of most maritime security programs. This carries with it both an increased role for economic considerations in the implementation of security measures and a lessened "price" (financial and convenience) customers are willing to pay when profit margins rather than lives are at stake. In addition, although the fear of terrorist use of maritime transport for deployment of weapons of mass destruction remains a major concern, the fact that actual terrorist incidents have been few and far between poses a particular problem in building and sustaining governmental (and public) support for maritime security measures.

Discussion Questions

1. Describe the objectives and major provisions of the International Ship and Port Facility Security Code.
2. What is the TWIC program, and what is its current status?
3. What is the global supply chain, and how does the SAFE Framework seek to secure it?
4. Describe the CSI and C-TPAT programs (including their major elements and current extent).
5. What are the Common Operational Picture, the Maritime Security Risk Analysis Model, the Long Range Identification and Tracking system, and the Maritime Intelligence Fusion Centers, and how do they contribute to maritime domain awareness?

References

Bennett, J.C., 2008. In Bragdon, C.R. (Ed.), Transportation security. Butterworth-Heinemann/Elsevier, Burlington, MA, pp. 149–181.

Chalk, P., 2008. The maritime dimension of international security: Terrorism, piracy, and challenges for the United States. Rand Corporation, Santa Monica, CA.

European Commission. 2012. Commission staff working document on transport security. Brussels.

European Union. n.d. How the EU works. <http://europa.eu/about-eu/index_en.htm> (accessed 10.26.14.)

FEMA. 2013. FY 2013 Port Security Grant Program (PSGP) funding opportunity announcement. Washington, DC.

Federal Register. Spring, 2014. Unified Agenda: Transportation Worker Identification Credential (TWIC); card reader requirements. <http://www.federalregister.gov/regulations/1625-AB21/transportation-worker-identification-credential> (accessed 10.31.14.)

GAO. February, 2008. Maritime security: Coast Guard inspections identify and correct facility deficiencies, but more analysis needed of program's staffing, practices, and data. Washington, DC.

GAO. July, 2010. Maritime security: DHS progress and challenges in key areas of port security. Washington, DC.

GAO. November, 2011. Port Security Grant Program: Risk model, grant management, and effectiveness measures could be strengthened. Washington, DC.

GAO. April, 2012a. Maritime security: Coast Guard efforts to address port recovery and salvage response. Washington, DC.

GAO. September, 2012b. Maritime security: Progress and challenges 10 years after the Maritime Transportation Security Act. Washington, DC.

GAO. October, 2012c. Supply chain security: CBP needs to conduct regular assessments of its cargo targeting system. Washington, DC.

GAO. October, 2012d. Combating nuclear smuggling: Megaports Initiative faces funding and sustainability challenges. Washington, DC.

GAO. June, 2013a. Transportation Worker Identification Credential: Card reader pilot results are unreliable; security benefits should be reassessed. Washington, DC.

GAO. July, 2013b. Coast Guard: Observations on progress made and challenges faced in developing and implementing a Common Operational Picture. Washington, DC.

GAO. September, 2013c. Supply chain security: DHS could improve cargo screening by periodically assessing risks from foreign ports. Washington, DC.

Government of the United Kingdom. 2013. Maritime security. <https://www.gov.uk/maritime-security> (accessed 10.26.14.)

International Maritime Organization. n.d.a. Summary of Status of Conventions. <http://www.imo.org/About/Conventions/StatusOfConventions/Pages/Default.aspx> (accessed 11.18.14.)

International Maritime Organization. n.d.b. What are the different security levels referred to in the ISPS Code? <http://www.imo.org/blast/mainframe.asp?topic_id=897> (accessed 10.26.14.)

International Maritime Organization. n.d.c. AIS transponders. <http://www.imo.org/OurWork/Safety/Navigation/Pages/AIS.aspx> (accessed 10.26.14.)

Khurana, G.S, 2007. India and the Container Security Initiative. Institute for Defence Studies and Analyses. July 17, 2007. <http://idsa.in/idsastrategiccomments/IndiaandtheContainerSecurityInitiative_GSKhurana_170707> (accessed 10.26.14.)

McNicholas, M., 2008. Maritime security: An introduction. Butterworth-Heinemann/Elsevier, Burlington, MA.

Peterman, D.R., Elias, B., Fritelli, J., January, 2013. Transportation security issues for the 113th Congress. Congressional Research Service, Washington, DC.

Randol, M.A., March, 2010. The Department of Homeland Security intelligence enterprise: Operational overview and oversight challenges for Congress. Congressional Research Service, Washington, DC.

SITPRO. 2008. A U.K. Review of Security Initiatives in International Trade. <http://webarchive.nationalarchives.gov.uk/20100918113753/http:/www.sitpro.org.uk/policy/security/initiatives0108.pdf> (accessed 10.26.14.)

U.S. Coast Guard. n.d.a. Area maritime security plan (AMSP). <http://www.uscg.mil/hq/cg5/cg544/amsp.asp> (accessed 10.26.14.)

U.S. Coast Guard. n.d.b. Area maritime security committee (AMSC). <http://www.uscg.mil/hq/cg5/cg544/amsc.asp> (accessed 10.26.14.)

U.S. Coast Guard. n.d.c. LRIT. <http://www.navcen.uscg.gov/?pageName=lritMain> (accessed 10.26.14.)

U.S. Coast Guard. n.d.d. Nationwide automatic identification system. <http://www.navcen.uscg.gov/?pageName=NAISMain> (accessed 10.26.14.)

U.S. Coast Guard. n.d.e. Interagency operations centers. <http://www.uscg.mil/acquisition/ioc/default.asp> (accessed 10.26.14.)

U.S. Coast Guard. 2013a. Port state control in the United States, 2012 annual report. Washington, DC.

U.S. Coast Guard. March, 2013b. Coast Guard transitioning interagency operations centers' software to sustainment. March 13, 2013. <http://www.uscg.mil/acquisition/newsroom/updates/ioc031313.asp> (accessed 10.26.14.)

U.S. Coast Guard. 2014. Port security advisory (2-14). Washington, DC.

U.S. Customs and Border Protection. n.d.a. Customs-trade partnership against terrorism mutual recognition. <http//www.cbp.gov/border-security/ports-entry/cargo-security/c-tpat-customs-trade-partnership-against-terrorism/mutual-recognition> (accessed 10.31.14.)

U.S. Customs and Border Protection. n.d.b. Office of Intelligence and Investigative Liaison. <http://www.cbp.gov/about/leadership/assistant-commissioners-office/intelligence-investigative-liaison> (accessed 10.31.14.)

U.S. Customs and Border Protection. August, 2013. Air and Marine Operations Center fact sheet. Washington, DC.

U.S. Customs and Border Protection. October, 2014. U.S., Mexico sign mutual recognition arrangement. October 17, 2014. <http//www.cbp.gov/newsroom/national-media-release/2014-10-17-000000/us-mexico-sign-mutual-recognition-arrangement> (accessed 10.27.14.)

U.S. Department of Homeland Security. September, 2005a. The national strategy for maritime security. Washington, DC.

U.S. Department of Homeland Security. October, 2005b. The national plan to achieve maritime domain awareness. Washington, DC.

U.S. Department of Homeland Security. April, 2008. Small Vessel Security Strategy. Washington, DC.

U.S. Department of Homeland Security. January, 2011. Small Vessel Security Implementation Plan report to the public. Washington, DC.

U.S. Department of Homeland Security. July, 2014. FY 2014 Port Security Grant Program (PSGP) fact sheet. Washington, DC.

White House. January, 2012. National Strategy for Global Supply Chain Security. Washington, DC.

White House. January, 2013. National Strategy for Global Supply Chain Security implementation update. Washington, DC.

World Customs Organization. June, 2012. SAFE Framework of standards to secure and facilitate global trade. www.wcoomd.org.

7

Implementing Land Transportation Security

CHAPTER OBJECTIVES:

In this chapter, you will learn about assessments, guidelines, training, and other programs for securing:

- Mass transit and passenger rail systems
- Freight rail
- Highway infrastructure and motor carriers
- Pipelines

Introduction

Land transportation systems continue to be the most frequent targets of terrorist attacks but also receive the lowest level of international and national security attention and resources. Furthermore, each of the four land transportation modes—mass transit and passenger rail; freight rail; highway infrastructure and motor carriers; and pipelines—present different security challenges and are subject to multiple government agency security efforts. In the United States, land transportation security has centered on federal risk assessments and guidelines coupled with voluntary industry participation and compliance, with far less mandatory regulation than in aviation or maritime transportation.

There is no equivalent in the land sector to the International Maritime Organization (IMO) or the International Civil Aviation Organization (ICAO). Nor is there a set of agreed-upon international standards to match those in the maritime and aviation modes.

In January 2006, an International Working Group on Land Transport Security was formed at a meeting in Tokyo and currently has 20 member countries (including Canada, France, Germany, Israel, Japan, the United Kingdom, and the United States) as well as the European Union (EU), the International Union of Railways (UIC), and the International Association of Public Transport (UITP). However, the group's mission is limited to the sharing of security information and the development of best practices while explicitly eschewing a role in producing international security standards such as those developed by IMO or ICAO. To date, its focus has been almost exclusively confined to mass transit and passenger and freight rail. For example, its major actions in 2013 included creating a framework of threat scenarios and key vulnerabilities in the rail sector to be used by national and local rail authorities in mitigating attacks and discussing how to organize freight rail services in the event of a crisis (Government of the United Kingdom, Department for Transport, 2013; International Union of Railways, 2013).

Earlier, in 2004, the UIC and UITP issued a joint "Declaration on Public Transport and Anti-Terrorism Security," which committed their members to organize "crisis drills" in consultation with national authorities and called on those national authorities to:

- Keep public transport operators informed "as rapidly as possible" about terrorist threats.
- Provide a "strong and prompt intervention in the areas with highest concentrations of people in and around transport systems in times of threat or effective crisis."
- Develop sensors and emergency plans for addressing attacks using bacteriological, radiologic, nuclear, and chemical weapons.
- Establish permanent collaboration with the organizations and their members at the national, European (EU), and international levels (International Union of Railways, n.d.).

The EU established an Advisory Group on Land Transport Security in 2012, and a staff working document was issued to assess "what can be done at the EU level to improve [land] transport security, particularly in areas where putting in place common security requirements would succeed in making Europe's transport systems more resilient to acts of unlawful interference." Among the areas recommended for policy development were intermodal interchange and mass transit security, rail security, training of staff, planning for the aftermath of an incident, technology and equipment, land transport security research, and better communication and sharing of information. The document concluded:

> *There are considerable merits in continuing the work already commenced in the aviation and maritime sectors in developing specific measures on [land] transport security at the EU level. The benefits could include: a higher overall level of security for citizens in the EU; lower levels of theft and other crimes—with consequential cost savings; simplification for transport operators by having common security requirements—with consequential cost savings; simplification for security providers—both equipment and personnel—by having common performance requirements; and having a stronger voice in international fora. Nonetheless, [land] transport security is a sensitive topic, and full account must be taken of the implications it can have for public authorities as well as for the fundamental rights of the individuals.*
>
> *European Commission, 2012, pp. 1, 4–7, 11*

The U.S. government has not produced a "national strategy" for land transportation to complement those it has established for the maritime and aviation modes. Instead, a more limited "Surface Transportation Security Priority Assessment" was issued by the Obama Administration in March 2010. The assessment represented "a collaborative process that produced recommendations compiled from participating stakeholders' individual recommendations for increasing the security of the surface transportation system" and involved "outreach across the spectrum of government and private stakeholders." The primary goals of the undertaking included (1) enhancing security and reducing risk; (2) improving the efficiency and effectiveness of the federal mission, organization, and program; (3) strengthening interactive stakeholder partnerships; and (4) using a systems management approach to surface transportation security. The report made 20 recommendations intended to provide guidance to federal agencies with land transportation security responsibilities (Table 7.1) (pp. i–iii, 13).

Table 7.1 Surface Transportation Security Priority Assessment Recommendations

1. Designate a lead agency to coordinate periodic modal and cross-modal security risk analyses.
2. Implement an integrated federal approach that consolidates capabilities in a unified effort for security assessments, audits, and inspections to produce more thorough evaluations and effective follow-up actions for reducing risk, enhancing security, and minimizing burdens on assessed surface transportation entities.
3. Identify appropriate methodologies to evaluate and rank surface transportation systems and infrastructure that are critical to the nation.
4. Implement a multi-year, multiphase grants program based on a long-term strategy for surface transportation security.
5. Establish a measurable evaluation system to determine the effectiveness of surface transportation security grants.
6. Establish an interagency process to inventory education and training requirements and programs, identify gaps and redundancies in surface transportation owner and operator education and training, and ensure that federal training requirements support counterterrorism and infrastructure protection.
7. Implement a unified environment for sharing transportation security information that provides all relevant threat information and improves the effectiveness of information flow.
8. Reemphasize National Infrastructure Protection Plan (NIPP) framework priorities with the Sector-Specific Agency, surface transportation owners and operators and state, local, tribal, and territorial (SLLT) partners to focus development and implementation of a relevant and representative model that enhances security of the Transportation Systems Sector partners.
9. Fully identify federal roles and responsibilities in surface transportation security, taking steps to efficiently leverage resources and ultimately lead to a budget "cross cut" that extends federal coordination to include both surface transportation safety and security.
10. Identify an interagency lead to establish a single data repository for all federally obtained security risk-related information and transportation systems and assets.
11. Coordinate data requests with the established single data repository to avoid redundant efforts, take advantage of existing data sets, and establish data access control.
12. Analyze the common features of existing analysis methods and tools, and then perform a gap analysis to identify additional characteristics that would ensure that analyses are more closely comparable and consistent with the risk assessment principles in the NIPP.
13. Define a process to assess and certify extant industry risk assessments for ranking risk remediation projects under the Transit Security Grant Program and other similar federal programs.
14. Establish a fee-based, centrally managed "clearing house" to validate new privately developed security technologies that meet federal standards.
15. Encourage the use of the SECURE and FutureTECH programs (that provide models for the Department of Homeland Security to work with the private sector in technology development) within appropriate directives.
16. Create a more efficient federal credentialing system by reducing credentialing redundancy, leveraging existing investments, and implementing the principle of "enroll once, use many" to reuse the information of individuals applying for multiple access privileges.
17. Collaborate with the Sector Coordinating Councils (SCCs) to develop a proposal for security threat assessment standards.
18. Incorporate formal and informal methods for surface transportation owner and operators, as well as SCCs that represent them, to provide direct input into setting surface transportation research and development priorities.
19. Develop a formal, recurring surface transportation security grant process for meeting with surface transportation SCCs, owners, operators, and SLLT governments, collecting and adjudicating recommendations, and making final decisions.
20. Review key policy issues and questions identified by the *Surface Transportation Security Priority Assessment* to address unresolved policy issues and provide solutions for resolving identified security gaps.

(*Source*: White House. March, 2010. Surface transportation security priority assessment. Washington, DC, pp. 14–23.)

Mass Transit and Passenger Rail

The mass transit and passenger rail subdivision stands out from all other land modes for several reasons. First, it has been subjected to terrorist attacks far more often than any other transportation mode (including aviation), and unlike the other land modes, its primary targets are people rather than goods or the transportation systems themselves.[1] This in turn has drawn higher levels of security attention than is typical for the land sector as a whole. Finally, most of the responsibility for these systems is at the local, rather than international, national, or state, level.

In the United States, the Transportation Security Administration (TSA) is the lead federal agency for mass transit and passenger rail security. Operating with relatively limited resources (especially compared with aviation security), it has established three principal strategic priorities:

- *Increasing visible deterrence through the deployment of canine teams, passenger screening teams, and antiterrorism teams.* The Visible Intermodal Protection and Response (VIPR) security teams, which "provide a random, announced, high-visibility surge into a transit agency," are an important element in fulfilling this objective.
- *Increasing infrastructure resilience by protecting the most critical tunnels, stations, and bridges.* A major component of this effort is the DHS/DOT interagency national tunnel security initiative, which has identified and assessed risk to underwater tunnels, prioritized tunnel risk mitigation efforts by targeting grant funding to the most pressing needs, developed funding strategies to help guide future technology research and development, and produced and disseminated recommended protective measures that transit agencies may take to enhance tunnel security.
- *Fully engaging the public and transit operators in the counterterrorism mission.* Examples include the TSA Mass Transit Security Training Program, the TSA-funded Land Transportation Anti-Terrorism Training Program at the Federal Law Enforcement Training Center, and the Connecting Communities program that brings together federal, state, local, and tribal security officials and first responders "to discuss security prevention and response efforts and ways to work together effectively to prepare and protect their communities" (Transportation Security Administration, 2014, p. 15; n.d.a).

In implementing mass transit and passenger rail security programs, TSA has used its regulatory authority sparingly. In May 2004, it issued the RAILPAX-04-01 and RAILPAX-04-02 security directives that required rail transportation operators to implement

[1]One exception to this finding concerns intercity passenger buses, which are considered by TSA (and in this work) to be part of the highways sector. Similar to their municipal transit bus counterparts, these vehicles have been frequent targets of attacks, particularly in India, Pakistan, and Israel.

FIGURE 7.1 Subway station. *(Source: Shutterstock, © Su Justen.)*

certain protective measures, report potential threats and security concerns to TSA, and designate primary and alternate security coordinators. The provisions concerning reporting of security incidents and designation of a security coordinator were also contained in the TSA's final rule on rail transportation security issued on November 26, 2008 (49 CFR Parts 1520 and 1580), which dealt primarily with freight rail. In addition, that rule formally established TSA's authority to inspect all covered rail systems (including mass transit and passenger rail) (Figure 7.1) (Transportation Security Administration, 2006; Government Printing Office, 2008a, p. 72131).

In December of 2006, TSA and the Federal Transit Administration (FTA) jointly issued a list of seventeen recommended "Security and Emergency Management Action Items for Transit Agencies," which, according to the federal agencies, "represent a comprehensive and systematic approach to elevate baseline security posture and enhance security program management and implementation:"

1. Establish written systems security programs and emergency management plans.
2. Define roles and responsibilities for security and emergency management.
3. Ensure that operations and maintenance supervisors, forepersons, and managers are held accountable for security issues under their control.
4. Coordinate Security and Emergency Management Plan(s) with local and regional agencies.
5. Establish and maintain a Security and Emergency Training Program.

6. Establish plans and protocols to respond to the DHS Homeland Security Advisory System (HSAS) threat levels.[2]
7. Implement and reinforce a public security and emergency awareness program.
8. Conduct tabletop and functional drills.
9. Establish and use a risk management process to assess and manage threats, vulnerabilities, and consequences.
10. Participate in an information-sharing process for threat and intelligence information.
11. Establish and use a reporting process for suspicious activity (internal and external).
12. Control access to security-critical facilities with ID badges for all visitors, employees, and contractors.
13. Conduct physical security inspections.
14. Conduct background investigations of employees and contractors.
15. Control access to documents of security-critical systems and facilities.
16. Establish process for handling and access to Sensitive Security Information (SSI).
17. Conduct security program audits at least annually (internal and external) (Transportation Security Administration, 2006).

DIGGING DEEPER
SECURITY SCREENING OF MASS TRANSIT AND RAIL PASSENGERS

Balancing Security with Civil Liberties and Passenger Convenience

"Faced with the virtual impossibility of imposing 100% passenger screening in the public mass transit environment, transit operators and government policymakers have been considering since at least 2002 whether selective passenger screening can provide an effective combination of defense and deterrence against terrorist attacks. If it can, how can such programs be managed to reduce the risks of terrorist attacks as much as possible while (1) remaining within the law, (2) maintaining legal, public, and passenger support, and (3) not altering transit systems so that they can neither move masses of people nor move them rapidly?" (Jenkins et al., 2010, p. 5).

The first attempt at passenger screening in a U.S. transit system was an experimental random passenger baggage inspection program carried out by the Massachusetts Bay Transit

[2]The HSAS was a color-coded warning system (green = low; blue = guarded; yellow = elevated; orange = high; red = severe) developed in 2002 by DHS of threat levels for the nation or specific sectors. After coming under criticism for being too general, it was replaced in April 2011 by the National Terrorism Advisory System (NTAS), which issues alerts containing "a concise summary of the potential threat including geographic region, mode of transportation, or critical infrastructure potentially affected by the threat, actions being taken to ensure public safety, as well as recommended steps that individuals, communities, business and governments can take to help prevent, mitigate or respond to a threat." The NTAS characterizes each alert as representing either an "elevated" (warning of a credible terrorist threat against the U.S.) or "imminent" threat (warning of a credible, specific, and impending terrorist threat against the U.S.) (U.S. Department of Homeland Security, 2011).

Authority in Boston during the Democratic National Convention in July 2004 (and resumed permanently in October 2006). This was followed by more systematic efforts in New York City. Immediately after the July 2005 attempted terrorist attack on the London subway system, the New York City Police Department (NYPD) and New York Metropolitan Transit Authority initiated a voluntary bag-screening program at transit stations in which individuals not wishing to be searched may exit the system without boarding. Under the New York program, which "is not comprehensive but designed to create uncertainty as to when and where inspection will occur," police officers located at the entrance to transit stations physically inspect any backpack, container, or other carry-on item, and the inspections are to be "limited to what is minimally necessary to ensure that the [item] does not contain an explosive device." NYPD added portable explosives detection equipment at select subway stations in November 2005. Amtrak and additional transit systems have instituted some form of passenger screening since that time (Jenkins et al., 2010, pp. 5–6; Sahm, 2006, pp. 10–11).

A 2006 analysis of the New York screening program reported, "critics ... argue that terrorists could simply detonate a device at the entrance to a subway station before they are searched or walk to another of the city's 468 subway stations where searches are not being conducted. NYPD terrorism officials, however, insist that random bag searches serve the dual purpose of keeping potential terrorists off balance and heightening public awareness" (Sahm, 2006, p. 11).

The U.S. DOT sponsored a study by the Mineta Transportation Institute (MTI) of selective passenger screening in transit systems. The study's findings and recommendations, first issued in 2007 and updated in 2010, included the following:

> *Selective passenger screening is a viable security option that can contribute to deterrence, oblige terrorists to take greater risks, complicate their planning, force them to use smaller quantities of explosives, and divert them to less lucrative targets. Selective searches must be carefully planned and closely managed to reduce the inevitable allegations of discrimination and profiling based upon race or ethnicity. An effective program of selective passenger screening must be based on clear policies and procedures; must combine random selection, behavioral profiling, and threat information; must maximize unpredictability; must allow for expansion, redeployment, and reduction; and must maximize interaction with riders, but not in a way that is perceived as harassment.*
>
> *Jenkins et al., 2010, p. 1*

Jenkins, Butterworth, and Gerston (2010) observed, "A primary factor—if not the primary factor—that transit authorities must address is whether they will be able to sustain legal authority in the face of the inevitable legal challenges that will be posed. Selective screening is a contentious issue that raises understandable concerns on the part of citizens and advocacy groups, which play an important role in our democracy. Although selective screening has been upheld in court, operators do anticipate challenges to any new program and even attempts to overturn current ones." The New York transit passenger screening program was challenged on privacy grounds almost immediately in the courts but was eventually upheld in an August 2006 decision by the Second Circuit Court of Appeals, which found, "We hold that the Program is reasonable, and therefore constitutional, because (1) preventing a terrorist attack on the subway is a special need; (2) that need is weighty; (3) the Program is a reasonable effective deterrent; and (4) even though the searches intrude on a full privacy interest, they do so to a minimal degree" (pp. 11–13).

How would you respond to the questions posed at the beginning of this segment on the effectiveness, constitutionality, public support, and passenger inconvenience of mass transit and rail passenger screening?

Transit Security Grant Program

Since most of the responsibility for implementing transit and passenger rail security rests with the local owners and operators of these systems, DHS has provided grant funding to transit agencies beginning in 2003, though administrative responsibility for the program within the department has shifted several times since then. In FY 2007, program management was moved to FEMA while TSA retained the lead role in setting grant priorities and making funding decisions (Government Accountability Office [GAO], 2009, p. 8).

Through December 2012, the Transit Security Grant Program (TSGP) had awarded $547 million to 60 U.S. mass transit and passenger rail systems in 25 states and the District of Columbia. In fulfilling its policy leadership responsibilities for the program, "TSA employs a risk-based prioritization in determining eligible passenger rail and transit agencies, funding allocations, and evaluations for award." TSA has established six funding priorities to assist transit systems in creating "the essential foundation for effective security programs":

- Protection of high-risk underwater and underground assets and systems
- Protection of other high-risk assets identified through system-wide risk assessments
- Use of visible, unpredictable deterrence (e.g., security patrols, canine explosives detection teams, mobile screening equipment)
- Targeted counterterrorism training for key front-line staff
- Emergency preparedness drills and exercises
- Public awareness and preparedness campaigns (Transportation Security Administration, n.d.b)

Owners and operators of transit systems (including mass transit, intra-city and commuter bus systems, ferries, and all forms of passenger rail) within high-risk urban areas are eligible for funding. A 2009 GAO report analyzed the program:

> *Although TSA allocated about 90 percent of funding to the highest-risk agencies, lower-risk agency awards were based on other factors in addition to risk. In addition, TSA has revised the TSGP's approach, methodology and funding priorities each year since 2006. These changes have raised predictability and flexibility concerns among transit agencies because they make engaging in long-term planning difficult.... The two agencies that manage the TSGP—TSA and FEMA—lack defined roles and responsibilities, and only 3 percent of the funds awarded for fiscal years 2006 through 2008 have been spent as of February 2009.... TSA and FEMA lack a plan and related milestones*

for developing measures specifically for the TSGP, and thus DHS does not have the capability to measure the effectiveness of the program or its investments.

GAO, 2009, Highlights

DHS generally concurred with the GAO analysis and recommendations and pledged to take action to address them. However, as of February 2012, GAO reported that performance measures for the TSGP had not yet been implemented, and congressional appropriators continue to express concern that the program has not focused sufficiently on areas of highest risk and that significant amounts of previously appropriated funds have still not been awarded to recipients (GAO, 2012a, p. 30; Peterman et al., 2013, p. 10).

In response to concerns about unnecessary duplication in federal homeland security preparedness grants, the fiscal year (FY) 2013 and FY 2014 DHS budget requests of the Obama Administration contained a proposal to create a National Preparedness Grant Program (NPGP) by consolidating 16 existing programs (including both the TSGP and the Port Security Grant Program) into a single, comprehensive program. According to the Administration, the NPGP would eliminate redundancies and requirements placed on both federal agencies and applicants in the current system of multiple, often disconnected, grant programs. However, the proposal has not been approved by Congress thus far because of concerns about the lack of congressional authorization and insufficient details on guidance to applicants and program implementation (GAO, 2013b, pp. 6, 6n10).

Surface Transportation Security Inspectors

TSA established the Surface Transportation Security Inspector (STSI) program (also referred to as Transportation Security Inspectors—Surface) in 2005 as a means of expanding its efforts with respect to mass transit and freight rail security. In the former role, STSIs are responsible for conducting security and risk assessments of mass transit and passenger rail systems through the Base Assessment and Security Enhancement (BASE) program, which is voluntary and concentrated on the 100 largest systems, accounting for more than 80% of public transportation passengers. The BASE process is designed to evaluate a system's security posture on the 17 action items developed jointly by TSA and FTA. The assessments are used by TSA "to establish a security profile and baseline posture for transit or passenger rail security programs, track improvements or diminutions from the baseline, and determine future program decisions and needs. They inform risk mitigation priority development and determine ... financial resource allocations, particularly in the transit security grants." The inspectors conduct follow-up visits to address any identified performance weaknesses and monitor the transit system's efforts to remove security vulnerabilities. As of March 2014, more than 120 transit and passenger rail agencies were assessed by the BASE program, representing 270 assessments and reassessments.TSA calculates that the process has produced a 19%

improvement in the largest systems' security results (Transportation Security Administration, 2013, pp. 10, 20; 2014, p. 9).

Other transit and passenger rail functions performed by STSIs include:

- Providing, in cooperation with FTA, assistance to state rail safety oversight agencies in completing security audits of the nation's 26 rail transit ("fixed guideway") systems
- Offering to transit and passenger rail systems the Security Analysis and Action Program, which uses several different tools to identify vulnerabilities based on specific threat scenarios
- Reviewing design plans of systems under construction to assess the adequacy of their security features and recommend improvements that can be accomplished in the final stages of construction
- Performing liaison functions with other governmental and non-governmental stakeholders
- Responding to security incidents to gather real-time, on-scene information provided to TSA leadership (Transportation Security Administration, n.d.b; 2013, pp. 18–19).

In a 2009 report, the DHS Office of Inspector General (OIG) found the STSI program "has been effective in its assessment and domain awareness initiatives … [but] appears understaffed for the long-term, and an aviation-focused command structure has reduced the quality and morale of the workforce" (U.S. Department of Homeland Security, 2009, p. 5).

After the OIG report, TSA moved rapidly to increase the number of STSIs, which grew from 175 in FY 2008 to 404 in FY 2011. In FY 2012 and FY 2013, there were a total of 379 STSIs, with 154 of these serving on VIPR teams. However, in FY 2014 the size of the STSI workforce was reduced to 300 via a cut of 79 positions from VIPR personnel (Peterman et al., 2013, p. 9; Transportation Security Administration, 2013, p. 18; 2014, p. 19).

The concerns expressed by the DHS OIG about the STSI program's command structure and workforce quality have continued. A 2012 Congressional Subcommittee hearing raised a number of issues, including "the lack of surface transportation expertise among the inspectors, many of whom were promoted from screening passengers at airports; the administrative challenge of having the surface inspectors managed by federal security directors who are located at airports, and who themselves typically have no surface transportation experience; and the security value of the tasks performed by surface inspectors" (Peterman et al., 2013, p. 9).

Visible Intermodal Protection and Response Teams

In the aftermath of the Madrid commuter train bombings of 2004, TSA developed the VIPR program, which was explicitly authorized by the Implementing Recommendations of the 9/11 Commission Act of 2007 (9/11 Act) to "augment the security of any mode of transportation at any location within the United States." The composition of VIPR teams varies, depending on the specific needs of a given deployment as determined by DHS and local officials but may include any DHS asset, including Federal Air Marshals, STSIs, explosive detection canine teams, and screening technology, as well as local law enforcement personnel. They may be deployed at random locations and times to deter and defeat terrorist activities or during specific alert periods or for special events. Their purpose is to "provide a visible presence to detect, deter, disrupt, and defeat suspicious activity while instilling confidence in the travelling public" (Transportation Security Administration, n.d.c; U.S. Department of Homeland Security, 2012, pp. 2–3).

The VIPR program is managed jointly by TSA's Federal Air Marshals Service and the Office of Security Operations. Funding was provided to establish and operate 10 multimodal VIPR teams in FY 2008, another 15 teams assigned primarily to land transportation systems in FY 2010, and 12 additional multimodal teams in FY 2012 (for a total of 37). VIPR teams conducted 12,845 operations in FY 2012 (8868, or 69%, involved land transportation systems, with the remainder in aviation). Well over half of this total was devoted to mass transit and passenger rail systems: in calendar year 2012, there were more than 7100 VIPR operations involving these systems. VIPR teams carried out over 10,000 operations in all land transportation modes in FY 2013, with 90% of these involving mass transit systems (U.S. Department of Homeland Security, 2012, p. 6; Transportation Security Administration, 2013, pp. 15, 21; 2014, p. 15).

A 2012 analysis by the DHS OIG reported:

> *The VIPR Program has improved its ability to establish effective partner and stakeholder relationships. However, organizational, programmatic, and operational challenges remain. For example, the VIPR program's placement within TSA hinders its ability to ensure coordinated field activities. Guidance is needed to clarify law enforcement activities, team member roles and responsibilities, and equipment use during VIPR operations. Additionally, the VIPR deployment methodology needs refinement, and resources are not allocated proportionately to team workloads across the Nation. Teams do not receive standardized training, and the length of VIPR team member assignments affects program effectiveness.*
>
> *U.S. Department of Homeland Security, 2012, p. 12*

National Canine Program

In 2005, TSA announced it was expanding the explosives detection canine team program—which was originally established by the Federal Aviation Administration in 1973 for explosives detection at airports—to cover mass transit and passenger rail systems. TSA Administrator Kip Hawley cited the particular utility of canine teams for the new mission: "These teams are a mobile and efficient method for identifying explosives materials and they can be quickly deployed to address a variety of situations" (Transportation Security Administration, 2005).

Initially, the National Canine Program (NCP) for mass transit and passenger rail systems provided (similar to its airport counterpart) for TSA to develop, train, deploy, and certify explosives detection canine teams composed of specially trained canines and state or local law enforcement officer (LEO) handlers, whose mission was to deter and detect the introduction of explosive devices into the transportation system. In 2011, the program was expanded by the creation of canine teams including a federal Transportation Security Inspector as the handler. As of September 2012, TSA had received funding for 111 LEO mass transit teams, 27 LEO multimodal teams (which may cover mass transit and passenger rail systems) and 46 TSI multimodal teams (which also cover mass transit and passenger rail) (GAO, 2013a, pp. 1–2, 6).

In the public version of a classified 2013 report on the overall NCP, GAO stated TSA "is collecting and using key data on its canine program, but could better analyze these data to identify program trends.... For example, GAO analysis ... showed that some canine teams were not in compliance with TSA's monthly training requirement, which is in place to ensure canine teams remain proficient in explosives detection" (GAO, 2013a).

Security Training and Exercises

As a result of its security assessments of mass transit systems, which identified "wide variations in the quality of transit agencies' security training programs and an inadequate level of refresher or follow-up training," TSA instituted its Mass Transit Security Training Program in 2007 to "elevate the level of training generally, bring greater consistency, and assist agencies in developing and implementing training programs." It provides curriculum guidelines for basic and follow-on security training and allows transit agencies to use funding from the TSGP to finance curriculum development and employee training costs (Transportation Security Administration, n.d.a; GAO, 2011, p. 10).

The DHS OIG reported in 2010 that the Mass Transit Security Training Program concentrated primarily "on law enforcement management of security incidents based on weapons of mass destruction scenarios," but "few training courses focus on passenger rail frontline employees, firefighter response efforts, or the threats posed specifically by improvised explosive devices." In response, TSA representatives indicated "they made the conscious decision to focus on security and law enforcement efforts because they believed that much of the training and exercises occurring in DHS, and at the state and local levels, was heavily focused on response and recovery efforts" (U.S. Department of Homeland Security, 2010b, pp. 8–9).

The 9/11 Act directed TSA to issue regulations requiring that operators of mass transit and passenger rail systems develop a comprehensive security training program for their front-line employees.[3] In keeping with the usual slow pace of regulatory proceedings, TSA did not develop a plan and milestones for carrying out this assignment until 2011 and issued a preliminary request for comments on June 14, 2013. The agency, which is currently in the process of preparing the proposed rule, indicated that part of the reason for the delays was the inherent difficulty in trying to address multiple transportation

[3]Intercity bus systems and freight rail were also included.

modes in one regulation (Table 7.2) (GAO, 2011, pp. 9–10; Federal Register, 2013, p. 35945; Transportation Security Administration, 2013, p. 7).

TSA's Intermodal Security Training Exercise Program (I-STEP) was an outgrowth of a 2002 pilot project undertaken by the agency in partnership with the Coast Guard to develop a port security training and exercise program. I-STEP seeks to improve the private sector's ability to respond to security incidents in all transportation modes (including

Table 7.2 Required Training Program Elements from the 9/11 Act

Mass Transit Systems	Passenger Rail Systems*
1. Determination of the seriousness of any occurrence or threat	1. Determination of the seriousness of any occurrence or threat
2. Crew and passenger communication and coordination	2. Crew and passenger communication and coordination
3. Appropriate responses to defend oneself, including using nonlethal defense devices.	3. Appropriate responses to defend or protect oneself
4. Use of personal protective devices and other protective equipment	4. Use of personal and other protective equipment
5. Evacuation procedures for passengers and employees, including individuals with disabilities and elderly adults	5. Evacuation procedures for passengers and railroad employees, including individuals with disabilities and elderly adults
6. Training related to behavioral and psychological understanding of, and responses to, terrorist incidents, including the ability to cope with hijacker behavior, and passenger responses	6. Psychology, behavior, and methods of terrorists, including observation and analysis
7. Live situational training exercises regarding various threat conditions, including tunnel evacuation procedures	7. Training related to psychological responses to terrorist incidents, including the ability to cope with hijacker behavior and passenger responses
8. Recognition and reporting of dangerous substances, and suspicious packages, persons, and situations	8. Live situational training exercises regarding various threat conditions, including tunnel evacuation procedures
9. Understanding security incident procedures, including procedures for communicating with governmental and nongovernmental emergency response providers and for on scene interaction with such emergency response providers	9. Recognition and reporting of dangerous substances, suspicious packages, and situations
10. Operation and maintenance of security equipment and systems	10. Understanding security incident procedures, including procedures for communicating with governmental and nongovernmental emergency response providers and for on scene interaction with such emergency response providers
11. Other security training activities that DHS deems appropriate	11. Operation and maintenance of security equipment and systems
	12. Other security training activities that DHS deems appropriate

DHS, Department of Homeland Security.

*Also applies to freight rail systems.

(*Source*: Federal Register. June 14, 2013. Transportation Security Administration: Request for comments on security training for surface mode employees. Washington, DC. p. 35947.)

intermodal systems) "by increasing awareness, improving processes, creating partnerships, and delivering transportation-sector network training exercises." The program provides participating stakeholders with facilitation of all planning meetings and exercise activities, objectives and scenarios to help drive exercise discussion, and documentation and software for exercise design, evaluation, and tracking for tabletop and functional exercises (Transportation Security Administration, n.d.d).

For the transit and passenger rail sector, I-STEP uses workshops, tabletop exercises, and working groups "to integrate mass transit and passenger rail agencies with law enforcement and emergency response partners to expand and enhance coordinated deterrent and incident management capabilities." However, in a 2010 report, the DHS OIG found TSA had not devoted sufficient staff and other resources to transit and rail I-STEP activities, only three I-STEP sessions had been conducted over the preceding 2 years, and the program primarily used tabletop exercises and did not include live drills or exercises for security or first responder emergency response. TSA officials responded they "decided to exclude a live drill component from the program because [they] believed that the component would hamper program implementation in its early stages" (U.S. Department of Homeland Security, 2010b, pp. 4, 8–9).

More recently, I-STEP has utilized operations-based exercises to test if operators are meeting identified desired security outcomes and "expanded its exercise planning capacity by introducing a public-facing, on-line exercise planning and information management system capable of supporting all modes in the transportation system" (Transportation Security Administration, 2014, p. 7).

Federal Transit Administration Security Activities

Before the creation of TSA, the FTA was the federal agency responsible for mass transit security, and it still retains a significant role, as spelled out in the Mass Transit and Passenger Rail Annex to the Transportation Systems Sector-Specific Plan (SSP):

> *FTA conducts a range of safety and security activities, including employee training, research, technical assistance, and demonstration projects. In addition, FTA promotes safety and security through its grant-making authority...FTA stipulates conditions of grants, such as certain safety and security statutory and regulatory requirements, and may withhold funds for noncompliance.... For formula-based grants ... transit agencies are required to spend at least one percent, and may spend more, of their annual allocations on security-related projects, or certify that they do not need to do so (based on criteria such as the availability of [other] funds for funding security needs or a record of assessments indicating no deficiencies).*
>
> *U.S. Department of Homeland Security, 2010a, p. 218*

In addition, FTA has certain security-related regulatory authority over "rail fixed guideway systems" (including subways, light rail, monorails, trolleys, and other passenger rail systems not regulated by the Federal Railroad Administration). Such systems are required to develop and maintain a security plan that meets specific requirements, including identifying the plan's security objectives; documenting the transit agency's process for managing threats and vulnerabilities during operations; identifying the controls in place that address the security of passengers and employees; and documenting the agency's process for conducting internal security reviews to evaluate compliance with and effectiveness of the security plan. FTA administers these requirements through State Safety Oversight Agencies (SSOA), which must ensure that transit agencies under their jurisdiction conduct annual reviews of their security plan and perform on-site reviews of the agencies' implementation of the security plan at least once every 3 years (U.S. Department of Homeland Security, 2010a, p. 219; Government Printing Office, 2012, pp. 525–526).

■ ■ Critical Thinking ■

After 9/11, there was debate on whether transportation security would be better served if responsibility was (1) retained in the various modal agencies, such as the FTA, where long-established connections with stakeholders and knowledge of transportation operations would—in theory—produce more informed decision-making and regulation, or (2) moved to a separate agency, where security considerations would always be paramount and not subordinated to economic or other factors (as was found to be the case, at least in part, with respect to FAA's pre-9/11 security regulation). With the creation of TSA, this debate was apparently resolved in favor of the latter viewpoint (although TSA is far from having comprehensive security authority over all transportation modes). What do you think?

Federal Railroad Administration Security Activities and Amtrak

As part of its regulatory responsibility for commuter rail and Amtrak, the Federal Railroad Administration (FRA) requires operators of such systems to "adopt and comply with a written emergency preparedness plan approved by FRA," with implementation monitored by several hundred FRA rail inspectors. The plan covers "security situations" in addition to other emergencies and must address:

- Crew member assessment of emergencies and prompt notification to the control center
- Control center notification to outside emergency responders
- Onboard emergency lighting, first aid kits, and other emergency equipment
- Passenger safety awareness of emergency procedures
- Conduct of passenger train emergency simulations to determine capabilities in executing the emergency preparedness plan, with after-action debriefing and critiques (U.S. Department of Homeland Security, 2010a, p. 219)

Amtrak carries out its own security program, which features the Amtrak Police Department, and includes the following elements, "some of which are conducted on an unpredictable or random basis:" presence of uniformed police officers and Special Operations

Units (e.g., VIPR teams), identification checks, checked baggage screening, use of explosives detection canine teams, onboard security checks, and "random passenger and carry-on baggage screening and inspection." The latter are to be conducted with "due respect to passengers' privacy" and "completed as quickly as possible—usually in less than a minute," with "passengers failing to consent [to the procedure] ... denied access to trains and refused carriage" while being offered a refund (Amtrak, n.d.).

Freight Rail

In contrast to passenger rail security, efforts to protect freight rail systems in the United States have received less attention (and resources) from the federal government and have been more reliant on voluntary measures undertaken by industry. Furthermore, security programs have been more narrowly focused, with particular concentration on the transportation of a certain class of hazardous materials.

As described in TSA's FY 2014 budget justification to Congress, "the overarching strategic security goal [of TSA's freight rail program] is to reduce the risk associated with the transportation of potentially dangerous cargoes by rail, and to increase the resiliency of the railroad network. The primary strategic objectives to achieve this goal are: (1) reduce the vulnerability of cargo, (2) reduce the vulnerability of the network, and (3) reduce the consequences of attack" (Transportation Security Administration, 2013, p. 15).

TSA's regulatory authority over freight rail systems was spelled out as the major portion of the final rule on rail transportation security issued by the agency in November 2008. The rule applies to freight railroad carriers, as well as shippers of rail security–sensitive materials (RSSM)[4] and receivers of RSSM located within a high threat urban area. It requires these entities to appoint Rail Security Coordinators and to report location and shipping information for RSSM rail cars, as well as significant security concerns, to TSA. In addition, the rule sets forth procedures for the secure transfer of custody of RSSM rail cars between railroads and at points of origin and delivery in high-threat urban areas. Finally, TSA is given inspection authority over the covered entities, with its STSIs monitoring compliance with the regulations (Government Printing Office, 2008a, p. 72131; Transportation Security Administration, 2013, p. 11).

The GAO critiqued TSA's freight rail security efforts in 2009, stating, to that point, the agency had focused almost exclusively on the transportation of hazardous materials and had not addressed a range of other identified threats. TSA concurred with this finding and expanded its activities to include assessment of risks posed by potential sabotage of critical infrastructure, especially freight rail bridges and tunnels (Figure 7.2) (GAO, 2011, pp. 5–6).

[4]Defined as the categories and quantities of certain hazardous materials determined by DHS to "pose a significant risk to national security while being transported in commerce by rail due to the potential use of one or more of these materials in an act of terrorism" (Government Printing Office, 2008a, p. 72134).

FIGURE 7.2 Freight rail tank car designed to carry liquid or gaseous commodities. *(Source: Harvey Henkelmann, http://en.wikipedia.org/wiki/File:TILX290344.JPG.)*

Toxic Inhalation Hazard Risk Reduction

TSA's initial focus with respect to freight rail security was on identifying and reducing the risk posed by the transportation of toxic inhalation hazard (TIH) materials (e.g., chlorine gas and anhydrous ammonia) through densely populated urban areas. These materials were singled out for special attention because of the serious harm or deaths through inhalation their release could cause over a widespread area. In 2006, DHS and the DOT jointly issued a total of 27 security action items for the rail transportation of TIH materials, which were developed in collaboration with industry and after the conduct of field reviews and vulnerability assessments. The 27 items covered system security, access control, en route security, and the movement of TIH rail cars through high-threat urban areas. Compliance with the measures is voluntary, but TSA's STSIs monitor implementation by the railroads (U.S. Department of Homeland Security, 2009, p. 7; Transportation Security Administration, n.d.e).

One of the primary objectives of the TIH risk reduction effort was to use the recommendations (e.g., the consideration of alternative routes "when they are economically practicable and result in reduced overall safety and security risks") to decrease the amount of time spent in high-threat urban areas by rail cars carrying these materials. By the end of FY 2012, this goal had been largely achieved, with the "dwell time" of TIH cars in such areas reduced by 98% compared with 2006 (Transportation Security Administration, 2013, p. 11).

Transportation Security Administration Security Assessments

A key part of the TSA approach for freight rail security is in assessing risks to various parts of the sector, and a number of risk assessment mechanisms have been developed toward that end:

Rail Corridor Assessments (RCAs) are carried out by teams composed of TSA, railroad, and state and local homeland security officials and assess the vulnerabilities of high-population areas to TIH materials transported by rail. They seek to identify the key security control points at each location and to develop mitigation strategies. Results of early RCAs were used in the development of the Security Action Items and the final rule for rail transportation security. RCAs have been undertaken for Washington, DC; Northern New Jersey; Cleveland; New Orleans; Houston; Buffalo; Oklahoma City; Sacramento; Baltimore; Denver; Charlotte; Las Vegas; Milwaukee; Memphis; Columbus, OH; and Atlanta.

Comprehensive Reviews (CRs) "are a larger-scale, more encompassing version of the RCA. CRs provide a thorough evaluation of the security of a specific rail corridor and a comparative analysis of risk across transportation modes and critical infrastructure sectors in the specific geographic areas." These review teams are expanded to include response and recovery officials from all levels of government. As of 2010, comprehensive reviews had been completed in Northern New Jersey, Los Angeles, Chicago, and Philadelphia.

Corporate Security Reviews (CSRs) involve a "review of a company's security plan and procedures, and ... provide the federal government with a general understanding of each company's ability to protect its critical assets and its methods for protection of hazardous materials under its control." The review may involve onsite visits to such critical locations as bridges, tunnels, operations centers, and rail yards and aims to "analyze the railroad's security plan for sufficiency, determine the degree to which mitigation measures are implemented throughout the company, and recommend additional mitigation measures" to the company (U.S. Department of Homeland Security, 2010a, pp. 294–296).

Critical Infrastructure Assessments were begun in 2009 and, in collaboration with the infrastructure owners, identify security vulnerabilities of potentially critical bridges and tunnels as well as recommendations for mitigating such vulnerabilities. These assessments also provide a means for TSA to rank the relative risk of the bridges and tunnels. Through the end of FY 2013, 286 railroad bridges and tunnels had been assessed (Transportation Security Administration, 2013, pp. 11,15; 2014, pp. 10–11).

■ ■ Critical Thinking Question ■

What are the pros and cons of TSA's freight security focus on hazardous materials?

Pipeline and Hazardous Materials Safety Administration Security Activities

The Pipeline and Hazardous Materials Safety Administration (PHMSA), which is housed within the DOT, is responsible for prescribing regulations "for the safe transportation, including security, of hazardous material in intrastate, interstate, and foreign commerce." Over the years, the agency has issued such regulations on several occasions, the most recent of which was in November 2008. The final rule (49 CFR Parts 172 and 174) was developed in coordination with TSA and FRA and was pursuant to a mandate in the 9/11

Act that rail carriers be required to "select the safest and most secure route to be used in transporting" certain hazardous materials. It directs freight rail carriers to analyze safety and security risks along rail routes where specified quantities of TIH, explosive, and high-level radioactive materials are transported; assess alternative routing options; and select practicable routes posing the lowest safety and security risks. In addition, the rule clarifies the rail carriers' responsibility to address en route storage and transit delays in their security plans, and to inspect rail cars carrying hazardous materials for signs of tampering or suspicious items (U.S. Department of Homeland Security, 2010a, p. 292; Government Printing Office, 2008b, p. 72182).

An August 9, 2006, annex to the 2004 Memorandum of Understanding (MOU) between DHS and DOT on transportation roles and responsibilities delineated lines of authority between PHMSA and TSA with respect to hazardous materials transportation security, and affirmed the latter's lead role. Furthermore, a September 28, 2006, annex to the same MOU concerning the roles of TSA and FRA specified that FRA has enforcement authority for PHMSA's hazardous materials regulations (Government Printing Office, 2008a, p. 72133).

Federal Railroad Administration Security Activities

The September 2006 annex concerning TSA and FRA "recognizes that TSA is the lead federal entity for transportation security in general and rail security in particular" and that "FRA has authority over every area of railroad safety *(including security)* [emphasis added]." In carrying out this somewhat ambiguous security role, FRA has focused primarily on providing information to railroads on terrorist activity and threats or acts against rail systems; monitoring the agency's accident and incident database for reported acts of vandalism, sabotage, criminal mischief, and other intentional acts of destruction within the rail system; and responding to bomb threats and other criminal acts against railroads (Government Printing Office, 2008a, p. 72133; U.S. Department of Transportation, 2008a, p. 29).

Association of American Railroads Terrorism Risk Analysis and Security Plan

One of the most influential elements of freight rail security in the United States has been the plan developed by the Association of American Railroads (AAR), which represents the major freight railroads in North America. The AAR security plan was adopted in December 2001 and last updated in 2009. It serves as a national plan for the industry as well as a template for security plans for individual carriers and uses threat and risk assessment to address five major functional areas: hazardous materials; operational security; physical infrastructure; military liaison; and information technology. Its specific components include:

- A database of critical railroad assets
- Railroad vulnerability assessments
- Analysis of the terrorist threat
- Risk calculations

- Identification of countermeasures to reduce risk
- Functions of the AAR operations center and railroad alert network (U.S. Department of Homeland Security, 2010a, pp. 294–296; Association of American Railroads, n.d.)

Highway Infrastructure and Motor Carriers

TSA's Highway and Motor Carrier (HMC) Branch serves as the lead federal agency for security in this sector, with DOT's Federal Highway Administration (FHWA) and Federal Motor Carrier Safety Administration (FMCSA) also playing important roles. The HMC Branch focuses on three strategic goals: reducing risk through enhanced preparedness to provide security and resiliency to the highway transportation system, enhancing security awareness and information sharing with public and private security partners, and evaluating and reporting on security efforts and resource management. The office has identified 16 objectives to guide its pursuit of these goals (Table 7.3) (U.S. Department of Homeland Security, 2010a, pp. 261–262; Transportation Security Administration, 2012, p. ii).

In January 2009, GAO questioned the extent to which TSA was fulfilling its leadership role with respect to the highway and motor carrier mode and reported further:

Table 7.3 TSA Highway and Motor Carrier (HMC) Branch Strategic Objectives

Objectives

1. Develop and provide guidance-based and other risk-reduction initiatives as appropriate.
2. Conduct periodic risk assessments to provide a landscape from which to influence decision making and planning at HMC.
3. Develop and maintain strategic-level planning documents that address prevention, protection, mitigation, response, and recovery.
4. Facilitate the development and implementation of a program to track shipments of Highway Security Sensitive Materials (HSSM).
5. Facilitate the development and implementation of a program for truck rental vetting.
6. Maintain HMC industry risk assessments to inform decision making.
7. Determine the most effective communication systems and processes.
8. Develop an effective process-based communication strategy.
9. Enhance and expand the use of the Transportation Security Administration Alert notification system.
10. Manage the HMC Private Industry Clearance Program.
11. Manage the HMC Stakeholder Database.
12. Enhance the use of intelligence and situational awareness internally and externally.
13. Develop evaluation tools for various HMC programs such as the Intermodal Security Training and Exercise Program (I-STEP), Baseline Assessment for Security Enhancements (BASE), Visible Intermodal Prevention and Response (VIPR) teams, and so on.
14. Establish and manage an HMC acquisitions process.
15. Develop HMC personnel training development process.
16. Coordinate and report administrative responses.

(*Source*: Transportation Security Administration. 2012. Office of Security Policy and Industry Engagement, Highway and Motor Carrier Branch 2012 Annual Report. Transportation Security Administration, Washington, DC, p. ii.)

[DHS risk assessments of highway infrastructure] varied considerably, were at various levels of completion, were not systematically coordinated, and the results had not been routinely shared among the [various DHS] entities or with another key stakeholder, the Federal Highway Administration...Without adequate coordination with federal partners.... TSA was unable to determine the extent to which specific critical assets had been assessed and whether potential adjustments in its methodology were necessary to target remaining critical infrastructure assets.

GAO, 2010a, p. 11

After the GAO report and pursuant to a series of provisions in the 9/11 Act of 2007 that related to the highway mode, TSA stepped up its security efforts in this sector. In 2010, TSA began using a Highway (HWY) BASE program similar to the one previously developed for mass transit and passenger rail. The HWY BASE program is "designed to assess and ultimately elevate the level of security across all transportation modes in the highway sector, including trucking, motor coach,[5] school bus and infrastructure." It uses a review process aimed at identifying high-risk assets within the highway mode and a "manageable list" of the largest owners and operators of those assets. The targeted stakeholders are asked to participate in a voluntary, on-site assessment by a TSA STSI who evaluates a company or facility utilizing a checklist of best security practices unique for each of the four highway modes. The participating stakeholder is given a report outlining its security strengths and weaknesses, as well as its overall security grade. Each company or facility is to be revisited at least once every 3 years. Results of the assessments are to aid TSA in identifying industry-wide security weaknesses; developing appropriate mitigation strategies; and making resource allocation decisions, such as the deployment of VIPR teams (Transportation Security Administration, 2012, p. 8; 2013, pp. 13–14).

In 2012, a total of 224 HWY BASE assessments were conducted: 41 trucking companies, 71 motor coach companies and terminals, 63 school districts and buses, 32 bridges, and 17 tunnels. TSA is in the process of completing its assessments for the trucking (non-HAZMAT [hazardous materials]), motor coach and school bus sectors, and of developing baseline security standards for all highway submodes while encouraging stakeholders to adopt the identified best security practices and have their front-line employees receive security training (Transportation Security Administration, 2013, p. 15; 2012, p. 8).

TSA started deploying VIPR teams to augment security in the highway mode in 2011. In FY 2012, 372 such VIPR operations were conducted, and more than half involved motor coach terminals (177) or companies (19), with the remainder broken down as follows: infrastructure, 32; trucking, 51; rest areas, 7; stadium or parking lots, 25; weigh stations, 23; and border crossings, 38. I-STEP is also used in the highway mode "to conduct security preparedness and program enhancement exercises [with stakeholders] throughout the country across all HMC sub-modes." In 2012, four tabletop exercises took place, two (in

[5]DHS defines the motor coach sector as bus companies operating "primarily in interstate operations that include wholly-owned bus terminals, shared terminals and other transportation modes" and excludes intracity and mass transit buses, as well as school buses. The terms "over-the-road buses" and "intercity buses" are sometimes used to refer to this sector (U.S. Department of Homeland Security, 2010a, p. 255).

Augusta, GA and Philadelphia, PA) involving the trucking industry, one (in Herndon, VA) involving motor coaches, and one (in Irvine, CA) involving highway infrastructure (Transportation Security Administration, 2012, pp. 3, 10–12).

Highway Infrastructure

The major component of TSA's activities in support of highway infrastructure security have been the bridge and tunnel assessments commissioned by TSA starting in 2009 and conducted by the U.S. Army Corps of Engineers. The assessments, which were developed in response to a requirement in the 9/11 Act, are "comprehensive structural and operational vulnerability assessments on selected significant highway structures." The Corps developed the risk-based bridge assessment methodology, which is "unique in that it is specifically designed to focus on a single structure and the risk associated with each of its many structural components." A similar assessment methodology for tunnels has been developed by the Corps in collaboration with the DHS Science and Technology Directorate. The assessment program seeks to identify and rank critical bridge and tunnel assets, identify vulnerabilities, and quantify costs for hardening or replacing structures. A total of 48 bridges and tunnels had been assessed as of the end of FY 2012, with five to 10 additional assessments planned for FY 2013 and 2014. In FY 2014, "TSA plans to complete aggregate reports on both bridge and tunnel assessments that will include best practices and risk mitigating action items" (Transportation Security Administration, 2012, pp. 15–16; 2013, pp. 13, 15). Figure 7.3 shows the Golden Gate Bridge.

FIGURE 7.3 Golden Gate Bridge, San Francisco, California. *(Source: Rich Niewiroski, Jr., http://en.wikipedia.org/wiki/File:GoldenGateBridge-001.jpg.)*

FHWA performs a number of security-related functions for highway infrastructure:

- Cooperates with TSA in assessing bridge and tunnel vulnerability and developing risk mitigation measures
- Ensures state and local highway departments are prepared to respond to attacks on the highway system
- Conducts security-related research
- Administers the Emergency Relief program to provide funds to repair and reconstruct highways and bridges damaged as a result of catastrophic failures
- Provides information to relevant stakeholders through a variety of programs (U.S. Department of Transportation, 2008a, p. 28; U.S. Department of Homeland Security, 2010a, p. 264).

Motor Coaches

As part of the same 9/11 Act–mandated rulemaking involving security training for mass transit and freight and passenger railroads, in June 2013, TSA requested comments on training for over-the-road bus (motor coach) employees. Per the 9/11 Act, the required training elements for front-line motor coach workers are virtually identical to those for freight and passenger railroad employees, with the covered employees including bus drivers, maintenance and maintenance support personnel, dispatchers, security personnel, ticket agents and other terminal employees, and "other employees of an over-the-road bus operator or terminal owner or operator that [DHS] determines should receive security training" (Federal Register, 2013, pp. 35945–35948).

TSA developed Operation Secure Transport, a voluntary, computer-based, interactive training resource designed to provide instruction for industry employees in how to recognize security threats and respond to security incidents. In collaboration with the motor coach industry, the agency also created a training DVD, titled "Operation Secure Transport—First Observer," which includes separate training modules on security awareness and crisis response for drivers, maintenance workers, terminal workers, and management. Approximately 4000 of the videos were distributed to commercial bus companies (Federal Register, 2013, p. 35948; Transportation Security Administration, 2012, p. 18).

School Transportation

In cooperation with national school transportation organizations, TSA developed the "School Transportation Security Awareness" video "to provide school bus drivers, administrators, and staff members with information that will enable them to effectively identify and report perceived security threats, as well as the skills to appropriately react and respond to a security incident should one occur." The video was distributed to all 14,755 public school districts in the United States. The National School Transportation Association worked with TSA in developing a voluntary list of best security practices for school transportation industry administrators and employees (U.S. Department of Homeland Security, 2010a, p. 265; Transportation Security Administration, 2012, p. 19).

Trucking

As with freight rail, much of the effort in seeking to secure the trucking sector has focused on the transportation of hazardous materials. PHMSA regulations (specifically HM-232):

> *Require persons who offer for transportation or transport HAZMAT to develop, implement, and maintain security plans, as well as provide in-depth, employee security training. Motor carrier security plans must include an assessment of the possible transportation security risks for shipments of covered HAZMAT and include the following elements: personnel security, facility security, and en route security. Mandatory HAZMAT employee training must provide an awareness of security risks associated with HAZMAT transportation and provide in-depth security training on the elements of the security plan and its implementation.*

Both PHMSA and FMCSA investigate industry compliance with the HM-232 security plans and training requirements, and their inspectors are given authority to issue citations for noncompliance (U.S. Department of Homeland Security, 2010a, p. 265).

After evaluating best industry practices for the secure transport of high-risk hazardous materials, TSA developed the voluntary Highway Security-Sensitive Materials Security Action Items for commercial motor carriers. The items fall into four categories: general security (including security threat assessments, security planning, and protection of critical information); motor carrier personnel security; unauthorized access to the motor carrier's facilities, equipment, vehicles and cargo; and en route security. TSA recommends that, if adopted, these practices be included in any security plans developed by the motor carrier and adds, "the security practices are voluntary to allow highway motor carriers to adopt measures best suited to their particular circumstances, provided the measures are consistent with existing regulations, laws, or directives" (Transportation Security Administration, n.d.f).

As required by the USA Patriot Act, TSA operates the Hazardous Materials Endorsement Threat Assessment Program, which conducts a security check for any driver seeking to obtain, renew, or transfer a hazardous materials endorsement (permitting the transport of hazardous materials) on a state-issued commercial driver's license (Transportation Security Administration, n.d.g).

TSA also offers voluntary training programs for motor carriers involved in HAZMAT transportation to assist the industry in developing a plan to address security risks. In addition, and as directed by the 9/11 Act, it is currently working on the development of a system to track HAZMAT shipments on a real-time basis. The effort is based on the TSA Hazmat Truck Security Pilot "where a prototype HAZMAT truck tracking system demonstrated that a hazardous material tracking center was feasible from a technological and systems perspective" (Transportation Security Administration, 2012, p. 19).

Pipelines

The August 2006 TSA/PHMSA annex to the DHS/DOT MOU on transportation security roles and responsibilities also covered pipeline security, affirming TSA's lead role but specifying PHMSA's responsibility for pipeline safety issues. Because the two matters are closely connected, considerable coordination is required between the agencies. However, as is generally the case in the land mode, implementation of security largely rests with the owners and operators of the systems. For example, in 2002 security guidelines were developed for the petroleum industry by the American Petroleum Institute (API) and for the natural gas industry by the Interstate Natural Gas Association of America (INGAA) and the American Gas Association (AGA). Each of these address planning, vulnerability assessment, and "physical security measures that operators can take to protect their critical facilities, but provide caveats explaining the general nature of the described security practices and the importance of each operator determining the security measures that are appropriate for each facility" (Figure 7.4) (GAO, 2010b, pp. 10–12).

In August 2010, GAO evaluated TSA's pipeline security efforts:

> *PSD[6] has taken actions to implement a risk management approach, including identifying the 100 pipeline systems it considers most critical and being the first of the surface*

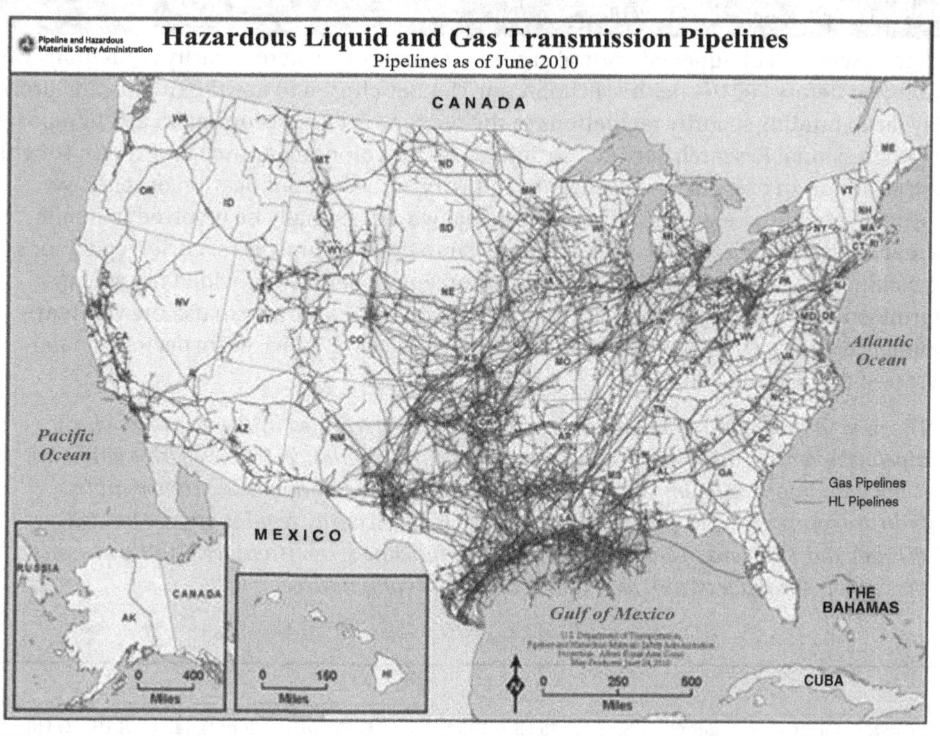

FIGURE 7.4 Hazardous liquid and natural gas pipelines in the United States. *(Source: PHMSA, https://opsweb.phmsa.dot.gov/pipelineforum/pipeline_safety_update/image_library.html#figure1.)*

[6]The TSA's Pipeline Security Division, currently called the Pipeline Security Branch.

transportation modes to develop a risk assessment model. Nevertheless, work remains to ensure that the highest risk pipeline systems are given the necessary scrutiny. PSD's risk assessment model is in its early stages of development; however, information is available or expected that could enhance the vulnerability and consequence components of the model.... PSD has taken actions to encourage private pipeline operators to employ security measures that will protect their pipeline systems, including critical facilities. While PSD officials have said that operators of the most critical pipeline systems are generally implementing voluntary security measures, [PSD security reviews] have identified shortcomings in operators' security programs and critical facilities that should be addressed to reduce vulnerabilities.... However, PSD is missing opportunities with respect to [its security recommendations to pipeline operators].

The GAO made a series of recommendations for addressing these and related problem areas, with which TSA concurred and has since begun implementing (GAO, 2010b, pp. 54–55; Parfomak, 2013, p. 12).

PIPELINE SECURITY ENFORCEMENT

Voluntary Compliance versus Regulation

With the exception of liquefied natural gas (LNG) facilities—where security regulations were established before 2001[7]—neither PHMSA nor TSA has chosen to use the rulemaking process to lay down binding security regulations in the pipeline sector. According to a 2013 report by the Congressional Research Service, "By initiating this voluntary approach, PHMSA sought to speed adoption of security measures by industry and avoid publication of sensitive security information (e.g., critical asset lists) that would normally be involved in public rulemaking." Although the 9/11 Act directed TSA to develop pipeline security regulations and conduct any necessary inspection and enforcement actions, provided the agency determined that such regulations are appropriate, it, too, continues to use the voluntary compliance approach, believing it to be adequate. The DOT Office of Inspector General addressed this issue in 2008:

The need for new security regulations will be partly determined by the degree to which pipeline operators are following TSA's current security guidance. However, [this guidance] is not mandatory and remains unenforceable unless a regulation is issued to require industry compliance. To adequately determine if new security regulations are needed, PHMSA and TSA need to conduct covert tests of pipeline systems' vulnerabilities to assess the current guidance as well as evaluate operators' compliance.

Parfomak, 2013, p. 29; U.S. Department of Transportation, 2008b, pp. 5–6

[7] These regulations cover general security (49 CFR 193 Subpart J), security training (49 CFR 193.2715), and facility security (49 CFR 195.436) and provide specific security requirements for LNG operators. PHMSA inspects the facilities to ensure they are in compliance (Pipeline and Hazardous Materials Safety Administration, n.d.; U.S. Department of Transportation, 2008b, p. 5).

Both agencies have continued to reject the imposition of regulations and the initiation of covert testing, with TSA indicating that most U.S. pipeline systems meet or exceed industry security guidelines and that the voluntary approach produces better security while maintaining a more cooperative and collaborative relationship with the pipeline industry. In contrast, in 2010, the National Energy Board of Canada promulgated enforceable security regulations for Canadian petroleum and natural gas pipelines because of "the critical importance of energy infrastructure protection" (Parfomak, 2013, p. 30).

■ ■ Critical Thinking ■

Do you believe land transportation security in the United States would be enhanced if the federal government established more mandatory regulations for this mode? Why?

Transportation Security Administration Security Assessments

Using a variety of methods, "TSA annually identifies and ranks the nation's highest risk pipeline systems based on analysis of total equivalent energy transported, system vulnerability and threat. TSA's risk reduction programs are guided by this ranking" (Transportation Security Administration, 2013, p. 16).

The most important of the pipeline assessment activities are the pipeline Corporate Security Review (CSR) program (under which TSA conducts onsite visits to the largest pipeline and natural gas distribution operators to review security plans, inspect facilities, and evaluate whether the company is following the intent of TSA security guidance) and the pipeline Critical Facility Inspection (CFI) program (under which TSA carries out in-depth inspections of all critical facilities of the 125 largest pipeline systems in the United States). As of the end of FY 2013, TSA had completed 144 Pipeline CSRs and planned to conduct 12 per year in FY 2014–2015. These reviews have revealed inadequacies in some company security programs (including not updating security plans, lack of management support, poor employee involvement, inadequate threat intelligence, and employee apathy or error), but in general, TSA reports a majority of the systems "do a good job with pipeline security" (Parfomak, 2013, p. 10; Transportation Security Administration, 2013, p. 16; 2014, p. 16).

The first phase of the pipeline CFI program was completed in May 2011, with 347 inspections having been completed. During FY 2012, pipeline operators updated their listings of essential facilities pursuant to TSA's revised Pipeline Security Guidelines. TSA completed 40 critical infrastructure reviews in FY 2013, and planned to conduct 90 to 100 such reviews annually starting in FY 2015 (Parfomak, 2013, pp. 10–11; Transportation Security Administration, 2013, pp. 13, 16; 2014, p. 16).

Transportation Security Administration Pipeline Security Guidelines

In April 2011, TSA issued *Pipeline Security Guidelines,* which replaced the guidance issued by DOT's Office of Pipeline Safety in 2002. The revised document was based on TSA's security assessments, as well as input from industry and other government agencies, and applies to natural gas and hazardous liquid transmission pipelines, natural gas distribution pipelines, liquefied natural gas facility operators, and pipelines carrying TIH materials. It proposes a series of security actions recommended by TSA but "does not impose mandatory requirements on any person." The major suggestions include the following (with additional details provided for each item

- *A risk-based corporate security program should be established and implemented by each pipeline operator to address and document the organization's policies and procedures for managing security related threats, incidents and responses.*
- *Operators should develop and implement a security plan ... [that is] comprehensive in scope, systematic in its development, and risk based, reflecting the security environment.*
- *The intent of these guidelines is to bring a risk-based approach to the application of the security measures throughout the pipeline industry.*
- *[Operators should determine] which pipeline facilities are critical ... to ensure that reasonable and appropriate security risk reduction measures are implemented to protect the most vital assets throughout the pipeline industry.*
- *Upon completion of the risk analysis process, operators should determine the appropriate mitigation measures for their assets. [This document] provides recommended measures for both critical and noncritical facilities.*
- *Developing and implementing appropriate [cyber] security measures reduces the risk to control systems.... To implement an effective cyber security strategy, pipeline operators should take advantage of industry and government efforts to develop methodologies, industry standards, and best practices for securing control systems*

Transportation Security Administration, 2011, pp. 1–16

Transportation Security Administration Training and Exercises

Three computer-based training programs have been developed by TSA for pipeline operators: "Pipeline Security Awareness;" "Pipelines: Countering IEDs;" and "Pipeline Infrastructure: The Law Enforcement Role." In FY 2012, the agency conducted nine I-STEP exercises for pipeline companies and established the goal of carrying out two such exercises per year thereafter (Transportation Security Administration, 2013, pp. 13, 16).

Pipeline Security and Incident Recovery Protocol Plan

In response to another provision in the 9/11 Act, TSA and PHMSA drafted the Pipeline Security and Incident Recovery Protocol Plan, which was completed in March 2010. "The objective of the Plan is to establish a comprehensive interagency approach to counter risks and minimize consequences of emergencies involving pipeline infrastructure, specifically focusing on actions the federal government can take to assist pipeline protection, response and recovery. The Plan identifies ways in which the federal government will provide increased security support to the most critical interstate and intrastate natural gas and hazardous liquid transmission pipeline infrastructure when threatened, and how the government will work to ensure continued transportation of product following an incident" (Transportation Security Administration, 2010, p. 2).

Information Sharing and Intelligence

The 2012 *National Strategy for Information Sharing and Safeguarding* proclaims, "It is a national priority to efficiently, effectively, and appropriately share and safeguard information so any authorized individual (federal, state, local, tribal, territorial, private sector or foreign partner) can prevent harm to the American people and protect national security." This is certainly the case in land transportation security, with its multiple and diverse stakeholders (White House, 2012, p. 3).

Homeland Security Information Network

The Homeland Security Information Network (HSIN) is one of the key information exchange systems established by DHS and provides a secure Internet portal that enables governmental and nongovernmental entities involved in homeland security activities to collaborate and share unclassified information. "The HSIN mission is to provide homeland security stakeholders with effective and efficient collaboration tools for decision making, secure access to data, and accurate, timely information sharing and situational awareness. To achieve this mission, HSIN provides a shared place for users to collaborate securely with features such as Connect (a Web conference tool) and Jabber (a chat tool)" (U.S. Department of Homeland Security, 2013, p. 2).

The HSIN is grouped into Communities of Interest, the most important of which for the land transportation modes is the HSIN Critical Sectors. This is used by private sector and governmental stakeholders across all critical infrastructure and key resource sectors, including transportation. In June 2013, the DHS Inspector General reported that the HSIN program had made progress in addressing previously identified planning and governance issues, including revalidating stakeholder requirements and realigning the program to address systemic challenges and concerns, but continued to face certain difficulties.

> *Migration from the legacy system to the new platform has been delayed because of contracting and technical challenges.... Although certain communities were using the system to share information successfully, the system was not routinely or widely used to share information throughout the homeland security enterprise. Specifically, the number of system account holders remained limited, and the extent to which those account holders were using the system was also constrained because of challenges with system content and performance.*

DHS concurred with the IG's recommendations for resolving these concerns and is in the process of implementing them. In September 2013, DHS announced upgrades to the HSIN platform that include enhanced security, new collaboration features and more advanced document management capabilities (U.S. Department of Homeland Security, 2013, p. 1; Roy, 2013).

Information Sharing and Analysis Centers

Information Sharing and Analysis Centers (ISACs) arose from a provision in the May 1998 Presidential Decision Directive-63 requesting that each critical infrastructure sector establish its own ISAC to facilitate information sharing about threats and vulnerabilities within the sector. In response, a number of such centers have been formed, including the Public Transit ISAC (for which the American Public Transportation Association was designated as the sector coordinator) and the Surface Transportation ISAC (which focuses on freight rail, with the Association of American Railroads as the designated coordinator).

ISACs are industry focused and carry out the following functions:

- Providing an around-the-clock secure operating capability that establishes specific information sharing and intelligence requirements for incidents, threats, and vulnerabilities
- Collecting, analyzing, and disseminating alerts and incident reports to members
- Assisting the government in understanding incident impacts in its sector
- Providing a trusted, electronic capability for members to exchange and share information on cyber, physical, and other threats
- Sharing with and providing analytical support to relevant government authorities and other ISACs regarding technical sector details and in mutual information sharing and assistance during actual or potential sector disruptions whether caused by intentional, accidental, or natural events (ISAC Council, 2009, pp. 4–5)

The Surface Transportation and Public Transit ISACs distribute the Transit and Rail Intelligence Awareness Daily (TRIAD), which provides security notifications within the mass transit, passenger rail and freight rail modes on suspicious activities, terrorism, and counterterrorism analysis. The Public Transit ISAC also disseminates Security Awareness Messages (SAMs) that include TSA-generated "voluntary protective measures, such as unpredictable inspections, surveillance and increased checks, and other methods of visible deterrence activities" (Transportation Security Administration, 2014, p. 10).

Rail Transportation Security Reporting

The one mandatory information sharing requirement in land transportation security is the reporting of significant security concerns contained as part of the November 2008 TSA rule on Rail Transportation Security. That rule applies to rail transit systems, commuter railroads, intercity passenger rail carriers, freight railroads, and rail hazardous materials shippers and receivers and requires them to report significant security concerns to TSA's Transportation Security Operations Center (a 24/7 facility serving as the agency's main contact point for monitoring security-related incidents in all transportation modes). The covered rail security concerns include, but are not limited to, interference with the rail crew; bomb threats; reports or discovery of suspicious items that disrupt rail operations; suspicious activity in a train or rail facility; discharge, discovery, or seizure of a firearm or other deadly weapon on a train or transit vehicle or at a rail or transit facility; indications of tampering with rail cars or rail transit vehicles; indications of possible unauthorized surveillance of a train or rail transit vehicle or facility; correspondence received by a rail carrier or rail transit system indicating a potential threat to rail transportation; and other incidents involving security breaches of rail carrier or rail transit systems. TSA is authorized to enforce the reporting requirement via viewing, inspecting, and copying rail agencies' records as necessary.

In December 2012, GAO issued a report on TSA's performance with respect to the passenger rail requirements:

> *TSA has inconsistently overseen and enforced its rail security incident reporting requirement because it does not have guidance and its oversight mechanisms are limited, leading to considerable variation in the number and type of incidents reported.... Local TSA inspection officials have provided rail agencies with inconsistent interpretations of the reporting requirement.... GAO also found inconsistency in TSA compliance inspections and enforcement actions because TSA has not utilized limited headquarters-level mechanisms as intended for ensuring consistency in these activities... TSA has not conducted trend analysis of rail security information, and weaknesses in TSA's rail security incident data management system, including data entry errors, inhibit TSA's ability to search and extract information.*

GAO recommended that to remedy these problems, TSA should develop guidance on what types of incidents should be reported, improve oversight of compliance inspections and enforcement actions, develop guidance to reduce data entry errors, and establish a process for trend analysis of incident data. "TSA concurred and is taking actions in response" (GAO, 2012b, pp. 8–9).

Department of Homeland Security Office of Intelligence and Analysis

The Office of Intelligence and Analysis (I&A) is the DHS departmental lead for intelligence. Its mission is to "equip the homeland security enterprise with the intelligence and information it needs to keep the homeland safe, secure, and resilient" by promoting the understanding of threats through intelligence analysis, collecting information and intelligence pertinent to homeland security, sharing information necessary for action, and managing homeland security intelligence. It "combines the unique information collected by DHS components as part of their operational activities with foreign intelligence from the intelligence community; law enforcement information from federal, state, local, and tribal sources; private sector data about critical infrastructure and key resources; and information from domestic open sources to develop homeland security intelligence." The products of its work are made available in unclassified form through the HSIN, among other methods (U.S. Department of Homeland Security, n.d.a.; Randol, 2010, pp. 5, 10).

I&A also has the overall lead within the federal government for sharing homeland security information and intelligence with the nonfederal entities (governmental and private sector) that generally have the primary role in preventing and responding to homeland security threats. To fulfill this role, the Office supports state and major urban area intelligence fusion centers (which are the focal point for two-way intelligence sharing between federal and nonfederal components of the homeland security effort) and manages DHS's participation in the Nationwide Suspicious Activity Reporting (SAR) Initiative (which established a standardized national process for gathering, documenting, processing, analyzing, and sharing SAR information while protecting civil liberties). These activities were, in part, a response to criticism by nonfederal partners that DHS's intelligence had become irrelevant to them "because that intelligence lacks timeliness and adds so little value to local terrorism efforts" (U.S. Department of Homeland Security, n.d.b.; Randol, 2010, p. 11).

Transportation Security Administration Office of Intelligence and Analysis

The TSA Office of Intelligence and Analysis (TSA-OIA) does not collect but receives, evaluates, and distributes information and is the only federal organization that analyzes threats specifically related to transportation. Among the principal sources of its information are reports from the intelligence community, other DHS components, law enforcement agencies, and the owners and operators of transportation systems. From these and other sources, TSA-OIA produces intelligence on current and emerging threats to all U.S. transportation modes (although its primary focus has been, and remains, on the aviation sector). It provides around-the-clock indications and warnings of threats to the transportation network and disseminates its reports—at the appropriate classification level—to other TSA components and other governmental and private transportation security stakeholders (Randol, 2010, pp. 37–38).

Department of Transportation Office of Intelligence, Security and Emergency Response

The Office of Intelligence, Security and Emergency Response provides all-source intelligence to the DOT secretary, as well as the Department's modal administrators, on "current developments and long range trends in international terrorism; [and] global and international topics concerning aviation, trade, transportation markets, trade agreements and related topics in international cooperation and facilitation." It also:

- Develops DOT policy and coordinates DOT participation in interagency policy development related to intelligence, security, and all aspects of all-hazards preparedness

- Operates the DOT Crisis Management Center
- Develops and participates in departmental training and exercise programs to ensure DOT personnel are adequately prepared for response to a disaster (U.S. Department of Transportation, n.d.)

Conclusion

Based on previous and continuing threats in the United States and ongoing security incidents abroad, the highest security priority in the land mode has been the protection of transit, rail, and bus passengers, with a number of federal initiatives—including voluntary security guidelines, risk assessments, enforcement mechanisms (including STSIs and VIPR teams), training efforts, and grant monies—developed primarily in support of that objective.

Following criticism by GAO and others, as well as directives in the 9/11 Act, TSA moved in recent years to beef up its security activities in the other land modes, with a principal focus on securing the transportation of hazardous materials by freight rail, trucks, and pipelines using similar tactics as had been developed for the passenger systems. The various DOT modal administrations have continued to play a significant role in land transportation security.

On the one hand, as measured by the lack of major security incidents involving land transportation systems in the United States, these efforts can be said to have been successful. And many vulnerability assessments have been performed, best practices adopted, and security training provided, all of which have undoubtedly improved security awareness throughout the sector. However, deficiencies have been detected in most federal land security programs. Such shortcomings are likely inevitable, given the inherent difficulties in securing these highly localized systems, the lack of resources and enforcement authority (in the form of binding laws and regulations) available to TSA and other federal agencies (with the concomitant need for industry cooperation and approval), and (as is also true of the maritime sector) the absence of major security incidents that could motivate and enable the adoption of stronger measures.

Discussion Questions

1. What are the "Security and Emergency Management Action Items for Transit Agencies?"
2. Compare and contrast TSA freight rail regulations and the AAR's security plan.
3. Describe the STSI program and provide examples of its role in mass transit, freight rail, and highway infrastructure protection.
4. What is the purpose and structure of the VIPR program?
5. Describe some examples of TSA training and exercise programs in the land mode (including I-STEP).

References

Amtrak. n.d. Safety and security. <http://www.amtrak.com/safety-security> (accessed 10.26.14.)

European Commission. May, 2012. Commission staff working document on transport security. Brussels.

Association of American Railroads. n.d. Railroad security. <https://www.aar.org/safety/Pages/Railroad-Security.aspx> (accessed 10.26.14.)

Federal Register. June, 2013. Transportation Security Administration: Request for comments on security training program for surface mode employees. Washington, DC.

GAO. June, 2009. Transit Security Grant Program: DHS allocates grants based on risk, but its risk methodology, management controls, and grant oversight can be strengthened. Washington, DC.

GAO. May, 2010a. Transportation security: Additional actions could strengthen the security of intermodal transportation facilities. Washington, DC.

GAO. August, 2010b. Pipeline security: TSA has taken actions to help strengthen security, but could improve priority-setting and assessment processes. Washington, DC.

GAO. June, 2011. Rail security: TSA improved risk assessment but could further improve training and information sharing. Washington, DC.

GAO. February, 2012a. DHS needs better project information and coordination among four overlapping grant programs. Washington, DC.

GAO. December, 2012b. Passenger rail security: Consistent incident reporting and analysis needed to achieve program objectives. Washington, DC.

GAO. January, 2013a. TSA explosives detection canine program: Actions needed to analyze data and ensure canine teams are effectively utilized. Washington, DC.

GAO. June, 2013b. FEMA has made progress, but additional steps are needed to improve grant management and assess capabilities. Washington, DC.

Government of the United Kingdom, Department for Transport. October, 2013. International Working Group Land Transport Security: Workshop on rail security, Geneva. October 22 , 2013. <http://www.unece.org/fileadmin/DAM/trans/doc/2013/sc2/SC2-Workshop-2013-Pres08e.pdf> (accessed 10.26.14.)

Government Printing Office. November, 2008a. 49 CFR Parts 1520 and 1580—Rail transportation security. <http://www.gpo.gov/fdsys/pkg/FR-2008-11-26/pdf/E8-27287.pdf> (accessed 10.26.14.)

Government Printing Office. November, 2008b. 49 CFR Parts 172 and 174—Hazardous materials: Enhancing rail transportation safety and security for hazardous materials shipments. <http://www.gpo.gov/fdsys/pkg/FR-2008-11-26/pdf/E8-27826.pdf> (accessed 10.26.14.)

Government Printing Office. October, 2012. 49 CFR 659—Rail fixed guideway systems; state safety oversight. <http://www.gpo.gov/fdsys/granule/CFR-2012-title49-vol7/CFR-2012-title49-vol7-part65> (accessed 10.26.14.)

International Union of Railways. n.d. Introduction. <http://www.uic.org/spip.php?article528&lang=en> (accessed 10.26.14.)

International Union of Railways. May, 2013. 10th session of the International Working Group on Land Transport Security. <http://www.uic.org/com/article/international-working-group-on?page=thickbox_enews> (accessed 10.26.14.)

ISAC Council. January, 2009. The role of Information Sharing and Analysis Centers (ISACs) in private/public sector critical infrastructure protection. <http://www.isaccouncil.org/images/ISAC_Role_in_CIP.pdf> (accessed 10.26.14.)

Jenkins, B.M., Butterworth, B.R., Gerston, L.N., January, 2010. Supplement to MTI study on selective passenger screening in the mass transit rail environment. Mineta Transportation Institute, San Jose, CA.

Parfomak, P.W., January, 2013. Keeping America's pipelines safe and secure: key issues for Congress. Congressional Research Service, Washington, DC.

Peterman, D.R., Elias, B., Fritelli, J., January, 2013. Transportation security issues for the 113th Congress. Congressional Research Service, Washington, DC.

Pipeline and Hazardous Materials Safety Administration. n.d. Security. <http://phmsa.dot.gov/pipeline/security> (accessed 10.26.14.)

Sahm, C., March, 2006. Hard won lessons: Transit security. Manhattan Institute for Policy Research, New York.

Randol, M.A., March, 2010. The Department of Homeland Security intelligence enterprise: Operational overview and oversight challenges for Congress. Congressional Research Service, Washington, DC.

Roy, D., September, 2013. Interoperability strengthened by upgrades and expansion to DHS's Homeland Security Information Network. <http://ise.gov/blog/donna-roy/interoperability-strengthened-upgrades-and-expansion-dhs%E2%80%99s-homeland-security> (accessed 10.26.14.)

Transportation Security Administration. n.d.a. Building security force multipliers. <http://www.tsa.gov/stakeholders/building-security-force-multipliers> (accessed 10.26.14.)

Transportation Security Administration. n.d.b. Advancing the security baseline. <http://www.tsa.gov/stakeholders/advancing-the-security-baseline> (accessed 10.26.14.)

Transportation Security Administration. n.d.c. Visible intermodal prevention and response (VIPR). <http://www.tsa.gov/about-tsa/visible-intermodal-prevention-and-response-vipr> (accessed 10.26.14.)

Transportation Security Administration. n.d.d. Intermodal Security Training Exercise Program (I-Step). <http://www.tsa.gov/sites/default/files/assets/pdf/i-step-flyer.pdf> (accessed 10.26.14.)

Transportation Security Administration. n.d.e. Rail transportation security rule—49 CFR 1580. <http://www.tsa.gov/stakeholders/rail-transportation-security-rule-%E2%80%93-49-cfr-1580> (accessed 10.26.14.)

Transportation Security Administration. n.d.f. Highway security-sensitive materials (HSSM) security action items (SAIs). <http://www.tsa.gov/highway-security-sensitive-materials-hssm-security-action-items-sais> (accessed 10.26.14.)

Transportation Security Administration. n.d.g. HAZMAT endorsement threat assessment program. <http://www.tsa.gov/stakeholders/hazmat-endorsement-threat-assessment-program> (accessed 10.26.14.)

Transportation Security Administration. September, 2005. TSA expanding national explosives detection canine teams to mass transit and commuter rail systems. News release. Washington, DC.

Transportation Security Administration. December, 2006. TSA/FTA security and emergency management action items for transit agencies. <http://www.tsa.gov/sites/default/files/assets/pdf/Intermodal/mass_transit_action_items.pdf> (accessed 10.26.14.)

Transportation Security Administration. March, 2010. Pipeline security and incident recovery protocol plan. Washington, DC.

Transportation Security Administration. April, 2011. Pipeline security guidelines. Washington, DC.

Transportation Security Administration. 2012. Office of Security Policy and Industry Engagement, Highway and Motor Carrier Branch 2012 Annual Report. Washington, DC.

Transportation Security Administration. April, 2013. Transportation Security Administration, surface transportation security fiscal year 2014 congressional [budget] justification. Washington, DC.

Transportation Security Administration. March, 2014. Transportation Security Administration, surface transportation security fiscal year 2015 congressional [budget] justification. Washington, DC.

U.S. Department of Homeland Security. n.d.a. About the Office of Intelligence and Analysis. <http://www.dhs.gov/about-office-intelligence-and-analysis> (accessed 10.26.14.)

U.S. Department of Homeland Security. n.d.b. More about the Office of Intelligence and Analysis mission. <http://www.dhs.gov/more-about-office-intelligence-and-analysis-mission> (accessed 10.26.14.)

U.S. Department of Homeland Security, Office of Inspector General. February, 2009. Effectiveness of TSA's surface transportation security inspectors. Washington, DC.

U.S. Department of Homeland Security. 2010a. Transportation systems sector-specific plan: an annex to the National Infrastructure Protection Plan. Washington, DC.

U.S. Department of Homeland Security, Office of Inspector General. March, 2010b. TSA's preparedness for mass transit and passenger rail emergencies. Washington, DC.

U.S. Department of Homeland Security. April, 2011. Secretary Napolitano announces implementation of National Terrorist Advisory System. <http://www.dhs.gov/news/2011/04/20/secretary-napolitano-announces-implementation-national-terrorism-advisory-system> (accessed 10.26.14.)

U.S. Department of Homeland Security, Office of Inspector General. August, 2012. Efficiency and effectiveness of TSA's Visible Intermodal Prevention and Response program within rail and mass transit systems. Washington, DC.

U.S. Department of Homeland Security, Office of Inspector General. June, 2013. Homeland Security Information Network improvements and challenges. Washington, DC.

U.S. Department of Transportation. n.d. Intelligence, security and emergency response. <http://www.dot.gov/mission/administrations/intelligence-security-emergency-response> (accessed 10.26.14.)

U.S. Department of Transportation. 2008a. Budget in brief for FY 2009. Washington, DC. February, 2008.

U.S. Department of Transportation, Office of Inspector General. 2008b. Actions needed to enhance pipeline security. Washington, DC.

White House. March, 2010. Surface transportation security priority assessment. Washington, DC.

White House. December, 2012. National strategy for information sharing and safeguarding. Washington, DC.

Implementing Aviation Security

CHAPTER OBJECTIVES:

In this chapter, you will learn about:

- Commercial aviation security layers, procedures, and technologies
- Air cargo security programs and regulations
- GA security measures

Introduction

Aviation systems, particularly those involved in transporting passengers via commercial aviation, continue to attract the greatest security attention of policymakers, the news media, and the general public internationally and in most developed countries.

The ICAO has updated and expanded its global security standards. In 2010, the ICAO Assembly unanimously adopted a Declaration on Aviation Security that "constitutes a very strong commitment by [member] States to strengthen aviation security worldwide, principally by enhancing international cooperation." In addition, the Assembly endorsed a new aviation security strategy, which, among other things, committed the ICAO to conduct aviation security audits to identify deficiencies and encourage their resolution by members, encourage the exchange of information among countries to promote mutual confidence in the level of aviation security in each country, assist members in the training of all security personnel, and assist members in addressing security-related deficiencies through technical cooperation programs (International Civil Aviation Organization, 2011, pp. 16–17).

The ICAO's 2011 annual report highlighted the organization's security activities over that year and is broadly representative of its current focus:

- Adoption of an amendment to Annex 17, which provided for more stringent air cargo security measures and the application of screening and other security controls to nonpassengers and emphasized the need for member states to use a risk-based approach toward security
- Provision for strengthened cooperation with the World Customs Organization to address threats to air cargo security while facilitating cargo movement
- Work with its Technical Advisory Group on Next Generation Screening to define concepts for a future passenger screening checkpoint that will achieve security objectives while minimizing the impact on operations
- Monitoring of compliance with the ICAO standard stipulating that member nations were to issue only ICAO-compliant machine-readable passports (MRPs) after April

1, 2010, and that all non-MRPs be out of circulation by November 24, 2015. As of January 2014, just one-third of members had responded to an ICAO survey on compliance with the deadlines, with five stating they would be unable to meet the 2015 requirement. Thus, ICAO reported, "the likelihood of universal compliance is not very high so far" but, at this stage—and given that "international law has no centralized enforcement authority"—"ICAO has no official position or information on the possible consequences of not meeting the deadline" (International Civil Aviation Organization, 2012, pp. 25–28; n.d.).

The European Union's (EU's) security standards for civil aviation are set forth in Regulation 300/2008 adopted by the European Parliament in 2008 and entering into full force in April 2010. The regulation applies to all nonmilitary airports in EU countries, air carriers using those airports, and all entities providing services to such airports. The basic standards cover airport security, demarcated areas of airports, aircraft security, passengers and carry-on baggage, checked baggage, air cargo and mail, in-flight and airport supplies, onboard security, staff recruitment and training, and security equipment. A committee of members assists in defining the measures for meeting the standards, and a European civil aviation Stakeholders' Advisory Group provides input. Each EU country is required to designate a single national authority to be responsible for implementation of the common basic standards, and covered airports, air carriers, and other entities are mandated to establish security programs that comply with the applicable national program (European Union, n.d.).

In the United States, the National Strategy for Aviation Security outlined in Homeland Security Presidential Directive-16 called for the development of a series of supporting plans to aid in its implementation, including the Aviation Transportation System Security Plan, which was issued by the Department of Homeland Security (DHS) on March 26, 2007, and is the element of the various strategic plans most relevant to ongoing security operations. It was designed to "ensure that efficient and effective aviation security is based on a system of shared responsibilities and costs, creating many interdependent, interlocking layers of security" and outlines specific security measures (and federal agency responsibilities) in the areas of passenger, employee, and crew security assurance; threat object detection and interdiction; and infrastructure protection. Among the plan's guiding principles are the following:

- *Effective aviation security is maintained through the inclusion of randomness and unpredictability to prevent terrorist identification of our measures and create a disruption of terrorist plots and criminal acts.*
- *The effectiveness of security measures must be continuously assessed and modified to reflect changes in the highly dynamic and adaptive terrorist threat. Terrorists closely study and actively attempt to defeat our security systems. Security measures cannot remain static in methodology, application, or technological approach. Instead, they must continually evolve, with the goal of being proactive rather than reactive.*

- *Any one of the current or recommended measures in our layered security system can potentially be compromised, but together provide greatly enhanced security. The United States government will address, enhance, and further strengthen all major layers and systems critical to risk reduction in aviation security.*
- *Cooperation in the implementation of the recommended multi-layered system across all federal departments and agencies, State, local, and tribal entities, and with our foreign partners, is essential and further enhances the strength of each measure pursued.*

U.S. Department of Homeland Security, 2007, pp. 3–15

Commercial Aviation

U.S. commercial aviation security—involving protection of the airports, aircraft, crew, and passengers involved in regularly scheduled flights within, from, to or over the United States—long predated the events of 9/11 and has continued to be the most elaborated, resourced, and scrutinized element of all transportation security efforts. Its post-9/11 evolution has been profoundly shaped by the requirements contained in the Aviation and Transportation Security Act (ATSA) of 2001.

TSA is the lead federal agency for commercial aviation security, and this is indeed the overwhelming priority of the agency, accounting for more than 93% of its workforce and more than three-quarters of its budget in recent years. It exercises the role formerly held by the Federal Aviation Administration (FAA) in regulating the security efforts of airports and airlines. However, it has also assumed direct responsibility for a wide range of security operations, most notably including the screening of passengers and their baggage (U.S. Department of Homeland Security, 2013a, p. 207; Transportation Security Administration, 2013a, p. 7).

In carrying out its mission, TSA employs Aviation Transportation Security Inspectors (TSI-As) who help to evaluate the security posture of the 448 commercial U.S. airports and more than 1500 domestic and international air carriers that operate in the United States. In addition to conducting annual inspections of these entities, TSI-As also:

- Review records and files to ensure compliance of airmen and aircrew with security requirements.
- Perform testing to determine compliance with transportation security regulations.
- Investigate incidents related to violations of TSA regulations, security directives, and approved security programs.
- Deliver technical briefings and provide assistance to the aviation industry in interpreting agency policies.
- Initiate enforcement actions against airports and air carriers when compliance violations have occurred or corrections have not been made.

In fiscal year (FY) 2014, there were 966 TSI-As, of whom 889 were assigned to domestic duties (including 190 canine handlers and 45 Visible Intermodal Prevention and Response [VIPR] inspectors), and 77 were stationed outside of the United States. Through October

2013, TSA had conducted 7725 airport inspections, 14,925 inspections of domestic air carriers, and 3368 inspections of foreign air carriers, with an emphasis on access control systems, security identification systems, surveillance systems, law enforcement response capabilities, and the physical security of aviation facilities and aircraft (Transportation Security Administration, 2014a, pp. 52–54).

Both the VIPR and National Canine Program teams employed by TSA in land security operations are also used in commercial aviation. The VIPR teams (which may be composed of TSI-As, TSA screening personnel, and Federal Air Marshals [FAMs}) screen passengers, look for suspicious behavior, and act as a visible deterrent against attacks. In FY 2012, TSA's 22 multimodal VIPR teams conducted 3977 aviation operations, representing 31% of all VIPR operations. In that same year, TSA had received funding for 684 canine explosives detection teams with commercial aviation security responsibilities:

- 491 aviation teams, which patrol airport terminals, curbside areas, and secured areas and respond to calls to search unattended items (including vehicles and baggage)
- 73 multimodal teams, which patrol and search aviation (as well as the other modes) in their geographic areas
- 120 passenger screening canine teams, which search for explosives odor on passengers in airport terminals (Transportation Security Administration, 2013b, p. 21; Government Accountability Office [GAO], 2013a, p. 6).

Fundamental to TSA's commercial aviation security measures is the concept of layering in which the individual security measures each add to the probability of successful prevention of terrorist or other attacks (Figure 8.1).

Airport Security

Of all of the major commercial aviation security layers, airport security is perhaps the least changed since 9/11, with the federal role exercised mainly through establishment of requirements for, and approval and oversight of Airport Security Programs (ASPs) that must be maintained by all commercial airports in the United States and airport operators primarily responsible for most security functions. ATSA did direct TSA to undertake certain efforts to improve the security of airport perimeters and access controls for secured areas and to reduce risks posed by airport workers.

The ASP serves as the "foundation for the entire airport security system" and "establishes the security areas and details how access to these areas will be controlled, defines the process for obtaining access/ID, explains how the airport will comply with the [regulation's] law enforcement requirements . . . and specifies the actual practices for airport compliance with federal regulations." Each program is unique to the individual airport, and the level of detail required varies, depending on the size and security risk of the airport, with larger (generally those servicing aircraft capable of carrying more than 60 passengers) or more at-risk airports required to operate "complete security programs" and smaller facilities facing fewer mandates. A complete ASP consists of the following elements:

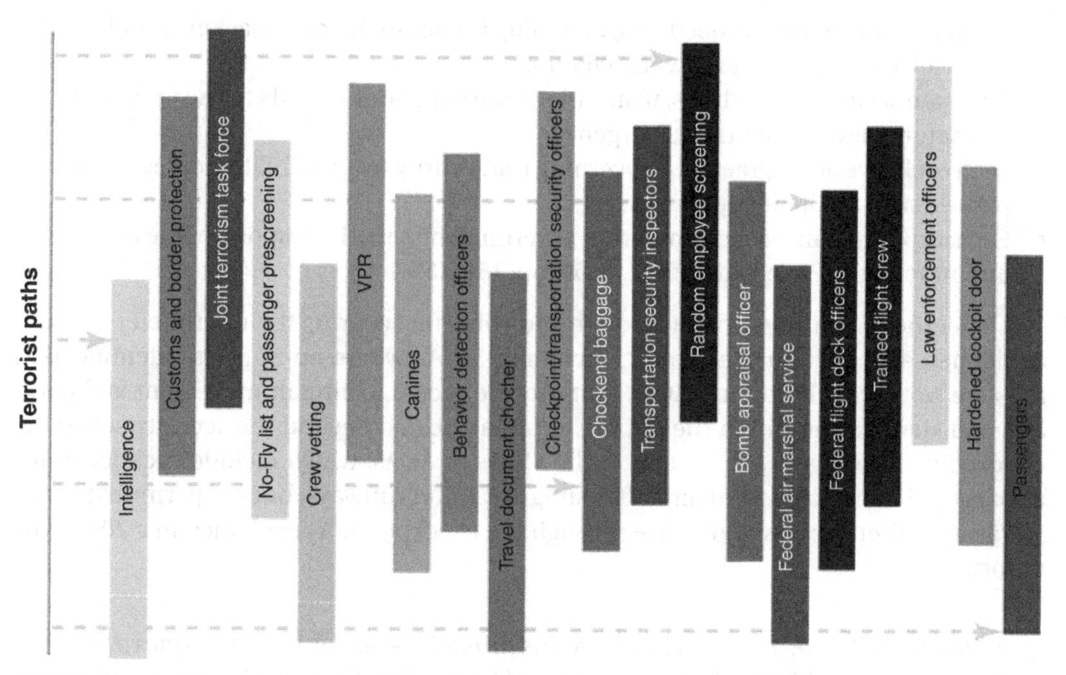

FIGURE 8.1 Layers of U.S. aviation security. VIPR, Visible Intermodal Prevention and Response. *(Source: Transportation Security Administration, http://www.tsa.gov/about-tsa/layers-security.)*

- The name, means of contact, duties, and training requirements of the airport security coordinator
- Descriptions of secured areas, air operations areas (AOAs), security identification display areas, (SIDA), and sterile areas
- Description of personnel identification systems, including fingerprint-based criminal history records checks
- Escort procedures for individuals who do not have unescorted access authority within the SIDA
- The content of the airport's SIDA training procedures
- A description of law enforcement personnel support requirements and training standards
- A system for maintaining security-related reports and forms
- Descriptions of the procedures, facilities, and equipment used to support TSA and aircraft operator screening of persons and property
- A contingency plan for complying with increased security measures
- Procedures for the secure distribution, storage, and disposal of ASPs; security directives; information circulars; implementing instructions; and classified information
- Procedures for posting public advisories, including warning notices

- Incident management procedures, including for bomb threats, hijackings, and other unlawful acts of interference with aviation
- Alternate security procedures, if any, the airport operator intends to use in the event of a natural disaster or other emergency
- Each exclusive area agreement governing the security responsibilities of regulated parties, such as aircraft operators
- Each airport tenant security program governing the security responsibilities of other airport tenants (Price and Forrest, 2009, pp. 151–155)

Under the ASP, airport operators are responsible for securing their perimeters, as well as defined areas within the airport, including SIDAs where appropriate identification must be worn for admittance, AOAs that provide access to aircraft movement and parking, and sterile areas within the airport terminal located beyond the security screening checkpoint. Passengers may not enter the SIDAs or AOAs, which include such locations as baggage loading areas and aircraft taxiing areas and runways, but are permitted in the sterile area after successful passage through the checkpoint. As explained in a 2009 GAO report:

> *Methods used by airports to control access through perimeters or into secured areas vary because of differences in the design and layout of individual airports, but all access controls must meet minimum performance standards in accordance with TSA requirements. These methods typically involve the use of one or more of the following: pedestrian and vehicle access codes using personal identification numbers, magnetic stripe cards and readers, turnstiles, locks and keys, and security personnel.*

According to that same report, in FY 2008, there were 2819 security breaches at U.S. airports, although TSA indicated that most of these were accidental and posed no threat (GAO, 2009, pp. 10–12).

TSA relies on its 120 Federal Security Directors (FSDs) located at major U.S. airports to oversee the implementation of and compliance with TSA security requirements at commercial airports (including TSA screening operations). They serve as "the central reference point on policy development, information technology, training, performance management, finance, and human resources to support the mission of TSA" in commercial aviation security (Figure 8.2) (Transportation Security Administration, 2013a, p. 68).

In response to provisions in ATSA, TSA has taken a variety of actions to strengthen airport security, including developing an inspection program to evaluate airport access controls, assisting airport operators in determining the effectiveness of access control technologies, requiring background checks (including fingerprint and name-based checks) on all workers with unescorted access to secure areas of airports, requiring airport operators to implement a security awareness plan to keep employees and contractors informed of the threat to airport security and their individual security responsibilities, providing reimbursement to local agencies for assignment of law enforcement officers to patrol airport

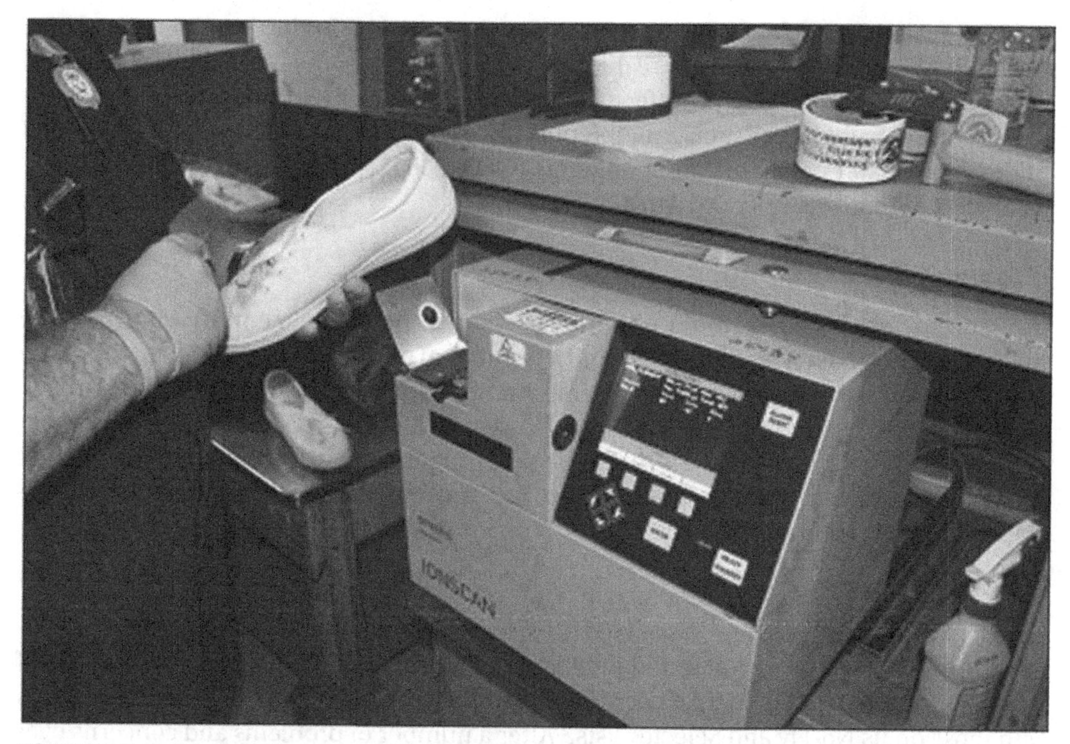

FIGURE 8.2 Explosives trace detection machine being used to manually resolve a suspect item. *(Source: GAO. 2012a. Checked bag screening: TSA has deployed optimal systems at the majority of TSA-regulated airports, but could strengthen cost estimates. GAO, Washington, DC, p. 10.)*

perimeters and be stationed at access points, and conducting a pilot program to assess several types of screening methods for airport workers[1] (GAO, 2009, pp. 26–29, 76–78).

Another TSA initiative relevant to airport security is the Security Playbook program that "employs security measures at direct access points and airport perimeters and uses a variety of resources and equipment to conduct screening of individuals and vehicles entering the AOA." Examples of such measures include vehicle inspection, explosives trace detection of individuals and property, enhanced screening, property searches, ID verification, and behavior detection (Transportation Security Administration, 2011).

In a 2009 assessment, GAO concluded: "Since 2004, TSA has taken efforts to strengthen airport security and implement new programs. . . . [These] efforts . . . have not been guided by a unifying national strategy that identifies key elements, such as goals, priorities, performance measures, and required resources." Although the agency and DHS concurred

[1]ATSA also required the development of a Transportation Worker Identification Credential (TWIC) that would apply to all transportation modes, including aviation. However, in response to concerns among aviation stakeholders about the cost and complexity of replacing the access control systems already in place in many large and medium airports and difficulties in implementing the TWIC program (including problems in developing biometric identifiers), TSA has not thus far extended it to the aviation sector (Price and Forrest, 2009, pp. 179–180).

with this evaluation and the accompanying recommendation for the development of such a strategy, it remains only partially realized to date. In addition, the DHS Inspector General (IG) conducts covert tests of the effectiveness of TSA's policies and procedures for airport access controls and issued a classified report in 2012 on the results, including identified vulnerabilities and policy recommendations (GAO, 2009, Highlights, p. 67; U.S. Department of Homeland Security, 2012a).

Passenger Prescreening

Assessing the risk posed by would-be airline passengers before their entrance into security checkpoints is a federal responsibility, and a number of agencies—in addition to TSA—are involved in fulfilling this role. Although representing an attempt to utilize risk management in allocating aviation security resources, as called for by GAO, the 9/11 Commission, and many other security analysts, prescreening programs have repeatedly raised issues about the appropriate balance among security and individual privacy, traveler convenience, and operational efficiency.

Secure Flight

Secure Flight is a TSA program developed pursuant to a recommendation by the 9/11 Commission and related provision in the Intelligence Reform and Terrorism Prevention Act of 2004 that the agency should take over from the airlines the matching of passenger names with those on its No-Fly and Selectee lists.[2] After a number of problems and concerns over privacy and other issues expressed in Congress and elsewhere, TSA began implementation of the program with the issuance of a final rule in October 2008, and by November 2010, it was fully operational, with the agency reporting that 100% of commercial passengers flying to, from, within, or over the United States were being vetted via Secure Flight against federal watchlists[3] (Johnstone, 2006, pp. 75–77; Transportation Security Administration, n.d.a).

The program is designed to support risk-based security by identifying high-risk passengers that are to receive appropriate security measures or actions. By having TSA solely in charge of the program's administration, it "decreases the chance for compromised watch list data by limiting its distribution; provides earlier identification of potential matches, allowing for expedited notification of law enforcement and threat management; provides a fair, equitable, and consistent matching process across all airlines; and offers consistent application of an integrated redress process for misidentified individuals through the Department of Homeland Security's Travel Redress Inquiry Program (DHS TRIP)" (Transportation Security Administration, n.d.a).

[2]Whereas those on the No-Fly list are to be prohibited from boarding the flight, those on the Selectee List are to be subjected to enhanced screening at the security checkpoint. These lists are subsets of the Terrorist Screening Database maintained by the federal government's Terrorist Screening Center (Transportation Security Administration, n.d.a).

[3]According to media reports, there were approximately 21,000 names on the No-Fly list as of the beginning of 2012, of whom about 500 were American citizens (Associated Press, "U.S. no-fly list doubles in one year," February 2, 2012.)

Airlines operating covered flights are required to obtain the full name (as it appears on government-issued ID), date of birth, and gender of individuals making ticket reservations (as well as a Redress number from DHS TRIP, if applicable) and then to transmit this information to TSA, which matches it to its No-Fly and Selectee lists and transmits the results back to the airlines for use in issuance of passenger boarding passes.

In response to privacy concerns, Secure Flight contains a number of elements designed to protect personal information, including a dedicated privacy officer and staff responsible for privacy compliance; specific privacy policies, procedures, standards, and rules of behavior for TSA personnel; redress and response systems; and the assurance that "TSA does not collect or use commercial data to conduct Secure Flight watch list matching" (Transportation Security Administration, n.d.a).

According to the DHS IG, "Since Secure Flight assumed responsibility for passenger prescreening from aircraft operators, the program has provided more consistent passenger prescreening. Secure Flight has a defined system and processes to conduct watch list matching. To ensure that aircraft operators follow established procedures, Secure Flight monitors records and uses its discretion to forward issues for compliance investigation. Secure Flight also includes privacy safeguards to protect passenger personal data and sensitive watch list records and information" (U.S. Department of Homeland Security, 2012c, Spotlight).

Pre√

ATSA authorized TSA to establish a "trusted traveler" program under which individuals who voluntarily submit to a background check process identifying themselves as low risks to aviation would receive expedited processing through security checkpoints. An initial attempt to create such a program ("Registered Traveler"), involving the use of private vendors who issued and scanned participants' biometric identifiers, was abandoned by TSA because it failed to demonstrate sufficient security benefits. In the view of the U.S. travel industry, its lack of success was the result of not offering participants (~250,000 at the program's height) significant advantages (e.g., allowing them to keep their shoes on during security screening) beyond shorter wait times because of TSA concerns that "trusted" travelers should be subject to the same level of scrutiny as others (U.S. Travel Association, 2011, pp. 10–11).

The Pre√ program was initiated by TSA in October 2011 and by the fall of 2014 was operational at 120 airports, making over 35% of all individuals passing through U.S. security checkpoints each day eligible for expedited physical screening.[4] Participation is voluntary, and eligible participants include members of Customs and Border Protection (CBP) Trusted Traveler programs, enrollees in certain airline frequent flyer programs, and members of the U.S. military. Individuals may also apply directly to TSA (including paying an $85 fee).

[4]These figures also include individuals processed through TSA's Known Crew Member program under which flight crews of participating airlines receive expedited screening (Peterman et al., 2013, p. 4).

Once TSA determines a passenger is eligible for TSA Pre√ expedited screening, their low-risk status is contained in the passenger's boarding pass... TSA reads the barcode at designated checkpoints and the passenger may be referred to a TSA Pre√ lane where they will undergo expedited screening. TSA Pre√ expedited screening procedures include options such as no longer removing shoes, leaving laptops in their bag, leaving on light jackets/outerwear and belts, and leaving compliant liquids/aerosols/gels in the carry-on bag.

Transportation Security Administration, 2014a, p. 3

More than 500,000 travellers had directly enrolled in the Pre-check program as of August 2014 (Transportation Security Administration, 2014b).

Those who participate via membership in airline frequent flyer programs (currently including Air Canada, Alaska Airlines, American, Delta, Hawaiian Airlines, Jet Blue Airways, Southwest, Sun Country Airlines, United, US Airways, and Virgin Atlantic) are only eligible on the airlines in which they are members. In addition, "TSA will always incorporate random and unpredictable security measures throughout the airport and no individual will be guaranteed expedited screening in order to retain a certain element of randomness" (Transportation Security Administration, n.d.b).

Critics have raised questions about the effectiveness of Pre√. Specific concerns have included the security of the boarding pass barcodes for expedited screening, the lack of biometric identity authentication, and the lack of detailed background checks (especially for those participating through frequent flyer memberships) (Peterman et al., 2013, pp. 3–4).

■ ■ Critical Thinking ■

Secure Flight and Pre√ represent two different approaches toward using risk management to allocate security resources. Discuss the pros and cons of each approach.

Customs and Border Protection Prescreening

In addition to its role in risk assessment of incoming cargo, CBP's Automated Targeting System (ATS) is used to prescreen passengers arriving in the United States. Specifically, its ATS-Passenger (ATS-P) system "is a web-based enforcement and decision support tool used to collect, analyze, and disseminate information for the identification of potential terrorists, transnational criminals, and, in some cases, other persons who pose a higher risk of violating U.S. law. ATS-P capabilities are used at ports of entry[5] to augment the CBP officer's decision-making about whether a passenger or crew member should receive additional screening." TSA officials may also access ATS-P information in augmenting

[5]Border crossings and seaports in addition to airports.

their own risk assessment of international travelers. Similar to other ATS systems, ATS-P "compares traveler, cargo, and conveyance information against law enforcement, intelligence, and other enforcement data." Because of the wide range of information sources potentially accessed by ATS, civil liberties groups have been especially concerned about its operations (U.S. Department of Homeland Security, 2012b, pp. 2, 6; Stellin, 2013).

CBP also operates a number of its own trusted traveler programs, including:

- Global Entry, which allows travelers who "undergo a rigorous background check and interview before enrollment in the program" and who pay an annual fee ($100 in 2014) to receive "expedited clearance" through Customs upon arrival in the United States, although similar to Pre√ members, they may be randomly selected for additional screening (Global Entry, n.d.).
- NEXUS, which serves as an alternative to a passport for air, land, and sea travel into the United States for U.S. and Canadian citizens. To qualify, an individual must submit to an interview by U.S. or Canadian customs authorities. Upon approval and the payment of a fee ($50 as of 2014), they are issued a special photo ID entitling them to expedited customs passage at air, land, and sea ports of entry (U.S. Customs and Border Protection, n.d.a).
- SENTRI, which provides for expedited customs processing at U.S.-Mexico border crossings. Applicants must "undergo a thorough biographical background check against criminal, law enforcement, customs, immigration, and terrorist indices; a 10-fingerprint law enforcement check; and a personal interview with a CBP officer." The total fee in 2014 was $122.25 per person. (U.S. Customs and Border Protection, n.d.b).

These CBP programs had enrolled over three million users as of August 2014 (Transportation Security Administration, 2014b).

Passenger and Carry-On Baggage Screening

If commercial aviation has been the transportation mode most attended to in terms of security efforts, the airport security checkpoint—where passengers and any items they seek to carry onboard are scrutinized—is the single security layer that has attracted the most funding, policy attention, and visibility. In FY 2014, more than 47,000 of the close to 53,000 TSA employees involved in aviation security worked in checkpoint operations,[6] and more than $3.4 billion of the $5.2 billion spent by the agency for all aviation security activities went to checkpoint-related functions (Transportation Security Administration, 2014a, p. 6).

At the checkpoints, passengers and their carry-on items are subjected to a variety of techniques aimed at detecting and removing dangerous items, with the goal being to "strike the appropriate balance between preventing security breaches and maintaining the efficient movement of law-abiding passengers through the security checkpoints."

[6]Some of these individuals are also assigned to operate checked bag screening equipment.

These techniques include validation of travel documents at the checkpoint's entrance, use of various types of electronic detection and imaging technologies, behavior recognition, and physical searches (Transportation Security Administration, 2013a, p. 13).

Travel Document Checkers

Travel Document Checkers are a specialized category of TSA's Transportation Security Officers (TSOs), who are the agency's "front-line workforce performing checkpoint security, document checking, airport employee screening and unpredictable security measures." The document checkers operate in front of the security checkpoints at U.S. commercial airports and are tasked with verifying an individual's identity and travel documents before allowing entry into the checkpoint. They use devices "to validate boarding passes and authenticate various forms of acceptable photo ID presented by passengers, airport/airline personnel, and law enforcement officers" and "make a visual comparison of the individual with the ID photograph, and ensure the boarding pass presented was issued to that individual, for that day's travel and from the correct airport." Any suspect cases are referred to the TSO Supervisor for that checkpoint. In FY 2014, TSA employed 2001 Travel Document Checkers (Transportation Security Administration, 2013a, pp. 14–17; 2014a, p. 15).

Checkpoint Screening Technologies

After being admitted into the checkpoint itself, travelers and their carry-on items are processed by one or more of the following currently deployed screening technologies (with the number of units deployed as of FY 2014 in parentheses).

- Enhanced Walk-through Metal Detectors (1505) that screen persons for metallic weapons, including guns and knives
- Advanced Technology x-ray systems (1647) that screen carry-on baggage and provide enhanced visual detection or automated explosives detection capabilities
- Explosives Trace Detection (2800) units used by TSOs to test selected carry-on items for explosives residue
- Bottled Liquids Scanners (1690) that screen bottles to determine if they contain explosives
- Chemical Analysis Devices (255), which are small devices used to identify suspect substances
- AIT (749), or full-body scanners, screens persons for metallic and nonmetallic threats, including weapons, explosives, and other concealed objects (Transportation Security Administration, 2014a, pp. 34, 48–49).

The AIT equipment was developed by TSA to replace the walkthrough metal detectors as the primary method for checkpoint screening of passengers to meet evolving threats to commercial aviation. Deployments began in 2007 and were accelerated after the December 2009 attempted bombing of Northwest Flight 253. Two types of AIT were deployed: backscatter technology (which projects low-level x-rays over the surface of the body that are converted into a computer image of the body) and millimeter wave technology (which uses electromagnetic waves to generate a three-dimensional computer image based on the energy reflected from the body). As of January 2013, 746 millimeter wave and 251 backscatter AITs had been purchased by TSA (U.S. Department of Homeland Security, 2013c, pp. 2–3).

The first AITs put into use raised significant health and privacy issues. The backscatter technology had been previously banned in Europe based on concerns that the x-ray radiation it generated could pose a health risk, although TSA has consistently maintained that both sets of AITs are "safe for all passengers" based on testing conducted by the National Institute of Standards and Technology (for backscatter technology) and the Food and Drug Administration (for millimeter wave technology). The privacy issues persisted, however (even though two national surveys conducted in 2010 indicated that two-thirds or more of the public supported use of the more intrusive AITs as a means of improving security). In response to concerns expressed in Congress and elsewhere, in FY 2011 and FY 2012, TSA developed and installed software for the millimeter wave equipment that displayed its body images as generic figures while auto-detecting potential threat objects. The FAA Modernization and Reform Act of 2012 mandated that TSA use this software on all AITs used for passenger screening by June 2012. TSA experienced difficulties in doing so for the backscatter AITs and in January 2013 announced it was removing all of those machines from airports and cancelling the contract calling for additional procurements (Halsey, 2013; Transportation Security Administration, 2013a, pp. 31–32; U.S. Department of Homeland Security, 2013c, pp. 2–3).

In a September 2013 audit, the DHS IG found that TSA "did not develop a comprehensive deployment strategy to ensure that all AIT units were effectively deployed and fully used for screening passengers" because the agency did not "have a policy or process requiring program officers to prepare strategic deployment plans for new technology that align with the overall goals of the passenger screening program, and have adequate internal controls to ensure accurate data on AIT utilization." TSA generally concurred with this analysis and was in the process of drafting a comprehensive deployment strategy at the time the IG report was released (U.S. Department of Homeland Security, 2013c, pp. 4–6, 9–10).

Checkpoint Screening Process

Checkpoint screening operations are handled by TSOs (or their private contractor counterparts at airports participating in the Screener Partnership Program). Primary screening involves a walk-through of an AIT or metal detector for the person and passage through the x-ray equipment for carry-on bags and other items required to be screened separately from persons, such as their shoes. For passengers who alarm the primary screening equipment or whose carry-on items are identified as potentially containing prohibited items, as well as those designated as selectees for additional screening, TSOs perform secondary screening (via physical searches of carry-on baggage, pat-down searches of travelers, or use of Explosives Trace Detection or other screening device).

TSA maintains a list of items that are prohibited from carriage on board an aircraft by a passenger, or placement in checked baggage, or both. Most sharp objects (including box cutters, knives, and scissors with blades longer than 4 inches) may not be carried on board but are allowed in checked bags. On the other hand, most explosive materials (e.g., blasting caps, flares, and plastic explosives) are prohibited altogether, as are most flammable items (e.g., fuels or gasoline). A complete and updated list is available at www.tsa.gov/traveler-information/prohibited-items. TSA reported approximately 2 billion items were screened at its checkpoints during FY 2013. Combined with the more than 425 million checked bags also examined, this resulted in the prevention of 111,000 dangerous prohibited items (including explosives, firearms, and other weapons) from being carried onto planes (U.S. Department of Homeland Security, 2013a, p. 133; 2014, p. 69).

TSA reported that, as a result of the Pre√ program and certain other exemptions (including for members of the military, children under the age of 12, and passengers over the age of 75), over half of all passengers passing through airport security checkpoints received expedited screening as of September 2014. This has helped reduce the occurrence of checkpoint wait times of 20 minutes or longer by 64% (Halsey, 2014, p. A2).

In FY 2014, there were 46,920 TSOs (including Travel Document Checkers and Behavior Detection Officers) involved in providing checkpoint security. Under ATSA, all security screening personnel, whether federal or private contractors, must be U.S. citizens; pass a background investigation; and possess a high school diploma, a general equivalency diploma, or sufficient experience. TSOs "must undergo extensive training and be certified to screen passengers and baggage through the use of detection equipment" and are "subject to ongoing testing and training requirements." Failure to pass any of such requirements is grounds for dismissal (Transportation Security Administration, 2014a, p. 14).

The performance of airport checkpoint screeners in detecting threat objects has been perhaps the most scrutinized, and criticized, element of transportation security in the United States, before 2001 and since. A 2009 report for Congress highlighted some of the key findings and issues involved:

> *Screener performance is a continuing concern as covert testing results have repeatedly demonstrated existing weaknesses in screening procedures and capabilities that could potentially be exploited by terrorists or criminals seeking to attack the aviation system. These weaknesses may reflect a combination of policies, procedures, technology capabilities, and screener human performance, although weakness in screener human performance has been emphasized as a particular concern. . . . While specific performance metrics for covert testing are considered security sensitive, various media reports of test results suggest that failure rates are often quite high, particularly with respect to screeners missing simulated improvised explosive devices and explosive components. For example, it has been reported that during tests conducted in 2006, TSA screeners missed fake bombs 75% of the time at Los Angeles International Airport, and 60% of the time at Chicago O'Hare Airport. The TSA contends that the results, on the surface, appear discouraging, but are a reflection of highly sophisticated*

concealment methods being used by testers to uncover specific system vulnerabilities so that corrective action can be taken.

Elias, 2009, pp. 6, 9

Screening Passengers by Observation Techniques

TSA began testing the SPOT concept in October 2003 and by FY 2012 had deployed more than 3000 Behavior Detection Officers (BDOs, who are TSOs with specialized training in SPOT) at 176 airports in the United States. TSA considers the SPOT program to be "essential" in adding "an important layer of security in all areas of an airport":

It provides a non-intrusive means of identifying potentially high-risk individuals who exhibit behaviors that deviate from an established environmental baseline (indicative of stress, fear and deception), which could possibly reflect intentions of terrorism. SPOT looks at involuntary physical and physiological reactions. A recent study sponsored by the DHS Science & Technology Directorate and conducted by the American Institutes for Research examined the SPOT indicators' effectiveness compared against a strict random protocol. The study confirmed that SPOT was significantly more effective at identifying persons of interest than random selection.

Transportation Security Administration, 2013a, pp. 17–18

BDOs primarily operate at airport screening checkpoints by engaging in brief verbal exchanges with passengers waiting in line. Those indentified as potential threats by BDOs are subjected to additional screening of their persons and carry-on baggage, and in certain cases, referred to law enforcement officers for further investigation. In cases when the observed behaviors are not reconciled satisfactorily, the passenger is prohibited from boarding the aircraft. SPOT referrals in FY 2012 resulted in 199 arrests (including outstanding warrants and drug and immigration violations) (U.S. Department of Homeland Security, 2013b, pp. 3–4).

DIGGING DEEPER

THE DEBATE OVER BEHAVIOR DETECTION

Conflicting Data and Interpretations

The effectiveness of the SPOT program has been questioned by both the DHS IG and GAO. In May 2013, the IG reported, "TSA has not implemented a strategic plan to ensure the program's success. For example, TSA did not (1) assess the effectiveness of the . . . program, (2) have a comprehensive training program, (3) ensure outreach to partners, or (4) have a financial plan. As a result, TSA cannot ensure that passengers at United States airports are screened objectively, show that the program is cost-effective, or reasonably justify the program's expansion." In response, TSA:

- Finalized strategic and performance measurement plans and began implementing them
- Instituted controls to enhance the completeness and accuracy of program data

- Implemented a mandatory recurrent and refresher training plan for all BDOs
- Developed and implemented an automated tool to help evaluate airports' use of BDOs (U.S. Department of Homeland Security, 2013b, pp. 1, 12–15; Transportation Security Administration, 2013c, p. 5).

In November 2013 testimony at a congressional hearing, GAO recommended that "TSA should limit future funding for behavior detection activities" because "our review of meta-analyses (studies that analyze other studies and synthesize their findings) . . . called into question the use of behavior observation techniques . . . as a means for reliably detecting deception. The meta-analyses we reviewed collectively found that the ability of human observers to accurately identify deceptive behavior based on behavioral cues or indicators is the same as or slightly better than chance (54 percent). We also reported on other studies that do not support the use of behavioral indicators to identify mal-intent or threats to aviation" (GAO, 2013c, p. 3).

At the same session, TSA Administrator Pistole addressed the GAO findings.

> *While TSA appreciates GAO's partnership in improving the [BDO] program, we are concerned that its most recent report relies heavily on academic literature regarding the detection of individuals who are lying. The report, however, fails to recognize all of the available research. . . It is important to note that TSA's behavior detection approach does not attempt to specifically identify persons engaging in lying; rather, it is designed to identify individuals who may be deemed high-risk based on objective behavioral indicators. . . . TSA believes the program should continue to be funded at current levels while [planned] improvements . . . are implemented.*
>
> *Transportation Security Administration, 2013c, p. 4*

In your view, how can, or should, this debate be resolved?

Screening Partnership Program

ATSA directed TSA to determine the feasibility of having qualified private screening companies provide airport security screening instead of federal screeners, and in late 2002, the agency began a 2-year pilot program at five airports[7] of varying sizes in which private companies were contracted to provide checkpoint and checked bag screening. At the conclusion of the pilot test, TSA created the Screening Partnership Program (SPP) under which any commercial airport authority can request to transition from federal to private contract screeners. If the airport's application for SPP participation is approved by TSA, a private screening workforce selected by the airport authority assumes screening responsibility but is subject to TSA regulations and oversight aimed at ensuring the contractor provides effective and efficient security.

The question of privatized airport screening has been controversial since the creation of TSA, and there have been regular disputes about the optimal degree of privatization. By

[7]San Francisco International Airport, Kansas City International Airport, Greater Rochester (NY) International Airport, Jackson Hole Airport, and Tupelo Regional Airport.

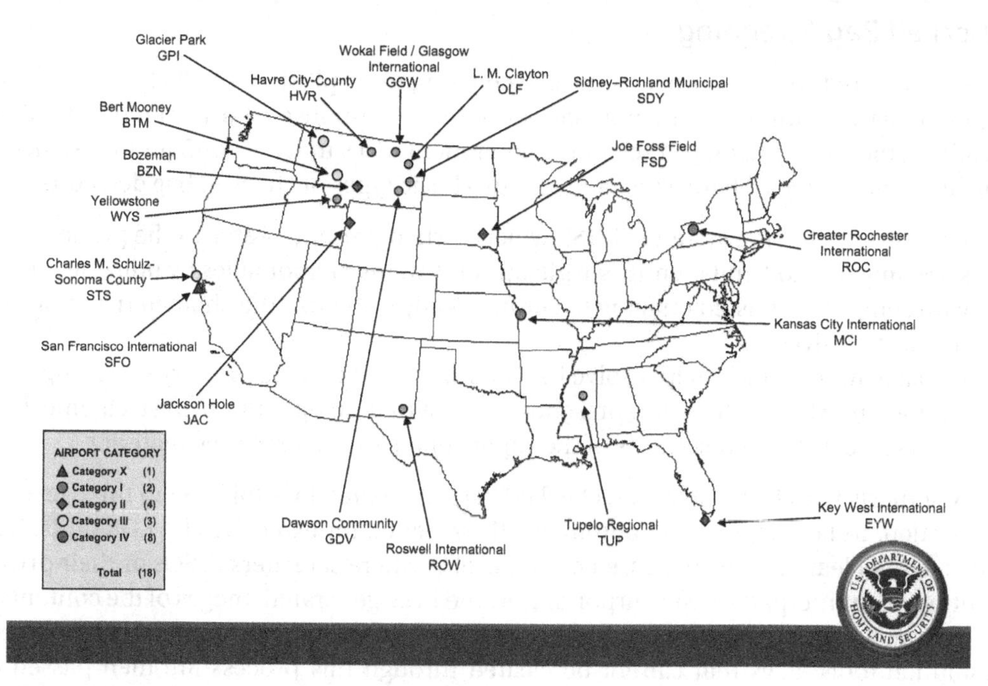

FIGURE 8.3 Airports currently participating in Screening Partnership Program. *(Source: Transportation Security Administration, http://www.tsa.gov/sites/default/files/images/stakeholders_IMG/spp_map_1.pdf.)*

January 2011, a total of 16 airports were in the SPP, and the TSA Administrator announced that no further expansion would take place "unless a clear and substantial advantage to do so emerges in the future." Supporters of expansion succeeded in gaining addition of a provision in 2012 legislation that changed the standard by which TSA evaluates SPP applications so that approval of such applications would be given unless the change would compromise security or detrimentally affect the cost efficiency or effectiveness of the airport's screening operations. As of mid-2014, there were 18 airports participating in the SPP, and a nineteenth (Orlando, FL) was added via the award of a private screening contract in September 2014 (Figure 8.3) (GAO, 2012d, pp. 1–3, 46; 2014, pp. 1–2; Transportation Security Administration, 2014c).

In a 2012 analysis of screener performance at similarly sized SPP and non-SPP airports, the GAO found that the privatized screeners at the former performed better than TSOs on certain measures and worse on others. However, the agency added, "The differences we observed in private and federal screener performance cannot be entirely attributed to the type of screeners (private or federal) at an airport, because, according to TSA officials and other subject matter experts we interviewed, many factors, some of which cannot be controlled for, affect screener performance. These factors include, but are not limited to, checkpoint layout, airline schedules, seasonal changes in travel volume, and type of traveler" (GAO, 2012d, pp. 26–29).

Checked Bag Screening

ATSA mandated that TSA provide for the screening for explosives of all checked bags on flights departing from U.S. commercial airports. In response to a December 31, 2003, deadline, the agency deployed screening equipment in a variety of configurations, including in airport lobbies. There were (and are) two basic types of checked bag detectors:

- Explosives detection systems (EDS), which—when present—serve as the primary screening method and identify suspicious bulk items or anomalies by utilizing x-rays with computer-assisted imaging to automatically recognize the characteristics of threat explosives
- ETD machines, which help resolve EDS alarms or serve as the primary screening equipment when EDS are not present and use a human operator using a chemical analysis device to detect traces of the vapors or residue of explosive material

At airports with EDS machines, checked bags are generally subjected to three levels of inspection. In Level 1, they are automatically screened for explosives by EDS units. Bags that are not cleared here are passed on to Level 2, where screeners (TSOs or their private contractor counterparts at SPP airports) view the EDS-generated images of the contents of the alarmed bags on a computer monitor and try to clear them through use of on-screen resolution tools. Bags that cannot be cleared through this process are then passed on to Level 3, where screeners manually inspect them using ETD machines. The generally smaller airports equipped only with ETDs use only this last level of checked bag screening (GAO, 2012a, p. 11).

The initial post-9/11 deployments produced higher than necessary operating costs and impaired the effectiveness and efficiency of the screening operations. Consequently, TSA instituted the Electronic Baggage Screening Program (EBSP) to (1) replace, reconfigure, and deploy screening equipment to increase throughput (processing speed), system capacity, and effectiveness while reducing staffing requirements and airport lobby installations; (2) increase equipment reliability, reduce equipment downtime, and extend its service life; and (3) develop improved capabilities to address evolving terrorist threats (GAO, 2012a, pp. 1–2).

As the terrorist threat evolved and new technologies became available, TSA revised the detection requirements for the EDS equipment. The machines in use on 9/11 and those deployed in the immediate aftermath of ATSA[8] were operating under standards established in 1998 by the FAA. In 2005 and again in 2010, TSA upgraded the requirements (including expanding the number and types of explosives that must be detected) and planned to

[8]These represent a "large portion" of the more than 1900 EDS units deployed as of January 2013 (Transportation Security Administration, 2013a, p. 38).

FIGURE 8.4 Stand-alone explosives detection system (EDS). *(Source: GAO. 2011b. Aviation security: TSA has enhanced its explosives detection requirements for checked baggage, but additional screening actions are needed. GAO, Washington, DC, p. 6.)*

implement the necessary improvements in a phased approach. A 2011 GAO report indicated that as of that time, "some number of EDS machines in TSA's checked baggage screening fleet were configured to detect explosives at the levels established in 2005 and that the remaining EDS machines are configured to detect explosives at levels established in 1998." GAO called for TSA to develop and implement a plan to ensure that all EDS machines, including existing and newly acquired units, meet the more stringent 2010 requirements. In March 2013, TSA announced it intended to complete upgrading deployed units by the end of FY 2013 (GAO, 2011b, pp. 9–11; 2013b, pp. 7–8).

In addition to the capabilities of the screening equipment, the other big issue in checked baggage screening concerns the manner in which they are deployed. The EDS machines are deployed in either stand-alone or in-line configurations. Whereas in the former, checked bags are manually loaded and unloaded by screeners, in the latter, the EDS units are integrated into a conveyor system that transports and sorts the bags from the ticket counter through the baggage screening system. Most of the pre- and immediately post-9/11 deployments were in the stand-alone mode (Figure 8.4).

The benefits of in-line systems have been widely recognized by TSA officials and others and include enhanced security, improved efficiency, and lowered costs (by reducing the number of screeners needed and decreasing work-related injuries)[9]. However, initial installation of such a system can be costly because it may require modification of an airport terminal as well as removal of the existing system and installation of the new one. In response to its own findings, plus recommendations of GAO and the 9/11 Commission, in February 2006, TSA released a strategic plan for achieving "optimal" checked bag screening systems at airports that were prioritized based on security risk and projected cost savings.

[9]TSA estimated the savings from deployment of in-line systems to be $189.4 million through the end of FY 2013 (Transportation Security Administration, 2014a, p. 16).

As of 2012, TSA was reporting that 76% (337 of 446) of all U.S. commercial airports had optimal baggage screening systems, with such a system defined by the agency as one in which installation and activation of "the in-line or stand-alone systems that best fit the airport's screening needs without relying on temporary stand-alone systems" had been completed. For example, use of an in-line EDS setup would be cost ineffective and thus not "optimal," for certain small, low-volume airports. According to a GAO analysis of the systems deemed by TSA to be optimal, 50% used ETDs only, another 27% used EDS in stand-alone mode, 16% had in-line EDS, and the remaining 6% had a mixture of systems. However, the proportion of optimal systems varied greatly, depending on the size of the airport, with just 36% (10 of 28) of the largest (Category X) commercial facilities having achieved optimal systems compared with 100% (all of the 157) of the smallest (Category III) airports. Ultimately, TSA plans on deploying in-line EDS at all of the largest (Category X and I) airports and in-line or stand-alone EDS in all medium-sized (Categories II and III) airports (GAO, 2012a, pp. 10, 12–13, 17–20).

In its FY 2015 budget justification submitted to Congress in March 2014, TSA reported that 407 of 448 (91%) U.S. commercial airports (including 18 of 28 Category X airports) have completed deployment of screening equipment "with the most efficient technologies for their screening configuration" (Transportation Security Administration, 2014a, pp. 40–41).

Aircraft and Onboard Security

The final layer of aviation security—protection of the aircraft and its crew and passengers on the ground and in flight—experienced the most immediate transformation after the 9/11 hijackings.

Hardened Cockpit Doors

Less than 1 month after those events, FAA issued a regulation to expedite the reinforcement of cockpit doors on U.S. airliners via the installation of steel bars and locking devices. By January 2002, these modifications had been voluntarily completed by 98% of U.S. airlines.

The enactment of ATSA in November 2001 mandated the further strengthening of cockpit doors and door locks to ensure that the doors could not be forced open from the passenger cabin, and in January of the following year, the FAA issued a rule requiring that approximately 7000 U.S. aircraft install reinforced cockpit doors by April 9, 2003. The rule established new design and performance standards for the doors, required that any modifications meet existing safety standards, and directed that the doors remain locked and access privileges be controlled. After leading a successful effort to have the ICAO set a global standard for hardened cockpit doors (with an international installation deadline of November 2003), in June 2002, the FAA directed that all foreign aircraft serving the United States meet the same time limit for deployment of reinforced cockpit doors previously set for U.S. air carriers. In an April 9, 2003, news release, the FAA reported that the deadline had been met, and more than 10,000 U.S. and foreign aircraft serving the United States were "now equipped with new, hardened cockpit doors, making air travel safer for passengers and crew" (FAA, 2002; 2003).

Federal Air Marshals

The number of FAMs was dramatically, and expeditiously, increased from the 33 agents in place on 9/11 to an estimated force of in excess of 3500 in 2012,[10] and their mission was enlarged to include deployment on domestic, as well as international, commercial flights (Johnstone, 2006, pp. 82–83; Ahiers, 2014).

According to TSA, "Federal Air Marshals must operate independently without backup, and rank among those federal law enforcement officers that hold the highest standard for handgun accuracy. They blend in with passengers and rely on their training, including investigative techniques, criminal terrorist behavior recognition, firearms proficiency, aircraft specific tactics, and close quarters self-defense measures to protect the flying public." FAMs are trained in basic aircraft operations in addition to the security-related capabilities. Their flight assignments are based on a risk-assessment of commercial flights incorporating intelligence and other classified information. FAMs may also be assigned to other aviation security-related duties, including serving as Assistant Federal Security Directors for law enforcement at certain airports and as members of VIPR teams and staffing positions at the National Counterterrorism Center, National Targeting Center, and FBI Joint Terrorism Task Forces (Transportation Security Administration, n.d.c; Price and Forrest, 2009, p. 138).

Some have questioned the expanded FAM force in view of other changes that occurred in onboard security after 9/11. One academic analysis found that hardened cockpit doors were more cost effective than air marshals in saving lives, and the March 2011 report by the U.S. Travel Association stated, "The FAMs program was originally expanded after 9/11 to provide a last line of defense when the other layers of aviation security were being improved and at a time when cockpit doors were not yet hardened against intrusion. The appropriate level of the FAMs program, both for international and domestic flights, is an example of a review that should be evaluated in a risk management context" (Rittgers, 2011; U.S. Travel Association, 2011, p. 29).

Recently, budgetary pressures have led to a reduction in the FAMs program, with a cut in funding between FY 2012 and FY 2014 leading to a closure of six of its 26 offices by June 2016 and an accompanying "reduction in [the number of] FAMs through attrition" (Ahiers, 2014).

Federal Flight Deck Officers Program

The Federal Flight Deck Officer (FFDO) program was established by the Arming Pilots Against Terrorism Act, which was incorporated into the Homeland Security Act and

[10]The precise number remains classified. The 3500 figure is based on media accounts and derived from reports of the number of pay disputes resolved by TSA and FAMs during 2012 (Ahiers, 2014).

signed into law in November 2002. The legislation authorized the deputation of qualified airline pilots and certain other cockpit personnel to act as federal law enforcement officers and to use firearms to defend aircraft against attempted hijackings or other criminal acts. Participation in the program is voluntary, and pilots, flight engineers, and navigators on commercial and cargo flights are eligible to apply. To be selected, an applicant must successfully complete certain assessments; meet all standards established by the FAM Service; and attend (and pay for) FFDO training, which includes instruction in use of firearms, use of force, legal issues, defensive tactics, the psychology of survival, and the standard operating procedures for the program. Upon successful completion of the training, the individual is admitted into the FFDO program but is required to successfully pass firearms requalification activities on a biannual basis (Transportation Security Administration, n.d.d; n.d.e).

The first 44 FFDOs were trained in April 2003, and there are currently "thousands" in the program, covering more than 100,000 flights per month, according to the Air Line Pilots Association (ALPA). In its FY 2014 budget, the Obama Administration proposed that the FFDO program be continued but funded by the airlines rather than the federal government. The proposal is "adamantly" opposed by ALPA, which represents a majority of FFDOs. The proposal was not adopted by Congress and did not appear in the Administration's FY 2015 budget request (Air Line Pilots Association, 2013; Transportation Security Administration, 2013a, p. 73; 2014a, pp. 72–73).

Crew Member Security Training

In the early 1980s, the FAA established guidelines for the security training of airline flight and cabin crews. Referred to as the Common Strategy, these standards (which were based on previous experiences in dealing with nonsuicidal hijackings) directed air carriers to develop security training programs instructing crew members to cooperate with threatening passengers or hijackers. In immediate response to the events of 9/11, ATSA was enacted, and among its provisions were requirements for FAA to develop new and detailed guidance for flight and cabin crew security training programs within 60 days of the law's enactment. Specifically, ATSA mandated that the new guidelines include training in determination of the seriousness of any occurrence, crew communication and cooperation, appropriate self-defense response, use of any protective devices assigned to crew members, psychology of terrorists to cope with hijacker behavior and passenger responses, live situational training exercises regarding various threat conditions, and flight deck procedures or aircraft maneuvers to defend the aircraft (GAO, 2005, pp. 11, 13).

In response, the FAA moved quickly to revise the Common Strategy guidelines, issuing new ones in January 2002 that represented "a shift in strategy from passive to active resistance by [aircraft] crew members" and included the following directions:

- *Any passenger disturbance, even those seemingly harmless, should be considered suspicious; it could be a diversion for other more serious acts.*

- *In any suspected or actual hijack attempt, the flight crew should land the airplane as soon as possible to minimize the time hijackers would have to commandeer the aircraft and use it as a weapon of mass destruction.*

FAA, 2002

The Vision 100—Century of Aviation Reauthorization Act of December 2003 made further changes in the requirements for air carrier crew member security training:

- All air carriers providing scheduled passenger air service must implement a training program that addresses the elements contained in ATSA plus instruction in recognition of suspicious activities, the proper commands to give passengers and attackers, and the proper conduct of a cabin search.
- TSA (which had assumed responsibility for oversight of air carrier security training programs from the FAA in February 2002) must approve the air carrier training programs.
- TSA (in consultation with FAA) must monitor and periodically review the training programs to ensure they are adequately preparing crew members for potential threat conditions and order air carriers to modify their training programs to reflect new or different security threats.
- TSA must develop and provide an advanced voluntary self-defense training program that provides both classroom and effective hands-on training in deterring a passenger who might present a threat; advanced control, striking, and restraint techniques; training to defend oneself against edged or contact weapons; methods to subdue and restrain an attacker; use of available items aboard the aircraft for self-defense; and appropriate and effective self-defense responses including the use of force against an attacker (GAO, 2005, pp. 12–14).

Questions have been raised over the years about the adequacy of the air carrier security training programs. In July 2012 testimony to a Congressional subcommittee, a representative of the Association of Flight Attendants (AFA) reiterated their concerns about the training provided to their membership:

> *Today basic security training provided by air carriers includes actual hands-on self-defense training that varies from 5 minutes to 30 minutes. . . . Despite repeated request by AFA and others for updated training that includes basic self defense maneuvers to allow flight attendants to defend themselves against a terrorist attack, we still do not receive mandatory training about how to effectively recognize suspect terrorist behavior and how to defend ourselves and others against terrorist attacks aboard the aircraft.*
>
> Association of Flight Attendants, 2012, p. 4

TSA's Crew Member Self Defense Training (CMSDT) program is voluntary and provides free instruction to all actively employed or temporarily furloughed U.S. passenger and cargo crew members at 22 sites nationwide. It currently consists of 4 hours of hands-on training developed by the FAM Service to help prepare crew members to face potential threat situations. As with the FFDO program, the President's FY 2014 budget proposed that the CMSDT program be continued but funded by the airlines rather than the federal government, but the proposal was rejected by Congress and not included in the Obama FY 2015 budget submission (Transportation Security Administration, n.d.f; 2013a, p. 73; 2014a, pp. 72–73).

■ ■ Critical Thinking ■

ATSA contained numerous mandates for the deployment of specific security measures within tight deadlines. The authors of these provisions were seeking to ensure the expeditious adoption of tangible security improvements while reassuring the American public that concrete steps were being taken to improve aviation security. Critics have questioned this approach as focusing security efforts on a fixed set of prescriptions that may or may not have represented optimal solutions. What do you think?

Aircraft Operator Security Role

The takeover by TSA of most commercial aviation prescreening and screening operations has reduced the security responsibilities of commercial air carriers, but they do retain a number of security obligations. Most important, they are required to create their own security program that complies with a standard program issued by TSA as a baseline for a particular category of operator. The most extensive of these is the full program (formally called the Aircraft Operator Standard Security Program, or AOSSP), which applies to operators of scheduled or public charter passenger flights that use aircraft with more than 60 passenger seats or that enplane from or deplane into an existing "sterile area" of an airport. Other standard security programs for scheduled or public charter passenger operators include the Partial Program Standard Security Program (PPSSP), which covers operators of scheduled or public charter flights with aircraft of 31 to 60 passenger seats not using sterile areas, and the Twelve-Five Standard Security Program (TFSSP), which applies to operators of aircraft weighing more than 12,500 lb and not covered by either the AOSSP or the PPSSP.

Under the full AOSSP, the aircraft operator is responsible for, among other obligations:

- Ensuring that TSA, or whatever entity is conducting such operations, properly screens passengers, carry-on and checked baggage, and cargo
- Designating a security coordinator
- Providing for law enforcement officer support to respond to security issues onboard or at a screening checkpoint
- Transporting FAMs
- Preventing unauthorized access to exclusive areas and aircraft
- Conducting criminal history record checks on all its personnel with access to checked baggage and cargo, any personnel with unescorted access authority, and all flight crew members
- Using airport-approved personnel identification systems
- Providing security training for ground security coordinators, flight crew members, and other employees with security-related duties

- Restricting access to the flight deck
- Evaluating threats and having plans to handle bomb and hijacking threats

The PPSSP and TFSSP contain some, but not all, of the same requirements (Price and Forrest, 2009, pp. 256–263).

Air Cargo

The first significant air cargo security effort in the United States resulted from the 1996 Gore Commission's recommendations for addressing the threat of bombs being placed in the baggage holds of passenger aircraft. In response, the Known Shipper Program was initiated through which the airlines developed a process for scrutinizing companies wishing to ship cargo via passenger aircraft. Generally, this involved advance inspection of a shipper's facilities by airline representatives, and after the company was approved as a trusted "known shipper," it was permitted to transport cargo on that airline.

In late 2001, ATSA directed the federal government to provide for the screening of cargo carried onboard passenger and all-cargo aircraft. In response, TSA established a program that required air carriers to ensure that all cargo representing an "elevated risk" was screened before loading onto passenger aircraft and prohibited the transport of cargo from unknown shippers on such aircraft. With a focus on preventing the placement of an explosive device on a passenger aircraft and the hijacking of an all-cargo aircraft for use as a weapon of mass destruction, the agency also initiated the development of an air cargo security regulation (with the final rule being promulgated in 2006) that included the following provisions:

- Created a new standard security program for all-cargo air operators using aircraft weighing more than 100,309.3 lb (45,500 kg) at takeoff and strengthened existing security requirements for all-cargo operators using aircraft weighing 12,500 to 100,309.3 lb
- Expanded screening requirements for persons boarding and cargo loaded onto all-cargo aircraft
- Directed airports to extend the areas requiring the display of security identification into cargo operations areas
- Expanded the definition of indirect air carriers (IACs)[11] to include all-cargo carriers and enhanced the security requirements for these entities by creating the IAC Standard Security Program and requiring IACs to have a TSA-approved security plan in place before they ship cargo via air
- Strengthened foreign air carrier cargo standards by requiring these operators to implement a level of security similar to that of U.S. operators using the same size aircraft
- Codified and strengthened the Known Shipper Program by having TSA take over the vetting process, maintain the program database, and conduct random screening of known shipper cargo agents

[11]IACs, also known as freight forwarders, act as intermediaries between shippers who supply the cargo and the air carriers who are to transport it. Typically, they collect and consolidate cargo into larger shipping units and then deliver those units to air carriers for transport. IACs are responsible for as much as 80% of all cargo shipped on passenger aircraft in the United States (Price and Forrest, 2009, pp. 330–331).

- Implemented methods to identify and screen high-risk cargo
- Required security threat assessments of individuals who have unescorted access to air cargo (Price and Forrest, 2009, pp. 330–332, 337–338)

To enforce these and other security regulations, TSA employs Transportation Security Inspectors-Cargo (TSI-Cs) for domestic and international inspections. The domestic inspectors (500 in FY 2014, located at 121 U.S. airports with high cargo volumes) conduct compliance inspections, investigations and tests of air carriers and IACs and "perform educational outreach to assist [them] in complying with air cargo security mandates." The foreign inspectors (61 in FY 2014) "verify compliance with cargo screening procedures" at foreign airport (Transportation Security Administration, 2014a, pp. 52–53, 78).

National Canine Program explosives detection teams also assist in the screening of cargo destined for transport on passenger aircraft. In FY 2014, 120 TSA-led canine teams were primarily assigned to screening cargo at high-volume air cargo facilities and 511 law enforcement officer-led teams at 79 airports spent approximately one-fourth of their time on air cargo security (Transportation Security Administration, 2014a, pp. 62–63).

Certified Cargo Screening Program

The Implementing Recommendations of the 9/11 Commission Act of 2007 (9/11 Act) further focused air cargo security efforts on the screening of such cargo with its mandate that 100% of all cargo on passenger aircraft be screened by August 2010. To accomplish this objective, TSA developed a two-part approach, one covering cargo departing from U.S. airports and the other for cargo inbound from foreign airports.

The domestic cargo screening requirement was implemented in part via the establishment of the Certified Cargo Screening Program (CCSP), which was begun as a pilot program in 2009 and formalized in a final rule published in September 2011.

> *Under CCSP TSA certifies cargo screening facilities located throughout the United States to screen cargo prior to providing it to airlines for shipment on passenger flights.... Participation in the program is voluntary and designed to enable vetted, validated and certified supply chain facilities to comply with the 100 percent screening requirement.... Certified Cargo Screening Facilities (CCSF) must carry out a TSA approved security program and adhere to strict chain of custody requirements. Cargo must be secured from the time it is screened until it is placed on passenger aircraft for shipment.*

To be approved, CCSP participants must implement procedures to evaluate the security risk posed by prospective and hired employees (including routine reviews of the latter); prevent unauthorized access to facilities where cargo is screened, prepared, or stored (including through the use of physical barriers); and maintain chain of custody standards for screened cargo (including proper documentation, methodology, and authentication). TSA reported that it met the August 1, 2010, deadline for screening 100% of cargo transported on passenger aircraft departing from U.S. airports (Transportation Security Administration, n.d.g; and TSA Office of Security Policy and Industry Engagement, 2013, p. 1).

In FY 2013, there were 1136 CCSP participants, and CCSP facilities accounted for more than 60% of cargo placed on passenger planes in the United States. Recertification of participants (required every 3 years) has begun, and more than 400 locations have been recertified (TSA Office of Security Policy and Industry Engagement, 2013, p. 4; Transportation Security Administration, 2014a, p. 75).

Screening of Inbound Air Cargo

TSA developed the National Cargo Security Program (NCSP) to assist in meeting the 9/11 Act's directive for screening of international cargo. Under this program, TSA assesses foreign cargo security measures, and if it determines that they provide a level of security commensurate with U.S. requirements, a recognition agreement is entered into through which air carriers operating in that country are allowed to implement its (rather than the United States') security program (Figure 8.5) (TSA Office of Security Policy and Industry Engagement, 2013, p. 2).

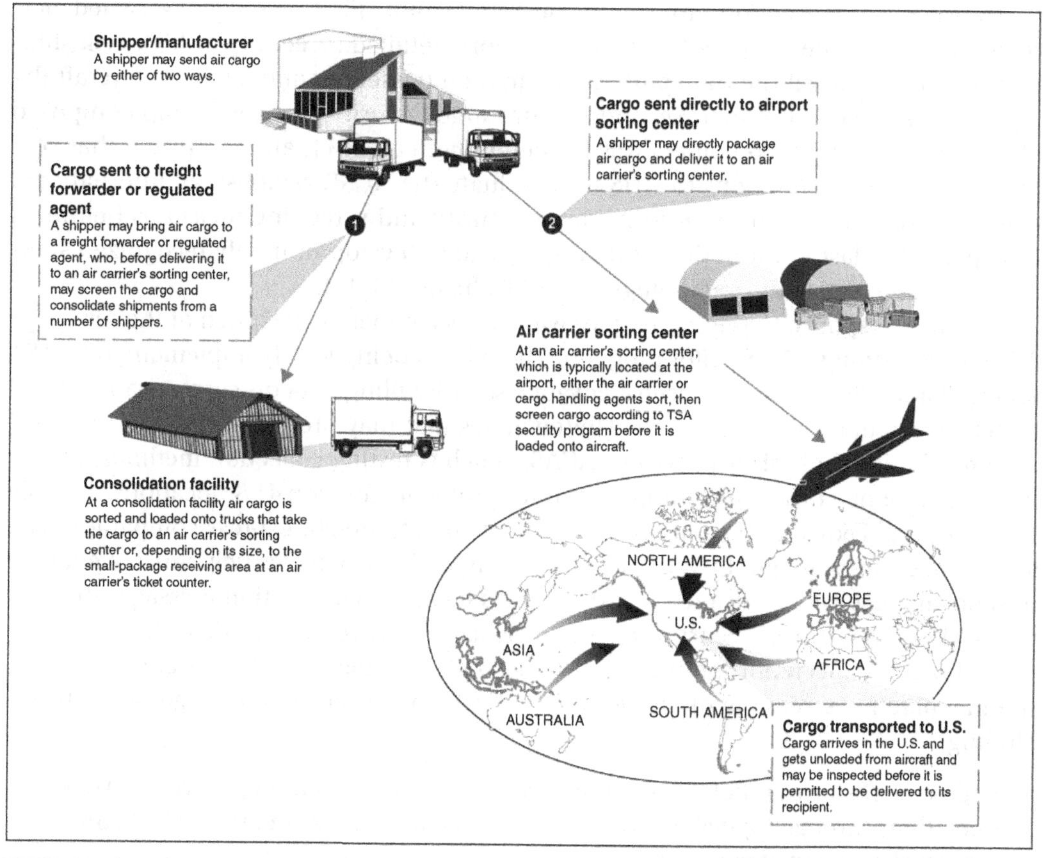

FIGURE 8.5 Flow of inbound air cargo transported to the United States. *(Source: GAO. 2012b. Aviation security: Actions needed to address challenges and potential vulnerabilities to securing inbound air cargo. GAO, Washington, DC, p. 9.)*

While TSA was still attempting to meet the 9/11 Act's mandate for 100% screening of international inbound cargo, the October 2010 Yemen air cargo plot raised concerns about the effectiveness of the overall air cargo screening system:

> *... because suspected packages were screened multiple times, using multiple methods, at various locations yet the threat items were detected only after foreign law enforcement officials opened the shipments based on a tip from an intelligence source. According to TSA, the threat item used in the incident likely would not have been detected by air carriers using TSA screening protocols in place at that time because screening requirements for all-cargo carriers focused on preventing and detecting stowaways or contraband items and not on detecting explosive devices and, for passenger air carriers, screening requirements primarily focused on detecting assembled explosive devices rather than on the types of specific components used to construct the explosive device associated with the Yemen incident.*
>
> *GAO, 2012b, p. 12*

TSA took several actions following the Yemen plot. First, the agency issued new screening requirements aimed at providing more detailed screening of high-risk shipments and prohibited the transport of air cargo on passenger and all-cargo aircraft flying from Somalia and Yemen. Second, an Air Cargo Security Working Group composed of air cargo security stakeholders was formed in January 2011, and in April of that year presented recommendations to DHS to reevaluate the NCSP, establish an international trusted shipper program to perform cargo security and screening measures before its transport from last point of departure airports, and develop mutually recognized standards for cargo screening technology (GAO, 2012b, pp. 13–14).

The third response to the Yemen plot was the December 2010 launch of the Air Cargo Advance Screening (ACAS) pilot program. The pilot is being jointly implemented by TSA and CBP and allows them to receive advance security filing data on cargo as a means of targeting inbound shipments to the United States that may present a high risk and thus require additional physical screening. ACAS, which is testing collection methods, targeting and response procedures, and screening protocols for non-U.S. locations, initially involved UPS, FedEx, DHL, and TNT but is now open to freight express companies, passenger air carriers, freight forwarders, and all-cargo air carriers. Participants are enabled to send and receive the advance security filing data and related action messages through CBP's ATS (TSA Office of Security Policy and Industry Engagement, 2013, p. 2).

A May 2012 GAO report pointed to several challenges facing TSA in its efforts to meet the mandate for screening of all inbound cargo transported on passenger aircraft, including:

- Logistical obstacles reported by a number of air carriers (including disruptions in operations through higher costs and slower screening times, and the lack of TSA-approved screening technologies)

- Problems in verifying self-reported screening data provided by passenger air carriers and used in determining the amount of inbound air cargo screened in accordance with TSA requirements and the absence of comparable data from all-cargo carriers
- Difficulties in evaluating foreign countries' cargo security programs (for purposes of the NCSP recognition program) because of a dependence on a country's willingness and ability to work with TSA in determining whether their program is commensurate with the U.S. program (GAO, 2012b, pp. 16–21).

TSA indicated to GAO that it had finalized a plan to achieve 100% screening of inbound air cargo transported on passenger aircraft by December 1, 2012, through use of the ACAS system. In its FY 2014 budget justification to Congress, the agency further reported that "as of December 3, 2012, TSA required 100 percent screening of international inbound air cargo on passenger aircraft using a risk-based approach to screening. . . . Screening totals reported by air carriers for December 2012 indicate that the 100 percent screening mandate has been attained, and there have been no reported issues or challenges associated with meeting this requirement" (GAO, 2012b, p. 28; Transportation Security Administration, 2013a, p. 76).

General Aviation

GA airports and aircraft—which include all civil aviation except for commercial passenger and cargo operations and account for the vast majority of air facilities and flights—have experienced fewer high-profile incidents and received far less security policy attention and resources than the other aviation sectors. Although TSA has the lead federal responsibility, the agency "has taken a less direct role in securing GA, in that it generally establishes standards that operators may voluntarily implement and provides recommendations and advice to general aviation owners and operators." Certain GA operations do fall under existing TSA security regulations, including many charter aircraft operations and airports that also serve commercial or air cargo flights (GAO, 2012c, p. 10).

General Aviation Airports and Flight Schools

One of the first actions taken by TSA in GA security was its issuance of "Security Guidelines for General Aviation Airports" in May 2004. This voluntary guidance, developed in consultation with industry stakeholders, outlined federally endorsed security enhancements in the fields of personnel controls, aircraft security, airport and associated facility security, surveillance, security plans and communications, and special provisions for agricultural aircraft operations (e.g., crop dusting). This was followed in December 2007 by the agency's initiation of the Secure Fixed Base Operator Program under which operators providing fueling, aircraft handling, and hangar storage services to GA aircraft are enabled (but not required) to check passenger and crew identification against flight manifests or Electronic Advance Passenger Information System filings to provide positive identification of passengers and crew onboard GA flights.

Two other security programs for airports in the Washington, DC, area were implemented in the aftermath of 9/11. The DCA Access Standard Security Program (DASSP) covers GA operations out of or into Ronald Reagan Washington National Airport. The DASSP for each operator must provide for TSA inspection of crew and passengers, TSA inspection of property and aircraft, the start and end dates of the flight, identification checks of passengers by TSA, submission of passenger and crew manifests 24 hours in advance of the flight, enhanced background checks for all passengers and fingerprint-based criminal checks for flight crew, and an Armed Security Officer on board each flight. The Maryland-Three Program applies to the three GA airports closest to Washington, DC, all of which are located in Maryland: College Park Airport, Potomac Airfield, and Hyde Executive Field. These facilities must have a security coordinator who administers TSA security requirements that are similar to those for commercial airports (Price and Forrest, 2009, pp. 307, 310–312).

The GA sector includes more than 7000 flight training providers and individual flight instructors. TSA's Alien Flight Student Program (AFSP) requires foreign flight school candidates to submit biographical information and fingerprints for TSA's use in conducting a security threat assessment of the student (including criminal record, immigration status, and presence of terrorist watch lists). The program also mandates that all flight school employees who have direct contact with students must receive initial and recurrent security awareness training (GAO, 2012c, p. 2; Transportation Security Administration, n.d.h).

In 2012, GAO found weaknesses in TSA's AFSP security threat assessments:

> *TSA has not ensured that all foreign nationals seeking flight training in the United States have been vetted through AFSP prior to beginning the training or established controls to help verify the identity of individuals seeking flight training who claim U.S. citizenship. TSA also faces challenges in obtaining criminal history information to conduct its security threat assessments as part of the vetting process, but is working to establish processes to identify foreign nationals with immigration violations.*
>
> *GAO, 2012c, p. 20*

The 9/11 Act directed TSA to develop and perform standardized threat and vulnerability assessments of GA airports. In 2010, TSA distributed an online security survey to approximately 3000 GA airports. The survey included questions on security measures implemented by the operator, such as securing hangar doors, closed-circuit TV deployments, and perimeter fencing. Results from the 1164 responses indicated that "while most general aviation airports had initiated some security measures, the extent to which different security measures had been implemented varied by airport." For example, 97% of the larger GA airports responding to the survey reported they had developed an emergency contact list, but less than 19% of them had established measures to positively identify passengers, cargo, and baggage (GAO, 2012c, p. 17).

GAO conducted its own security assessment in 2011 at 13 airports with GA operations by determining (through onsite observation) what physical security measures were in place to prevent unauthorized access (Table 8.1):

Table 8.1 Security Measures in Place at Selected Airports

	General Aviation										Commercial and General Aviation		
Security Measure	**1**	**2**	**3**	**4**	**5**	**6**	**7**	**8**	**9**	**10**	**11**	**12**	**13**
Perimeter fencing or natural barrier	✓	✓	✓	✓	✓	✓	✓	✓	✓	x	✓	✓	✓
Controls at designated access points	✓	✓	✓	✓	✓	✓	x	✓	✓	x	✓	✓	✓
Lighting around perimeter	x	x	x	x	x	x	x	x	x	x	✓	✓	✓
Lighting at designated access points	✓	✓	✓	✓	✓	x	✓	x	✓	x	✓	✓	✓
Lighting around hangars	✓	✓	✓	✓	✓	✓	✓	✓	✓	✓	✓	✓	✓
Hangars locked and secured	✓	✓	✓	✓	✓	✓	✓	✓	✓	✓	✓	✓	✓
Aircraft locked and secured	✓	✓	✓	✓	✓	✓	✓	✓	✓	✓	✓	✓	✓
On-site law enforcement or security	x	x	✓	✓	✓	✓	x	✓	✓	✓	✓	✓	✓
Transient pilot sign-in/sign-out procedures	x	✓	✓	✓	✓	x	✓	x	✓	x	✓	✓	n/a
Intrusion detection system	x	x	x	x	x	x	x	x	✓	x	✓	✓	✓
CCTV cameras in areas related to unauthorized access	✓	✓	✓	✓	✓	x	✓	✓	✓	✓	✓	✓	✓
Passenger and baggage screening*	x	✓	x	x	x	x	x	x	✓	x	✓	✓	✓
Package and cargo screening*	x	x	x	x	x	x	x	x	x	x	✓	x	✓
Back-up generator or power supply	x	✓	✓	✓	✓	✓	✓	✓	✓	x	✓	✓	✓

✓ Security measure partially in place at time of Government Accountability Office (GAO) visit.
✓ Security measure in place at time of GAO visit.
x, Security measure not in place at time of GAO visit.
CCTV, closed-circuit television.
*TSA suggested guidelines for general aviation airports do not discuss physical screening of passengers, their baggage, packages, or cargo. GAO included these security measures based on their experience in conducting physical security reviews.
(*Source*: GAO. 2011a. General aviation: Security assessments at selected airports. GAO, Washington, DC, p. 8.)

> *The 13 airports GAO visited had multiple security measures in place to protect against unauthorized access, although the specific measures and potential vulnerabilities varied across the airports. The 3 airports also supporting commercial aviation had generally implemented all the security measures GAO assessed, whereas GAO identified potential vulnerabilities at most of the 10 general aviation airports that could allow unauthorized access to aircraft or airport grounds, facilities, or equipment.*
>
> *GAO, 2011a, Highlights*

General Aviation Aircraft

At present, most GA aircraft operators are not required to have security programs. Exceptions are private charters (i.e., do not offer service to the general public), which must meet standards under TSA's Private Charter Standard Security Program (PCSSP), and GA aircraft weighing more than 12,500 lb at takeoff and not covered under any other standard security program, which are subject to the Twelve-Five Standard Security Program (TFSSP). To enhance general aviation security and to provide greater standardization of security requirements for GA and certain other operators of large (more than 12,500 lb) aircraft, TSA published a notice of proposed rulemaking (NPRM) in the October 30, 2008, *Federal Register* for the establishment of the Large Aircraft Security Program (LASP).

The LASP would supplant both the PCSSP and TFSSP for GA aircraft of more than 12,500 lb[12] and impose the following major requirements on operators:

- Implement a TSA-approved and audited security program.
- Ensure flight crew members have undergone a fingerprint-based criminal history records check.
- Conduct watchlist matching of their passengers through TSA-approved watchlist matching service providers.
- Undergo a biennial compliance audit by a TSA-approved third-party auditor.
- Screen passengers and their accessible property for aircraft with takeoff weights of more than 100,309.3 lb (45,500 kg) and operated for compensation or hire.
- Check onboard property for unauthorized persons (Federal Register, 2008, pp. 64791–64792).

TSA received considerable push-back from the GA industry in response to the proposed rule. The National Air Transportation Association (NATA), which represents much of that industry, summarized those concerns in 2012:

> *The NPRM demonstrates the TSA's lack of knowledge and understanding of the general aviation community. . . . After receiving 7,400 comments, almost all negative, the TSA met with the general aviation industry to come up with a plan to draft a supplemental rulemaking that would be more practical and manageable for the industry. The supplemental NPRM was due in late 2011 and it remains uncertain . . . when it will be released. NATA's concern with the NPRM is vast, and such expansion of regulation without providing the public with any justification of the necessity for increased security on general aviation aircraft is disconcerting.*
>
> *National Air Transportation Association, 2012*

In November 2013, TSA announced it is preparing a revised NPRM for the LASP, with a planned publication date of August 2014. However, as of November 2014 the revision had not been published (Office of Information and Regulatory Affairs, 2013).

TSA Transportation Security Inspectors currently conduct periodic compliance inspections of GA operators covered under the TFSSP and PCSSP. Among other things, the inspections are designed to examine whether the operator has emergency procedures in

[12]It would also cover some non-GA aircraft operations, including certain midsize passenger and all-cargo operators.

place, ensures individuals are denied boarding if they lack valid identification, ensures passenger identification documents are checked against flight manifests, and has adequate procedures for addressing incidents of tampering or unauthorized access. According to TSA data, aircraft operator compliance with security requirements has been "well over 90 percent" in the 2007 to 2011 period (GAO, 2012c, pp. 15–16).

■ ■ Critical Thinking ■

Compare and contrast U.S. air cargo and general aviation security policies. Consider legislative mandates, resources, and regulations.

Intelligence and Information Sharing

Intelligence is considered the first layer of aviation security and has been key to foiling terrorist plots against aviation. Much of the intelligence activity within DHS, and especially TSA, is devoted to the aviation mode. In addition, significant efforts have been made to develop mechanisms for the sharing of security information between federal and nonfederal stakeholders in aviation security.

Transportation Security Administration Office of Intelligence and Analysis

The DHS Office of Intelligence and Analysis (OIA) is formally the departmental lead for both intelligence and information sharing for all transportation modes, including aviation, and it plays a part in aviation security, including managing DHS participation in the Nationwide Suspicious Activity Reporting (SAR) Initiative. However, the deep involvement of TSA in aviation security has meant, in practice, that the agency's own Intelligence and Analysis unit has performed the principal role in aviation security activities.

TSA-OIA does receive and evaluate information from other DHS components, law enforcement agencies, and the owners and operators of transportation systems, but it is uniquely positioned to review suspicious activity reports generated within the agency, including from TSOs and FAMs. In addition, the office deploys Field Intelligence Officers at major airports throughout the United States to provide intelligence support to the Federal Security Directors at those airports and to serve as TSA liaisons with law enforcement officials and intelligence fusion centers.

TSA-OIA disseminates intelligence via (1) Transportation Intelligence Notes, which are regularly distributed to TSA officials and select transportation security partners, and provide (at the classified or unclassified level) information or analysis on a single topic or situational awareness of an ongoing incident; (2) the Global and Regional Intelligence Digest, which is produced monthly for select transportation security officials, and analyzes reporting on suspicious activities and surveillance directed against all transportation modes; (3) annual classified or unclassified assessments of the threat to each transportation mode; and (4) semi-annual classified or unclassified assessments of current threats to various classes of U.S. airports.

The intelligence from TSA-OIA and other TSA aviation security–related information is shared through a variety of mechanisms, including the Homeland Security Information Network and the Transportation Security Information and Analysis Center (TS-ISAC) created by TSA-OIA in March 2010 and now called TSA Intel.

The 9/11 Act directed GAO to carry out a survey to measure the satisfaction of recipients of certain transportation security–related information disseminated by TSA. The survey was conducted in April and May 2011 and included responses from 275 stakeholders. GAO summarized the results:

> *Transportation stakeholders who GAO surveyed were generally satisfied with TSA's security-related information products, but identified opportunities to improve the quality and availability of the disseminated information. . . . Fifty-seven percent of stakeholders indicated they were satisfied with the products they receive. However, stakeholders who receive these products were least satisfied with the actionability of the information—the degree to which the products enabled stakeholders to adjust their security measures.*

More specifically, the survey found mode-based variations in the level of satisfaction with airport (64% satisfied) and passenger air stakeholders (62%) far more satisfied than those in the air cargo sector (38%) (The survey also covered short line and regional freight rail—63% satisfied, highways—56% satisfied, and class I freight rail—29% satisfied.) (GAO, 2011c, pp. Highlights, 3, 6–9, 13).

Other Aviation Security Information Sharing Programs

TSA has partnered with the air cargo industry to create Air Cargo Watch, which aims to "increase security domain awareness so that individuals are empowered to detect, deter, and report potential or actual security threats" by improving the ability of the public and air cargo industry to report suspicious activity. Toward that end, the program has developed an online presentation, posters, and a two-page guide "to encourage increased attention to potential security threats among several audiences" (Transportation Security Administration, n.d.i).

The Aircraft Owners and Pilots Association (AOPA), which is the largest trade group representing the general aviation sector, developed the nationwide Airport Watch Program that uses the more than 600,000 pilots "as eyes and ears for observing and reporting suspicious activity." The program "includes warning signs for airports, informational literature, and a training video to teach pilots and airport employees how to enhance security at their airports." TSA established the General Aviation Secure Program to work in tandem with Airport Watch in "encouraging everyone to be vigilant about general aviation security and report any unusual activities to TSA." Examples of the types of suspicious activities that both of these programs seek to have reported are aircraft with unusual modifications or activity, pilots appearing to be under the control of others, unfamiliar persons loitering around the airfield, suspicious aircraft lease or rental requests, and anyone making threats (Aircraft Owners and Pilots Association, n.d.; Transportation Security Administration, n.d.j).

Conclusion

Commercial aviation security has long occupied the preeminent position in the field of transportation security, and that continues to be the case in the second decade of the 21st century. Indications of ongoing terrorist intention to target this sector—in the form of intelligence reports and foiled plots—still provide an impetus for such prioritization. Furthermore, these large investments have produced improvements in the capabilities of virtually every layer of passenger aviation security, including prescreening procedures and screening technologies. However, in the absence of successful attacks and with the reassertion of other societal priorities, including privacy, economic efficiency, and budgetary constraints, it is not clear that the current level of commercial aviation security is sustainable in the long run.

The situation with respect to both air cargo and general aviation is more akin to the experiences in the other modes, with limited security systems in place before 9/11 and post-9/11 developments heavily shaped by legislative directives, such as the 100% air cargo screening mandate in the 9/11 Act or the absence thereof. Federal security programs for these sectors continue to be relatively small, and TSA has been moving to enhance security through the rulemaking process (including the Certified Cargo Screening Program and the Large Aircraft Security Program), but that process remains cumbersome and subject to outside pressure.

Discussion Questions

1. What are the main elements of U.S. commercial airport security?
2. Describe the purposes and operations of Secure Flight and Pre√.
3. Describe the procedures and technologies used at commercial airport security checkpoints.
4. According to TSA, what is an "optimal" checked bag screening system?
5. How has TSA moved to implement the 9/11 Act's mandate for 100% screening of cargo on passenger aircraft?

References

Ahiers, M., February, 2014. Homeland security thins air marshal ranks. CNN 26, 2014.

Air Line Pilots Association. April, 2013. Fact sheet on the Federal Flight Deck Officer Program. Washington, DC.

Association of Flight Attendants. July, 2012. Testimony of Colby Alonso before House Committee on Homeland Security Subcommittee on Transportation Security. Washington, DC.

Elias, B., April, 2009. Airport passenger screening: background issues for Congress. Congressional Research Service, Washington, DC.

European Union. n.d. Civil aviation security: Common rules. <http://europa.eu/legislation_summaries/transport/air_transport/tr0028_en.htm> (accessed 10.27.14.)

FAA. 2002. Fact sheet: Aircraft security accomplishments since Sept. 11. news release.

FAA. 2003. Press release: Airlines meet FAA's hardened cockpit door deadline. news release.

Federal Register. 2008. Transportation Security Administration: Large Aircraft Security Program, Other Aircraft Operator Security Program, and Airport Operator Security Program. Washington, DC.

GAO. September, 2005. Aviation security: Flight and cabin crew member security training strengthened, but better planning and internal controls needed. Washington, DC.

GAO. September, 2009. Aviation security: A national strategy and other actions would strengthen TSA's efforts to secure commercial airport perimeters and access controls. Washington, DC.

GAO. May, 2011a. General aviation: Security assessments at selected airports. Washington, DC.

GAO. July, 2011b. Aviation security: TSA has enhanced its explosives detection requirements for checked baggage, but additional screening actions are needed. Washington, DC.

GAO. November, 2011c. Transportation security information sharing: Stakeholders generally satisfied but TSA could improve analysis, awareness, and accountability. Washington, DC.

GAO. April, 2012a. Checked bag screening: TSA has deployed optimal systems at the majority of TSA-regulated airports, but could strengthen cost estimates. Washington, DC.

GAO. May, 2012b. Aviation security: Actions needed to address challenges and potential vulnerabilities to securing inbound air cargo. Washington, DC.

GAO. July, 2012. General aviation security: Weaknesses exist in TSA's process for ensuring foreign flight students do not pose a security threat. Washington, DC.

GAO. December, 2012d. Screening Partnership Program: TSA should issue more guidance to airports and monitor private versus federal screener performance. Washington, DC.

GAO. January, 2013a. TSA explosives detection canine program: Actions needed to analyze data and ensure canine teams are effectively utilized. Washington, DC.

GAO. May, 2013b. Homeland security. DHS and TSA continue to face challenges developing and acquiring screening technologies. Washington, DC.

GAO. November, 2013c. Aviation security: TSA should limit future funding for behavior detection activities. Washington, DC.

GAO. January, 2014. Screening Partnership Program: TSA issued application guidance and developed a mechanism to monitor private versus federal screener performance. Washington, DC.

Global Entry. n.d. About Global Entry. <http://www.globalentry.gov/about.html> (accessed 10.27.14.)

Halsey, A., 2013. TSA to pull revealing scanners from airports. Washington Post. January 18, 2013.

Halsey, A., 2014. TSA says wait times are shorter. Washington Post. September 5, 2014.

International Civil Aviation Organization. n.d. 24 November 2015 deadline. <http://www.icao.int/Security/mrtd/Pages/24-NOV-2015.aspx> (accessed 10.27.14.)

International Civil Aviation Organization. 2011. Annual report of the Council—2010. <http://www.icao.int> (accessed 10.27.14.)

International Civil Aviation Organization. 2012. Annual report of the Council—2011. <http://www.icao.int> (accessed 10.27.14.)

Johnstone, R.W., 2006. 9/11 and the Future of Aviation Security. Praeger, Westport, CT.

National Air Transportation Association. 2012. Large Aircraft Security Program. Alexandria, VA.

Office of Information and Regulatory Affairs. November, 2013. Unified agenda of regulatory and deregulatory actions: General aviation security and other aircraft operator security. <http://www.reginfo.gov/public/do/e/AgendaViewRule> (accessed 10.27.14.)

Peterman, D.R., Elias, B., Fritelli, J., January, 2013. Transportation security issues for the 113th Congress. Congressional Research Service, Washington, DC.

Price, J.C., Forrest, J.S., 2009. Practical Aviation Security. Butterworth-Heinemann/Elsevier, Burlington, MA.

Rittgers, D., September, 2011. Abolish the Air Marshals. Cato Institute.

Stellin, S., October, 2013. Security check now starts long before you fly. New York Times.

Transportation Security Administration. n.d.a. Secure Flight program. <http://www.tsa.gov/stakeholders/secure-flight-program> (accessed 10.27.14.)

Transportation Security Administration. n.d.b. Participation in TSA Pre√. <http://www.tsa.gov/tsa-precheck/participation-tsa-precheck> (accessed 10.27.14.)

Transportation Security Administration. n.d.c. Federal Air Marshals. <http://www.tsa.gov/about-tsa/federal-air-marshals> (accessed 10.27.14.)

Transportation Security Administration. n.d.d. Federal Flight Deck Officers. <http://www.tsa.gov/about-tsa/federal-flight-deck-officers> (accessed 10.27.14.)

Transportation Security Administration. n.d.e. Selection and training: Federal Flight Deck Officers. <http://www.tsa.gov/about-TSA/selection-and-training> (accessed 10.27.14.)

Transportation Security Administration. n.d.f. Crew member self defense training program. <http://www.tsa.gov/stakeholders/crew-member-self-defense-training-program> (accessed 10.27.14.)

Transportation Security Administration. n.d.g. Certified Cargo Screening Program. <http://www.tsa.gov/certified-cargo-screening-program> (accessed 10.27.14.)

Transportation Security Administration. n.d.h. Alien Flight Student Program. <http://www.tsa.gov/stakeholders/training-and-exercises-0> (accessed 10.27.14.)

Transportation Security Administration. n.d.i. Air cargo programs and initiatives. <http://www.tsa.gov/stakeholders/programs-and-initiatives-1> (accessed 10.27.14.)

Transportation Security Administration. n.d.j. General aviation security programs and initiatives. <http://www.tsa.gov/stakeholders/security-programs-and-initiatives> (accessed 10.27.14.)

Transportation Security Administration. July, 2011. Statement of John Sammon before U.S. House of Representatives Committee on Oversight and Government Reform Subcommittee on National Security, Homeland Defense, and Foreign Operations. Washington, DC.

Transportation Security Administration. April, 2013a. Transportation Security Administration, aviation security fiscal year 2014 congressional [budget] justification. Washington, DC.

Transportation Security Administration. April, 2013b. Transportation Security Administration, surface transportation security fiscal year 2014 congressional [budget] justification. Washington, DC.

Transportation Security Administration. November, 2013c. Statement of Administrator John S. Pistole before the United States House of Representatives Committee on Homeland Security, Subcommittee on Transportation Security. Washington, DC.

Transportation Security Administration. 2014a. Transportation Security Administration, aviation security fiscal year 2015 congressional [budget] justification. Washington DC, March.

Transportation Security Administration. 2014b. DHS achieves trusted traveler program milestones. August 26. <http://www.tsa.gov/press/releases/2014/08/26/dhs-achieves-trusted-traveler-program-milestones>. (accessed 10.27.14).

Transportation Security Administration. 2014c. TSA awards private screening contract at Orlando Sanford International Airport. September 19. <http://www.tsa.gov/press/releases/2014/09/19/tsa-awards-private-screening-contract-orlando>. (accessed 10.27.14).

TSA Office of Security Policy and Industry Engagement. January, 2013. Air Cargo Newsletter. Vol. II: Issue 1. Washington, DC.

U.S. Customs and Border Protection. n.d.a. NEXUS program description. <http://www.cbp.gov/travel/trusted-traveler-programs/nexus/nexus-overview> (accessed 07.11.14.)

U.S. Customs and Border Protection. n.d.b. SENTRI program description. <http://www.cbp.gov/travel/trusted-traveler-programs/sentri/sentri-overview>. (accessed 07.11.14).

U.S. Department of Homeland Security. March, 2007. Aviation transportation system security plan. Washington, DC.

U.S. Department of Homeland Security, Office of Inspector General. January, 2012a. Covert testing of access controls to secured airport areas: Unclassified summary. Washington, DC.

U.S Department of Homeland Security. June, 2012b. Privacy impact assessment for the Automated Targeting System. Washington, DC.

U.S. Department of Homeland Security, Office of Inspector General. July, 2012c. Implementation and coordination of TSA's Secure Flight program. Washington, DC.

U.S. Department of Homeland Security. April, 2013a. Budget-in-brief fiscal year 2014. Washington, DC.

U.S. Department of Homeland Security, Office of the Inspector General. May, 2013b. TSA's Screening of Passengers by Observation Techniques. Washington, DC.

U.S. Department of Homeland Security, Office of Inspector General. September, 2013c. Transportation Security Administration's deployment and use of Advanced Imaging Technology. Washington, DC.

U.S. Department of Homeland Security. 2014. Budget-in-brief fiscal year 2015. Washington DC, March.

U.S. Travel Association. 2011. A better way: Building a world class system for aviation security. Washington, DC.

9

Evaluating Transportation Security

CHAPTER OBJECTIVES:

In this chapter, you will learn about evaluations of the effectiveness of U.S. transportation security policies from several distinct perspectives, including:

- DHS performance measurements
- Independent performance assessments
- Congressional oversight
- The DHS workforce
- Public opinion

Introduction

Security analyst Brian Jenkins described the essence of the challenge presented in attempting to objectively evaluate transportation security measures:

> *The United States spends $200 billion a year on homeland security. This includes physical barriers, guards, closed-circuit TV, explosives detection, body scanners, security software and other technology and services intended to keep the nation safe from terrorists and other non-military adversaries. Does it work? And how do we measure the results? At a glance, those seem to be easy questions. The country has invested heavily in homeland security and is safer now. In terms of terrorist activity in the United States, the years since the September 11 attacks have been the most tranquil since the 1960s, when terrorism in its contemporary form first emerged as a threat. . . . If, however, we ask whether the visible security measures that have become so prevalent in the landscape prevented more terrorist attacks, what security measures are most effective against terrorists, or whether the difference can be measured in substantially reduced risk, then hard proof is much harder to come by.*
>
> *Jenkins, 2014*

In addition to performance measurement and assessment—whether by the agencies carrying out security measures or by other entities—there are other important viewpoints to consider in evaluating transportation security efforts in the United States, including Congressional oversight, the security workforce itself, and public opinion.

Department of Homeland Security Performance Measurement

Each year, the DHS issues its *Annual Performance Report* (U.S. Department of Homeland Security, 2013a), which "presents the Department's performance measures and applicable results aligned to our missions, provides the planned targets [for the two succeeding fiscal years], and includes information on the Department's Priority Goals. . . . DHS has created a robust performance framework that drives performance management and enables the implementation of performance initiatives and the reporting of results within the Department for a comprehensive set of measures that are aligned with the mission outcomes articulated in the Department's Strategic Plan" (foreword, p. 8).

The development of the performance measures and targets generally begins with an effort to improve upon the previous year's measures through feedback from senior departmental leadership, Office of Management and Budget (OMB) examiners, departmental performance analysts "working to fill gaps and improve quality," and component agency leadership and program managers "wishing to better characterize the results of their efforts." Proposed changes are first submitted to DHS leadership, with the approved proposals then sent to OMB for final review and concurrence. DHS seeks to ensure that the data used in the performance measures are "complete, accurate, and reliable" and uses a two-part approach to verify and validate the information:

- A Performance Measure Checklist for Completeness and Reliability used by component agencies to "self-evaluate key controls over . . . performance measure planning and reporting actions"
- An independent assessment of a sample of performance measures conducted under the auspices of the DHS's Office of Program Analysis and Evaluation (which evaluates the selected measurements for completeness and accuracy, makes recommendations for improvement, and performs a follow-up review of the implementation of the recommendations) (pp. 9–10).

Quarterly performance reports are provided within DHS and include assessments by the individual program managers of the likelihood of their meeting prescribed targets by the end of the fiscal year. When it appears that the targets may not be met, the managers are required to take corrective action (pp. 11–12).

During fiscal year (FY) 2012, there were a total of 84 performance measurements across all DHS missions, and for FY 2013 and 2014, 26 new measures were added, and 27 were "retired" (deemed to be "less informative for leadership"), leaving a total of 83 metrics. Within these totals, 19 addressed transportation security in FY 2012, with one added and five retired for FY 2013 and 2014, yielding a total of 15 in effect for the latter periods. All 20 of the transportation security statistics are described as follows:

1. *Percentage of DHS intelligence reports rated in customer satisfaction surveys as "very satisfactory" or "somewhat satisfactory" in enabling customers to understand the threat*

FY 2012	Target: 80%	Result: 90%
FY 2013	Target: 90%	Results: not available
FY 2014	Target: 90%	

2. *Percentage of DHS intelligence reports rated in customer satisfaction surveys as "very satisfactory" or "somewhat satisfactory" in enabling customers to anticipate emerging threats.* (retired)

FY 2012	Target: 80%	Result: 89%

3. *Flights conducted by foreign passenger airlines arriving in, departing from, or flying over the United States that are vetted by TSA against the terrorist watch list through Secure Flight*

FY 2012	Target: 100%	Result: 100%
FY 2013	Target: 100%	Result: 100%[1]
FY 2014	Target: 100%	

4. *Flights operated by U.S. air carriers required to have a full security program that are vetted by TSA against the terrorist watch list through Secure Flight*

FY 2012	Target: 100%	Result: 100%
FY 2013	Target: 100%	Results: not available
FY 2014	Target: 100%	

5. *U.S.-flagged air carriers operating from domestic airports that are in compliance with "leading security indicators," which "are derived from security laws, rules, regulations and standards" and "may be predictive of the overall security posture of an air carrier"*

FY 2012	Target: 100%	Result: 98.1%
FY 2013	Target: 100%	Results: not available
FY 2014	Target: 100%	

6. *Air carriers operating flights from foreign airports that serve as last point of departure to the United States in compliance with "leading security indicators," which "are*

[1]Where available, FY 2013 results are from U.S. Department of Homeland Security, 2013, *Agency Financial Report, Fiscal Year 2013*, pp. 13, 16.

derived from security laws, rules, regulations, and standards and are applied to both U.S.-flagged aircraft operators (operating from foreign airports to any destination) and foreign air carriers operating from foreign airports serving as last point of departure." The indicators "may be predictive of the overall security posture of an air carrier" (retired).

FY 2012	Target: 100%	Result: 94.1%

7. *Foreign airports serving as last point of departure to the United States in compliance with leading security indicators derived from "critical ICAO [International Civil Aviation Organization] aviation and airport security standards"* (retired)

FY 2012	Target: 100%	Result: 94%

(Note: DHS states: "TSA is engaged with counterpart agencies' governments whose airports are last point of departure to the United States in order to track the implementation of security improvements; however, host governments of these sovereign nations are occasionally unwilling or unable to implement the required improvements.")

8. *Foreign airports that serve as last points of departure and air carriers involved in international operations to the United States advised by TSA of necessary actions to mitigate identified vulnerabilities to ensure compliance with critical ICAO aviation and airport security standards* (new measure).

FY 2013	Target: 100%
FY 2014	Target: 100%

9. *Overall compliance of domestic commercial airports with established security practices and standards that "are key indicators and may be predictive of the overall security posture of an airport"*

FY 2012	Target: 100%	Result: 95%
FY 2013	Target: 100%	Result: 94.4%
FY 2014	Target: 100%	

10. *Average number of days for DHS to process Traveler Redress Inquiry Program (TRIP) forms after all required documents are submitted based on a sampling of 15% of closed cases for each month*

FY 2012	Target: <97	Result: 93
FY 2013	Target: <93	Results: not available
FY 2014	Target: <91	

11. *Air cargo on commercial passenger flights originating from the United States and its territories screened by physical search, x-ray systems, explosives trace detection, explosives detection systems, canine teams, or other approved detection equipment*

FY 2012	Target: 100%	Result: 100%
FY 2013	Target: 100%	Results: not available
FY 2014	Target: 100%	

12. *Inbound air cargo on international passenger flights originating from outside the United States and its territories screened by physical search (with manifest verification), x-ray systems, explosives trace detection, explosives detection systems, canine teams, or "additional methods approved by the TSA Administrator pursuant to" the 9/11 Act*

FY 2012	Target: 85%	Result: 93%
FY 2013	Target: 100%	Result: 99.5%
FY 2014	Target: 100%	

13. *Security compliance rate for high-risk maritime facilities as determined by whether Coast Guard inspections discover a "major" security problem at such facilities*

FY 2012	Target: 100%	Result: 98.7%
FY 2013	Target: 100%	Result: 99.3%
FY 2014	Target: 100%	

14. *Containerized cargo conveyances that pass through fixed radiation portal monitors at sea ports of entry*
    Targets and results are considered "for official use only" and are not released.
15. *Compliance rates for Customs-Trade Partnership Against Terrorism (C-TPAT) members with the established C-TPAT security guidelines* (retired).

FY 2012	Target: 100%	Result: 94.5%

(Note: DHS reported that "the overall compliance rate decreased after a number of companies were suspended or removed due to the implementation of strengthened C-TPAT security criteria.")

16. *Proportion of Customs and Border Protection (CBP)–requested examinations of higher risk cargo at foreign ports of origin "resolved or conducted by foreign customs officials" meeting Container Security Initiative (CSI) standards and requirements* (retired)

FY 2012	Target: 100%	Result: 98%

17. *Inbound cargo coming to the United States via air or sea (or land, beginning in FY 2013) identified by the Automated Targeting System as potentially high risk that is assessed or scanned before departure or at arrival at a U.S. port of entry*
Targets and results are considered "for official use only" and are not released.

18. *Proportion of all cargo (by value) imported to the U.S. by participants in CBP trade partnership programs, including C-TPAT and Importer Self Assessment programs*

FY 2012	Target: 45%	Result: 54.7%
FY 2013	Target: 57%	Results: not available
FY 2014	Target: 59%	

19. *Overall level of implementation of industry agreed upon Security and Emergency Management action items (including defined security responsibilities, background checks on employees and contractors, security training, exercises and drills, risk management, and public awareness activities) by the largest mass transit and passenger rail agencies (average weekday ridership > 60,000)*

FY 2012	Target: 75%	Result: 39%
FY 2013	Target: 75%	Results: not available
FY 2014	Target: 77%	

(Note: DHS indicated that, for this measure, "implementation of recommended security enhancements has been impacted by budgetary constraints primarily at state and local governments.")

20. *Cargo conveyances that pass through radiation detection systems upon entering the United States via land border and international rail ports of entry*

Targets and results are considered "for official use only" and are not released. (U.S. Department of Homeland Security, 2013a, pp. 14–16, 18, 22–23, App. A. pp. 4–11, 16–18, 24–31).

The FY 2013 DHS performance report also outlined four agency priority goals with associated performance measures. One of these, "strengthen aviation security counterterrorism capabilities by using intelligence-driven information and risk-based decisions," was in the field of transportation security. In addition to the metrics discussed above for vetting passengers against the terrorist watch list via Secure Flight (items 3 and 4) and the time required for the processing of TRIP forms (item 10), several additional measurements were included.

- *Number of daily passengers who have qualified for expedited physical screening based on assessed low risk (via TSA Pre✓ and other programs)*

FY 2012	Target: 89,250	Result: 122,684 (of an average total of 1.7 million)
FY2013	Target: 255,000	Results: not available

- *Level of passenger security screening assessment results (including for TSA Pre✓)*
  All targets and results are classified.
- *Level of baggage security screening assessment results.*
  All targets and results are classified.
- *Passengers satisfied with TSA Pre✓ security screening, calculated via checkpoint kiosk surveys.*

FY 2012	Target: 90%	Result: 93%
FY 2013	Target: 95%	Results: not available

- *Nationwide airport operational hours with security checkpoint wait times of less than 20 minutes*

FY 2012	Target: 99%	Result: 99.05%
FY 2013	Target: 99%	Results: not available

(U.S. Department of Homeland Security, 2013a, pp. 47–48).

These DHS performance indicators reveal many of the limitations in evaluating transportation security. First, very few of them actually measure effectiveness or outcomes. Most focus on tallying outputs (e.g., percentage of cargo screened or passengers vetted), inspection-based compliance rates with "leading security indicators" (which were the basis for evaluating the pre-9/11 security system), or "customer satisfaction surveys." Second, reflective of the allocation of departmental resources, most of the measures pertain to commercial aviation and air cargo, with few addressing maritime or land transportation security. Finally, when effectiveness is directly measured (e.g., for airline passenger and baggage screening), the results are classified. As Jenkins pointed out, "hard proof" of effectiveness is "hard to come by," and there are reasons for this, including difficulties in quantifying certain aspects of effectiveness (e.g., deterrence) and the understandable worry about disclosing vulnerabilities to those who pose a threat to transportation systems. However, the shortcomings in current performance measurements do impose costs on U.S. transportation security efforts. In its 2013 analysis evaluating DHS after its 10th year in operation, the Government Accountability Office (GAO) stated, "DHS continues to miss opportunities to optimize performance across its missions due to a lack of reliable performance information or assessment of existing information; evaluation among possible alternatives; and, as appropriate, adjustment of programs or operations that are not meeting mission needs" (GAO, 2013a, p. 13).

In addition to the publicly reported DHS performance measures, TSA's Office of Inspection (OOI) is responsible for evaluating the effectiveness of transportation security systems through unannounced covert testing and audits "designed to identify system vulnerabilities and provide mitigation strategies" and conducting inspections of TSA operations "to ensure all offices and airports are in full compliance with federal laws, regulations, and current policies." (The OOI is also responsible for investigating allegations of criminal and administrative misconduct of the agency's employees and contractors, and, as of FY 2011, 60% of its employees were criminal investigators.) In FY 2013, the OOI conducted more than 400 covert tests (with the results being classified), "which focused on potential vulnerabilities in existing policies, procedures, supervision and training," and "implemented risk-based initiatives through the development and implementation of tools, conducted risk-based analysis of information for program development and execution, and collaborated with internal and external stakeholders" (Transportation Security Administration, 2014, pp. 2–3; U.S. Department of Homeland Security, 2013b, p. 5).

In September 2013, the DHS OIG issued a report that was critical of the OOI's operations. Although most of the document addressed the work of the OOI's criminal investigators, some of its findings dealt with covert testing and inspections:

> *OOI did not operate efficiently. Specifically, it did not use its staff and resources efficiently to conduct cost-effective inspections, internal reviews, and covert testing.... Quality controls were not sufficient to ensure that inspections, internal reviews, and covert testing complied with accepted standards; staff members were properly trained; and work was adequately reviewed. Finally, the office could not always ensure that other TSA components took action on its recommendations to improve TSA's operations.... OOI did not establish adequate performance measures or set standards to demonstrate improvement over time. The office also did not create outcome-based performance measures, which would compare the results of its activities with the intended purpose, to assess its operations.*
>
> *U.S. Department of Homeland Security, 2013b, pp. 6, 12*

TSA concurred with most of the OIG's findings and recommendations, including for the development of outcome-based performance measures, and is working to address them (Transportation Security Administration, 2014, p. 4).

The U.S. Coast Guard maintains its own, quite different internal performance metrics, but its "summary" measures have been made available to the DHS Office of Inspector General (OIG) and included in that office's annual review of the Coast Guard's mission performance. As described in the most recent such annual review, four of the Coast Guard assessments for maritime security are:

> *risk-based outcome measures that begin with an assessment of likely high-consequence maritime terrorist attack scenarios. Threat, vulnerability, and consequence levels are estimated for each scenario, which generates a proxy (index) value of "raw risk" that exists in the maritime domain. Next, the USCG [U.S. Coast Guard] interventions (security and response operations, regime and awareness activities) for the fiscal year are scored against the scenarios with regard to the decreases in threat, vulnerability, and consequence that each has been estimated to have afforded. The resulting measures are proxy measures of performance.*

In FY 2013, the Coast Guard met five of its six targets for maritime security enhancement and missed the sixth by less than 1 percentage point.

1. *Overall reduction of all maritime security risk subject to Coast Guard influence*

FY 2013	Target: 36%	Result: 36%

2. *Reduction of maritime security risk resulting from Coast Guard consequence management*

FY 2013	Target: 4%	Result: 4%

3. *Reduction of maritime security risk resulting from Coast Guard efforts to prevent a terrorist entering the United States via maritime means*

FY 2013	Target: 34%	Result: 34%

4. *Reduction of maritime security risk resulting from Coast Guard efforts to prevent a weapon of mass destruction from entering the United States via maritime means*

FY 2013	Target: 24%	Result: 24%

5. *Annual Maritime Transportation Security Act (MTSA) facility compliance rate with Transportation Worker Identification Credential regulations*

FY 2013	Target: 99%	Result: 99.9%

6. *Security compliance rate for high risk maritime facilities*

FY 2013	Target: 100%	Result: 99.3%

(U.S. Department of Homeland Security, 2014, pp. 23–26).

The four risk-based metrics are a significant attempt to use outcome-based measures that relate directly to the effectiveness of security efforts. However, their reliability is critically dependent on the validity of the scenarios and risk estimates used in their calculation, which cannot be ascertained through publicly available information.

■ ■ Critical Thinking Question ■

Taking into account the need to assess a "measurable" trait and to not disclose information about vulnerabilities that could prove usable by potential threats to transportation systems, describe a new outcome-based performance measure that would enhance our ability to evaluate the performance of transportation security systems. Discuss why such a measure should be adopted and how it will aid transportation security oversight and performance.

Independent Assessments

The DHS, TSA, and Coast Guard performance measurement efforts are largely aimed at an internal audience within the agencies and among other transportation security stakeholders. Evaluations from outside DHS and its components play the key role in informing Congress and the general public about the effectiveness of various transportation security programs and "provide vital input to DHS as they offer insight to the performance of our programs and identify areas for improvement" (U.S. Department of Homeland Security, 2013a, App. B. p. 2).

The DHS OIG and the GAO are the primary sources of independent assessments of U.S. transportation security activities, and their analyses of various security measures have been used throughout this work. Nongovernmental research organizations also contribute to the discourse on transportation security policies and programs.

Department of Homeland Security Office of Inspector General

The Inspector General Act of 1978 established such an office in all existing federal departments and most independent federal agencies "in order to create independent and objective units":

1. *to conduct and supervise audits and investigations relating to the programs and operations of [their department or agency];*
2. *to provide leadership and coordination and recommend policies for activities designed (A) to promote economy, efficiency, and effectiveness in the administration of, and (B) to prevent and detect fraud and abuse in, such programs and operations; and*
3. *to provide a means for keeping the head of [the department or agency] and the Congress fully and currently informed about problems and deficiencies relating to the administration of such programs and operations and the necessity for and progress of corrective action.*

The law provided that these offices would be lead by an inspector general appointed by the president and confirmed by the Senate, who would report to and be directly under the general supervision of the departmental or agency head. However, the agency head was prohibited from preventing the Inspector General "from initiating, carrying out, or completing any audit or investigation, or from issuing any subpoena during the course of any audit or investigation."

The DHS OIG was created, along with DHS itself, in the Homeland Security Act of 2002. Its mission is to conduct and supervise independent audits, investigations, and inspections of DHS programs and operations; make recommendations on "ways for DHS to carry out its responsibilities in the most effective, efficient, and economical manner possible;" and "deter, detect and address fraud, abuse, mismanagement, and waste of taxpayer funds involved in Homeland Security." In FY 2013, the office made 532 recommendations to improve DHS operations; produced 177 arrests, 130 indictments, and 70 personnel actions; questioned $383.7 million in departmental costs; and achieved $28.3 million in recoveries, restitution, fines, and cost savings (U.S. Department of Homeland Security, n.d.).

In addition to producing reports on individual DHS programs and operations, the DHS OIG also provides an annual assessment summarizing the most serious management and performance challenges facing the department and an evaluation of the progress being made in addressing those challenges. The FY 2012 report identified the following as major challenges facing TSA:

- *Although [our] test results are classified, [airport] access control and checkpoint screening vulnerabilities were identified at the domestic airports tested. Although Transportation Security Officers (TSO) were ultimately responsible for not fully screening checked baggage, our audit identified additional improvements TSA can make in the evaluation of new or changed procedures, and improvements in supervision of TSOs that could have mitigated the situation.*
- *We analyzed vetting data from airport badging offices [which provide credentials for entrance into secured airport areas] and identified badge holder records with omissions or inaccuracies in security threat assessment status, birthdates, and birthplaces. These problems existed because TSA did not: (1) ensure that airport operators had quality assurance procedures for the badging application process; (2) ensure that airport operators provided training and tools to designated badge officer employees; and (3) require Transportation Security Inspectors to verify the airport data during their reviews.*
- *Through covert testing we identified vulnerabilities in [air] cargo screening procedures employed by air carriers and cargo screening facilities to detect and prevent explosives from being shipped in air cargo transported on passenger aircraft. Although TSA has taken steps to address air cargo security vulnerabilities, the agency did not have assurance that cargo screening methods always detected and prevented explosives from being shipped in air cargo transported on passenger aircraft.*
- *We determined that TSA does not have guidance for and oversight of the reporting process [for security breaches at airports]. This need for guidance resulted in the agency missing opportunities to strengthen airport security. TSA agreed with the recommendations in our report, and as a first step, is developing a standard definition of a security breach. In addition, TSA is also updating its airport performance metrics to track security breaches and airport checkpoint closures at the national, regional, and local levels.*

- *Amtrak did not mitigate critical vulnerabilities reported in risk assessments. These vulnerabilities remain because TSA: (1) did not require Amtrak to develop a corrective action plan addressing its highest ranked vulnerabilities; (2) approved Amtrak investment justifications for lower risk vulnerabilities; and (3) did not document roles and responsibilities for the [rail and transit security] grant award process.*

The Inspector General report also highlighted several significant accomplishments by TSA over the preceding year, including:

- Initial development of detailed utilization reports for Advanced Imaging Technology (AIT) passenger screening equipment to ensure that the units are being used efficiently
- Provision of additional training for TSOs, "which should help their performance"
- Deployment of the Secure Flight Program, which "has provided more consistent passenger prescreening . . . includes privacy safeguards to protect passenger personal data and sensitive watch list records . . . [and] focuses on addressing emerging threats through multiple initiatives."

More generally, the OIG indicated "TSA continues to work on improving operations, keeping us informed of the progress made in response to our work" (U.S. Department of Homeland Security, 2012, pp. 4–7).

The FY 2013 OIG performance assessment reported that the covert testing of airport access controls and passenger and baggage screening continued to identify vulnerabilities and that the office had initiated covert testing of the effectiveness of AIT units. Other problem areas noted were TSA's temporary permission for airports to issue ID badges to address a backlog in the processing of ID applications, which allowed some individuals with criminal records to receive the credential; the absence of a strategic plan for the Screening of Passengers by Observation Techniques (SPOT) program, which prevented TSA for ensuring the program is screening passengers objectively and that it is cost-effective; and inefficiencies in the TSA Office of Inspections with respect to inspections, internal reviews, and covert testing. Again, however, the report indicated TSA was taking action to comply with the OIG recommendations, specifically citing continued advances in the effective utilization of AIT units and improvements in the SPOT program's planning and training of Behavior Detection Officers (Table 9.1) (U.S. Department of Homeland Security, 2013c, pp. 14–16).

Government Accountability Office

The General Accounting Office (GAO) was established by the Budget and Accounting Act of 1921, which was aimed at improving federal financial management and required the president to submit an annual budget. The Office was made independent of the executive branch and given broad authority to investigate federal expenditures. In July 2004, its name was changed to Government Accountability Office to better reflect its current organization and mission. That mission is "to support the Congress in meeting its constitutional responsibilities and to help improve the performance and ensure the accountability of the federal government for the benefit of the American people . . . [by providing] Congress

Table 9.1 Select Recent DHS OIG Reports on Transportation Security (as of November 13, 2014)

Name	Number	Date
Vulnerabilities Exist in TSA's Checked Bag Screening Operations (unclassified version)	OIG-14-142	09/16/14
Annual Review of the United States Coast Guard's Mission Performance (FY 2013)	OIG-14-140	09/05/14
Transportation Security Administration's Deployment and Use of Advanced Imaging Technology (revised)	OIG-13-120	03/26/14
Transportation Security Administration's Office of Inspection's Efforts to Enhance Transportation Security	OIG-13-123	09/24/13
Research and Development Efforts to Secure Rail Transit Systems	OIG-13-111	08/27/13
Transportation Security Administration's Screening Partnership Program	OIG-13-99	06/20/13
Transportation Security Administration's Screening of Passengers by Observation Techniques (redacted)	OIG-13-91	05/29/13
United States Customs and Border Protection's Radiation Portal Monitors at Seaports	OIG-13-26	01/29/13
Annual Review of the United States Coast Guard's Mission Performance	OIG-12-119	09/13/12
Efficiency and Effectiveness of TSA's Visible Intermodal Prevention and Response Program Within Rail and Mass Transit Systems (redacted)	OIG-12-103	08/15/12
Implementation and Coordination of TSA's Secure Flight Program (redacted)	OIG-12-94	07/04/12
Improvements Needed to Strengthen the Customs-Trade Partnership Against Terrorism Initial Validation Process for Highway Carriers	OIG-12-86	06/01/12
Transportation Security Administration's Efforts to Identify and Track Security Breaches at Our Nation's Airports (redacted)	OIG-12-80	05/03/12
Transportation Security Administration's Use of Back Scatter Units	OIG-12-38	02/16/12
Covert Testing of Access Controls to Secured Airport Areas (unclassified summary)	OIG-12-26	01/06/12

(*Source*: DHS Office of Inspector General. <http://www.oig.dhs.gov/> (accessed 11.13.14.))

with timely information that is objective, fact-based, nonpartisan, non-ideological, fair, and balanced."

The GAO's work assignments are made by Congress through either a request by a committee or subcommittee or a mandate in enacted legislation or committee reports. The head of GAO, the Comptroller General of the United States (who is appointed to a 15-year term by the president from a slate of candidates proposed by Congress) may also authorize GAO research. GAO assessments may involve:

- Auditing agency operations to determine whether federal funds are being spent efficiently and effectively
- Investigating allegations of illegal and improper activities
- Reporting on how well government programs and policies are meeting their objectives
- Performing policy analyses and outlining options for congressional consideration
- Issuing legal decisions and opinions in certain instances, such as bid protest rulings and reports on agency rules
- Making recommendations to Congress and the heads of federal agencies about ways to make government more efficient, effective, ethical, equitable, and responsive

In FY 2013, GAO submitted testimony to Congress on 114 occasions, made 1430 recommendations to Congress and the executive branch (calculating that 75% of its recommendations over the preceding 5 years had been implemented), and reported $51.5 billion in savings as a result of its work (GAO, n.d.).

In February 2013, testimony before a House subcommittee, GAO's managing director for homeland security programs, Cathleen Berrick, summarized GAO evaluations of DHS during the department's first 10 years in operation and indicated that in that time GAO had issued more than 1300 reports and congressional testimonies concerning the department and its components and made approximately 1800 recommendations, of which more than 60% had been implemented (Table 9.2).

> *The department has implemented key homeland security operations and achieved important goals in many areas. These included developing strategic and operational plans across its range of missions; hiring, deploying, and training workforces; establishing new, or expanding existing offices and programs; and developing and issuing policies, procedures, and regulations to govern its homeland security operations.... But more work remains for DHS to address gaps and weaknesses in its current operational and implementation efforts, and to strengthen the efficiency and effectiveness of those efforts.... Forming a new department while working to implement statutorily mandated and department-initiated programs and responding to evolving threats, was, and is, a significant challenge facing DHS. Key threats, such as attempted attacks against the aviation sector, have impacted and altered DHS's approaches and investments, such as changes DHS made to its processes and technology investments for screening passengers and baggage at airports. It is understandable that these threats had to be addressed immediately as they arose. However, limited strategic and performance planning by DHS, as well as assessment to inform approaches and investment decisions, has contributed to programs not meeting strategic needs or not doing so in an efficient manner. Further, DHS has made important progress in analyzing risk across sectors, but it has more work to do in using this information to inform planning and resource-allocation decisions.*
>
> *GAO, 2013a, pp. 1, 3, 5, 12–13*

Other Independent Assessments

A number of nongovernmental organizations have also carried out evaluations of U.S. transportation security efforts. Among those that have been most active in issuing assessments in recent years are the Homeland Security Project of the Bipartisan Policy Center and the Rand Corporation.

Homeland Security Project

The Homeland Security Project's "core mission is to be an active, bipartisan voice on homeland and national security issues. . . . [It] works to foster public discourse, evaluate reform, provide expert analysis, and develop proactive policy solutions on how to best address emerging security challenges." At present, the work of the project is focused on cybersecurity and intelligence and information sharing, but its key role with respect to

Table 9.2 Select Recent GAO Reports on Transportation Security (as of November 13, 2014)

Name	Number	Date
Secure Flight: Additional Actions Needed to Determine Program Effectiveness and Strengthen Privacy Oversight Mechanisms	GAO-14-796T	09/18/14
Screening Partnership Program: TSA Has Improved Application Guidance and Monitoring of Screener Performance, and Continues to Improve Cost Comparison Methods	GAO-14-787T	07/29/14
Explosives Detection Canines: TSA Has Taken Steps to Analyze Canine Team Data and Assess the Effectiveness of Passenger Screening Canines	GAO-14-695T	06/24/14
Maritime Security: Ongoing U.S. Counterpiracy Efforts Would Benefit from Agency Assessments	GAO-14-422	06/19/14
Maritime Security: Progress and Challenges with Selected Port Security Programs	GAO-14-636T	06/04/14
Trusted Travelers: Programs Provide Benefits, but Enrollment Processes Could Be Strengthened	GAO-14-483	05/30/14
Cruise Vessels: Most Required Security and Safety Measures Have Been Implemented, but Concerns Remain About Crime Reporting	GAO-14-43	12/20/13
Maritime Security: Progress and Challenges in Key DHS Programs to Secure the Maritime Borders	GAO-14-196T	11/19/13
Aviation Security: TSA Should Limit Future Funding for Behavior Detection Activities	GAO-14-158T	11/14/13
Maritime Security: DHS Could Benefit from Tracking Progress in Implementing the Small Vessel Security Strategy	GAO-14-32	10/31/13
Supply Chain Security: DHS Could Improve Cargo Security by Periodically Assessing Risks from Foreign Ports	GAO-13-764	09/16/13
Coast Guard: Clarifying the Application of Guidance for Common Operational Picture Development Would Strengthen Program	GAO-13-321	04/25/13
Transportation Security: Action Needed to Strengthen TSA's Security Threat Assessment Process	GAO-13-629	07/19/13
Homeland Security: DHS and TSA Continue to Face Challenges Developing and Acquiring Screening Technologies	GAO-13-469T	05/08/13
Transportation Worker Identification Credential: Card Reader Pilot Results Are Unreliable; Security Benefits Need to Be Reassessed	GAO-13-198	05/08/13
Passenger Rail Security: Consistent Incident Reporting and Analysis Needed to Achieve Program Objectives	GAO-13-20	12/19/12
Air Passenger Screening: Transportation Security Administration Needs to Improve Complaint Processes	GAO-13-186T	11/29/12
Aviation Security: 9/11 Anniversary Observations on TSA's Progress and Challenges in Strengthening Aviation Security	GAO-12-1024T	09/11/12
Maritime Security: Progress and Challenges 10 Years After the Maritime Transportation Security Act	GAO-12-1009T	09/11/12
Aviation Security: Actions Needed to Address Challenges and Potential Vulnerabilities Related to Securing Inbound Air Cargo	GAO-12-632	05/10/12

(*Source*: GAO. <http://www.gao.gov/docsearch/repandtest.html> (accessed 11.13.14.))

transportation security has been in serving as the successor to the 9/11 Commission and the 9/11 Public Discourse Project in monitoring compliance with the 9/11 Commission's transportation security recommendations. Under the leadership of former 9/11 Commission Chairman Thomas Kean and Vice Chairman Lee Hamilton, in September 2011, the project (under its former name as the National Security Preparedness Group) issued its "Tenth Anniversary Report Card: The Status of the 9/11 Commission Recommendations." After noting the implementation of TSA's Secure Flight program as a "major success," the report indicated "we are still highly vulnerable to aviation security threats" and singled out explosives screening capabilities and DHS's research and development efforts as areas of particular concern (Bipartisan Policy Center, n.d.; 2011, pp. 17–18)

Rand Corporation

The Rand Corporation was originally part of the Douglas Aircraft Company but in 1948 became an independent, nonprofit organization that seeks to "help improve policy and decision making through research and analysis. . . . Rand's research is commissioned by a global clientele that includes government agencies, foundations, and private-sector firms." In the homeland security field, Rand has been particularly involved in research on terrorism and aviation security. For example, in August 2012, the organization published a paper titled "Efficient Aviation Security: Strengthening the Analytic Foundation for Making Air Transportation Security Decisions." In an effort "to perform analyses that define key tradeoffs, map out the major sources of uncertainty, and make more informed security decisions," the Rand report considers historical and projected threats to aviation, estimates of the intangible costs of security measures, the interactive impact of security layers, inclusion of the effects of deterrence in cost–benefit analyses, evaluation of the benefits of trusted traveler programs, and the use of modeling to assess the terrorist threat (Jackson et al., 2012, pp. xviii–xx).

DIGGING DEEPER

THE DEBATE OVER CHECKPOINT SCREENING

"Security Theater?"

The growth in transportation security measures after 9/11 has produced an ongoing debate about the cost effectiveness of those measures, a debate that increases in intensity in the absence of major security incidents. Given the size, expense, and visibility of passenger aviation security checkpoints, most of the contention has centered on that one layer of one transportation mode.

In March 2012, *The Economist* organized an online debate on the question, "Have changes made to airport security since 9/11 done more harm than good?" Cryptographer, security analyst, and persistent TSA critic Bruce Schneier represented the affirmative view, and former TSA Administrator Kip Hawley argued against the notion.

> Schneier: *In the entire decade or so of airport security since the attacks on America on September 11th 2001, the Transportation Security Administration (TSA) has not foiled a single terrorist plot or caught a single terrorist. Its own "Top 10 Good Catches of 2011" does not have a single terrorist on the list. The "good catches" are forbidden items carried by mostly forgetful, and entirely innocent, people—the sort of guns and knives that would have been just*

as easily caught by pre-9/11 screening procedures. Not that the TSA is expert at that. It regularly misses guns and bombs in tests and real life. Even its top "good catch"—a passenger with C4 explosives—was caught on his return flight; TSA agents missed it the first time through.

Hawley: *More than six billion consecutive safe arrivals of airline passengers since the attacks on September 11th 2001 mean that whatever the annoying and seemingly obtuse airport security measures may have been, they have been ultimately successful. However one measures the value of our resilient society careening through ten tumultuous years without the added drag of one or more industry-crushing and national psyche-devastating catastrophic 9/11-scale attacks, the sum of all that is more than its cost.... TSA was created six weeks after 9/11 with the mantra "never again" resonating throughout much of the world. Within two years, it had built an aviation security system that did, in fact, protect travelers from hijackings or suit case bombings, the major methods of attack that had been experienced to that point.*

Schneier: *Airport security is the last line of defense, and it is not a very good one. If there were only a dozen potential terrorist tactics and a hundred possible targets, then protecting against particular plots might make us safer. But there are hundreds of possible tactics and millions of possible targets. Spending billions to force the terrorists to alter their plans in one particular way does not make us safer. It is far more cost-effective to concentrate our defenses in ways that work regardless of tactic and target: intelligence, investigation and emergency response. That being said, aircraft require a special level of security for several reasons: they are a favored terrorist target; their failure characteristics mean more deaths than a comparable bomb on a bus or train; they tend to be national symbols; and they often fly to foreign countries where terrorists can operate with more impunity. But all that can be handled with pre-9/11 security. Exactly two things have made air travel safer since 9/11: reinforcing the cockpit door, and convincing passengers that they need to fight back. Everything else has been a waste of money. Add screening of checked bags and airport workers and we are done. All the rest is security theater. If we truly want to be safer, we should return airport security to pre-9/11 levels and spend the savings on intelligence, investigation and emergency response.*

Hawley: *A steady stream of al-Qaeda threats came in during 2006, 2007 and 2008.... On an average day during this period, I, as TSA Administrator, had threat discussions about half a dozen to a dozen specific, separate, serious plots with intelligence analysts to consider security operations that would counter threats targeting transport.... The original, and I believe outdated, pre-9/11 risk model that relies on regulation, compliance and enforcement is dangerously static and rigid in the face of highly adaptive enemies. An enemy like al-Qaeda incurs trivial cost by changing attack methods to get around regulation-based security, but defensive forces have to spend disproportionately large amounts of money and effort to close off increasing numbers of new types of attack. A better risk model against al-Qaeda-like attackers is to employ many changeable, flexible layers and make it simple for the defense to change measures while inflicting a dangerously high cost on would-be attackers who could never be sure what defense they were going to face.... Undercover air marshals, canine teams, unpredictable patrols, behavior detection specialists, and integrated watch list and checkpoint operations are all examples of lower-cost, more flexible security options than the old model of digging in at the checkpoint with a checklist of prohibited items. The transition from the old risk model to current risk strategy has not been smooth, nor is it complete.*

Economist, 2012

Schneier elaborated on the concept of "security theater" in a 2009 essay:

Security theater refers to security measures that make people feel more secure without doing anything to actually improve their security. An example: the photo ID checks that have sprung up in office buildings. No one has ever explained why verifying that someone has a photo ID provides any actual security, but it looks like security to have a uniformed guard-for-hire looking at ID cards.... Security is both a feeling and a reality. The propensity for security theater comes from the interplay between the public and its leaders. When people are scared, they need something done that will make them feel safe, even if it doesn't truly make them safer. Politicians naturally want to do something in response to crisis, even if that something doesn't make any sense.... Any terrorist attack is a series of events: something like planning, recruiting, funding, practicing, executing, aftermath. Our most effective defenses are the beginning and end of that process—intelligence, investigation, and emergency response..... Unfortunately for politicians, the security measures that work are largely invisible.... These security measures don't make good television, and they don't help come re-election time. But they work, addressing the reality of security instead of the feeling.

Schneier, 2009

Terrorism and transportation security analyst Brian Jenkins addressed the application of "security theater" to aviation security, adding the concept of deterrence to the debate.

Many criticize security as being "just for show." However, illusion is an important component of security. The objective is to convince would-be attackers that they will fail. We tend to focus on detection and prevention. Judging by the evidence, the most important effect of security is deterrence. There are very few instances where terrorists are caught trying to smuggle weapons or bombs on board airliners. If deterrence is working, that means fewer attempts, but it is difficult to count things that don't occur.... While quantifiable preventions of terrorist attacks are rare, we do have indirect indicators of their effects.... Airline security measures have increased over the last four decades since 100 percent passenger screening was imposed in response to the increase in hijackings during the late 1960s and early 1970s. Each decade since then has seen fewer attempts to hijack or sabotage commercial airliners, although it appears that terrorists remain obsessed with attacking aviation targets. This is not simply because the security measures chased away the less-determined non-terrorist adversaries, although that contributed to the overall decline. Even terrorist attempts declined.... Whatever we may think of aviation security, terrorists attempting to smuggle bombs aboard airliners take security seriously. They attempt to build smaller, more concealable devices with undetectable ingredients.... The security measures did not prevent [certain] attempts [after 9/11], but they persuaded the terrorists to trade reliability for concealment—an achievement nonetheless.

Jenkins, 2014

Consider the arguments made by Schneier, Hawley, and Jenkins. What evidence do they use to support their major points? What other information is available that might shed further light on the question of the cost-effectiveness of checkpoint security measures? What argument or arguments are most convincing? Why?

Congressional Oversight

In addition to its duties of passing legislation and appropriating funds, Congress also conducts oversight of executive branch operations, which involves "the review, monitoring, and supervision of federal agencies, programs, activities, and policy implementation." This function is carried out by a variety of means, including "authorization, appropriations, investigative and legislative hearings by standing committees; specialized investigations by select committees; and reviews and studies by congressional support agencies and staff." The authority for congressional oversight is not spelled out in the Constitution but derives from the "implied" powers of the Congress in the Constitution, various public laws, and the rules of the House and the Senate. Its objectives include:

- Improving the efficiency, economy, and effectiveness of governmental operations
- Evaluating programs and performance
- Protecting civil liberties and constitutional rights
- Informing the public and ensuring that executive policies reflect the public interest
- Gathering information to assist in the development of new legislation or amendments to existing statutes
- Ensuring administrative compliance with legislative intent
- Preventing executive branch encroachment on legislative branch authority and prerogatives (Kaiser, 2006, pp. 1–2)

In an interview, Lee Hamilton described the importance of congressional oversight from his perspective as a 12-term member of the U.S. House of Representatives, which included serving as chairman of both the Foreign Affairs and Intelligence Committees:

> *Congress' job is to look into every nook and cranny of the executive branch to see that the laws are being properly executed, to make suggestions [about] where improvements can be made, to understand what the policy of the executive branch is, [and] to try to be constructive and to be a critic as well if they don't like what the executive is doing. If it is properly done, if the right questions are asked, it can greatly strengthen the operation of a department.... Proper, tough, robust oversight can put the bureaucracy on its toes, can make sure that the law is being implemented, can see that there's not a lot of hanky-panky going on—corruption, and [can] make sure that the people are being well-served.*
>
> *Annenberg Foundation Trust and Aspen Institute, 2013, p. 6*

The complexity and challenge of creating the DHS out of 22 existing federal agencies was mirrored in the Congress as that body attempted to adapt its existing committee system to accommodate the new department and its components. Eventually, both houses established principal authorizing committees (House Committee on Homeland Security and Senate Committee on Homeland Security and Governmental Affairs) and appropriations subcommittees for homeland security. However, the jurisdiction of these panels is essentially confined to DHS, and thus they have little role in non-DHS homeland security programs. In addition, they do not exercise exclusive oversight over many DHS programs, sharing jurisdiction with a number of other congressional panels (e.g., with the House Committee on Transportation & Infrastructure and the Senate Committee on Commerce, Science & Transportation on many transportation security issues). The 9/11 Commission took note of this fragmentation in is 2004 final report and called for the creation of "a single, principal point of oversight and review for homeland security" in each chamber.

> *Of all our recommendations, strengthening congressional oversight may be among the most difficult and important. So long as oversight is governed by current congressional rules and regulations, we believe the American people will not get the security they want and need. . . . The leaders of the Department of Homeland Security now appear before 88 committees and subcommittees of Congress. One expert witness (not a member of the administration) told us that this is perhaps the single largest obstacle impeding the department's successful development.*
>
> *The 9/11 Commission Report, 2004, pp. 419, 421*

Little improvement had occurred as of the 10th anniversary of the 9/11 hijackings in 2011, when former 9/11 Commission leaders Kean and Hamilton participated in the development of a "report card" on the status of the Commission's recommendations:

> *When we issued our 2004 report, we believed that congressional oversight of the homeland security and intelligence functions of government was dysfunctional. It still is. . . . The homeland security committees in the House and Senate do not have sufficient jurisdiction over important agencies within the Department of Homeland Security. Instead, jurisdiction has been carved up to accommodate antiquated committee structures. As a result, too many committees have concurrent and overlapping jurisdiction. This is a recipe for confusion. This is not just a theoretical problem; it has already produced unclear security policies. The Senate Commerce Committee has jurisdiction over the TSA and has used this authority to set security standards for screening cargo shipped from abroad on airplanes. But cargo shipped on maritime vessels is governed by the security policies of U.S. Customs and Border Protection, which falls under the jurisdiction of the Senate Homeland Security Committee. . . . The security of cargo should not depend on whether it moves by air or sea and the committee that has jurisdiction over the agency that regulates that method of transit. . . . The unwieldy jurisdictional divisions result in the inefficient allocation of limited resources needed to secure our nation. . . . The result is that DHS receives conflicting guidance and Congress lacks one picture of how that enormous organization is functioning. Congress should be helping integrate the sprawling DHS; a fragmented oversight approach defeats that purpose.*
>
> *Bipartisan Policy Center, 2011, p. 16*

This fragmentation in congressional oversight is illustrated by figures on the department's interactions with Congress in 2011 and 2012. More than 100 committees and subcommittees asserted jurisdiction over the department, representing approximately three times the number of Congressional panels—36—that provided oversight of the far larger Department of Defense in the same period. More than 400 DHS representatives were called on to provide testimony at 289 formal committee, subcommittee, caucus, and commission hearings, and DHS participated in an additional 4300 briefings and other nonhearing engagements with Congress. Fewer than half of these interactions were with the homeland security authorizing committees (Figures 9.1 and 9.2) (Annenberg Foundation Trust and Aspen Institute, 2013, pp. 10–11).

In April 2013, the Annenberg Foundation Trust and the Aspen Institute convened a task force of former government officials (including Thomas Kean and Lee Hamilton) and members of Congress to address the issue of congressional oversight of DHS. The *Task Force Report on Streamlining and Consolidating Congressional Oversight of the U.S. Department of Homeland Security* (2013) identified three major ways in which faulty oversight has negatively affected national security:

1. *A drain on resources.* "Every request for a briefing or invitation to attend a hearing requires a commitment of [DHS] resources. By one estimate, no other agency

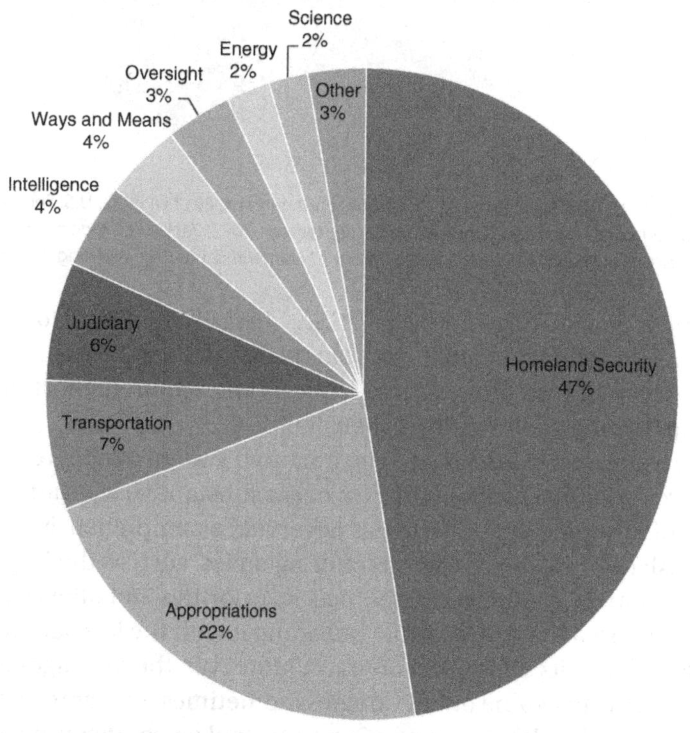

FIGURE 9.1 Department of Homeland Security congressional engagement, 2011 to 2012, U.S. House of Representatives, *(Source: Annenberg Foundation Trust and Aspen Institute. September, 2013. Task force report on streamlining and consolidating congressional oversight of the U.S. Department of Homeland Security Washington, DC. p. 11.)*

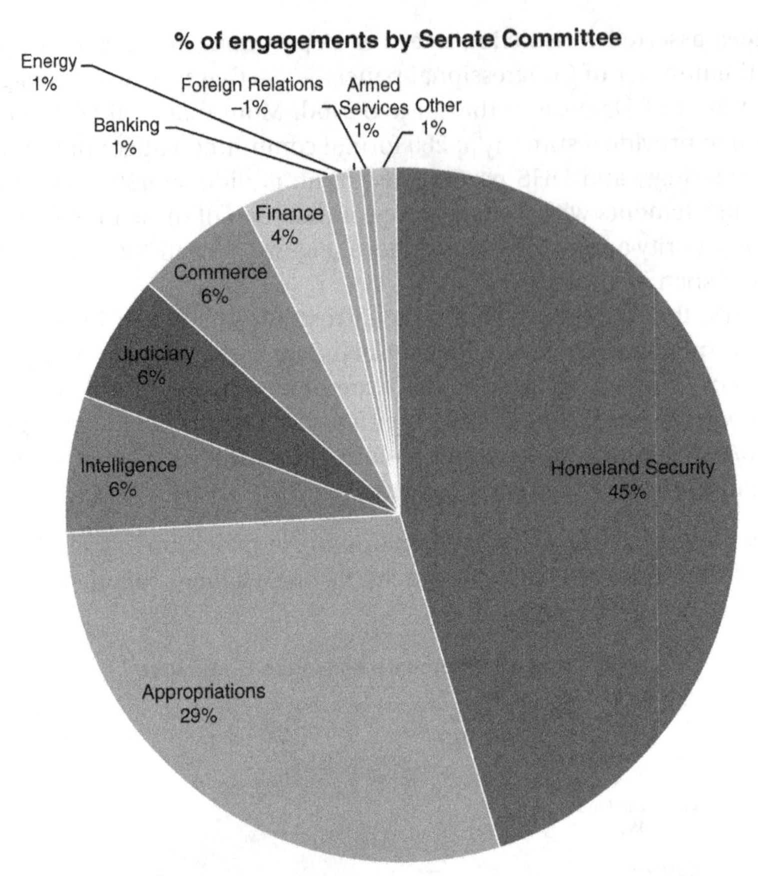

FIGURE 9.2 Department of Homeland Security congressional engagement, 2011 to 2012, U.S. Senate. *(Source: Annenberg Foundation Trust and Aspen Institute. September, 2013. Task force report on streamlining and consolidating congressional oversight of the U.S. Department of Homeland Security. Washington, DC. p. 11.)*

spends as much time on Capitol Hill as DHS. . . . In 2010 . . . then-Homeland Security Secretary Janet Napolitano [stated] that, 'Our principals and their staff [are] spending more time responding to Congressional requests and requirements than executing their mandated homeland security responsibilities'."

2. *Diminished congressional influence.* "The fractured system of congressional oversight makes it difficult for Congress to enact substantive legislation guiding DHS. Emblematic of this difficulty . . . DHS has never had a comprehensive authorization bill. Such legislation, routine for comparable agencies, such as the Department of Defense, is the forum in which Congress sets its priorities and offers comprehensive policy direction to a department, while providing it with the legislation necessary to effectively perform its daily operations. . . . Moreover, the messages regarding homeland security that come out of Congress sometimes appear to conflict or are drowned out altogether. With so many Congressional voices dictating to DHS, there is little cost to the department in ignoring the messages it dislikes or the policies it

wishes not to implement. . . . Among the problematic results is a reduced rather than enhanced Congressional role in protecting the homeland."

3. *Delayed response to pressing concerns.* "In a fragmented structure, no one committee is tasked with—and as a result accountable for—seeing the big picture. At the same time, getting legislation passed is complicated by competing demands from multiple committees and by a process that is filled with opportunities for intervention by those whose interests are not served by passage of the bill. . . . Task force members identified vulnerabilities that highlight the need to consolidate oversight as soon as possible: unregulated small aircraft and boats, cybersecurity, and biological threats" (pp. 9–15).

The Task Force pointed to resistance from the leadership of committees that would have to surrender some of their current jurisdiction and the lack of media and public attention to homeland security issues (other than after a significant incident) as key obstacles to reform of homeland security oversight. It called upon the news media to provide more information about homeland security concerns in order to build public support for reform and convince the congressional leadership that such reform is in the national (and their own) interest. The Task Force further recommended:

- *Congress should significantly reduce the number of committees with jurisdiction over homeland security and consolidate primary oversight of the key DHS component agencies under one committee in the House and one in the Senate, with coordinated jurisdiction.*
- *The oversight structure for DHS should resemble the one governing other departments, such as the Departments of Defense and Justice.*
- *Committees claiming common jurisdiction should have some overlapping membership to encourage the sharing of information and curtail redundant [information] requests [to DHS].*
- *[Congress needs to pass] an authorization bill for DHS, giving the department clear direction from Congress.*
- *Congress should limit the time for action of sequential referrals [of homeland security bills coming under the jurisdiction of multiple committees] to another committee, ensuring that if committees fail to act on what has been sent to them within a set period of time their jurisdiction would lapse, with the matter returning to the primary [homeland security] committee* (pp. 3–4, 18–21).

■ ■ Critical Thinking ■

Name some specific ways in which homeland security policy would be improved if congressional oversight of DHS were to be consolidated into a single committee in each house. Are there any disadvantages? In your opinion, how likely is it that such a reform will occur within the next 5 years? Why?

Workforce Morale

Since its beginnings, DHS has faced challenges with its workforce, including low job satisfaction, limited job engagement, and low regard for departmental leadership. Such difficulties were to be expected given the circumstances of DHS's creation as an amalgamation of multiple existing federal departments and agencies, each with its own history and work culture. In addition, even before DHS began operations, its TSA component was under legislative mandate to hire, train, and deploy more than 55,000 airport checkpoint and checked bag screeners by the end of 2002, producing significant personnel issues in what remains the largest single employment group within DHS. However, that the problems have largely persisted throughout the first decade of DHS's existence has raised serious concerns.

There are two primary measurements of federal employee morale: the Federal Employee Viewpoint Survey (FEVS) conducted by the Office of Personnel Management (OPM) and the Partnership for Public Service's "Best Places to Work in the Federal Government":

- The FEVS is a survey of federal workers that was conducted for the first time in 2002, with subsequent surveys in 2004, 2006, 2008, 2010, 2011, 2012, 2013, and 2014. It "measures employees' perceptions of whether, and to what extent, conditions that categorize successful organizations are present in their agencies" and "provides general indicators of how well the federal government is managing its human resources management systems." Employees are asked a series of questions relating to their work experience, work unit, agency, supervisor, leadership, and job satisfaction, and these responses are used in compiling index measures on leadership and knowledge management ("the extent to which employees hold their leadership in high regard, both overall and on specific facets of leadership"), results-oriented performance culture ("the extent to which employees believe their organizational culture promotes improvement in processes, products and services, and organizational outcomes"), talent management index ("the extent to which employees think the organization has the talent necessary to achieve organizational goals"), job satisfaction index ("the extent to which employees are satisfied with their jobs and various aspects thereof"), employee engagement index (the extent to which employees are immersed in the content of the job and energized to spend extra effort in job performance as indicated by leadership, supervision and intrinsic work experience), and global satisfaction index ("a more comprehensive indicator of employees' overall work satisfaction"). The index scores are calculated by averaging the percentage of positive responses on the questions comprising each index.
- The "Best Places to Work in the Federal Government" is compiled by the Partnership for Public Service, a nonprofit, nonpartisan organization that "works to revitalize the federal government by inspiring a new generation to serve and by transforming the way government works." The "Best Places" reports—issued in 2003, 2005, 2007, 2009, 2010, 2011, 2012, and 2013—are developed using data from the FEVS, which is arranged into a number of workplace category indices (including effective leadership,

employee skills-mission match, pay, and teamwork, among others) that allow for the ranking of federal departments and agencies on each measure. The aim is to enable readers of the report "to conduct side-by-side comparisons of how agencies or their subcomponents rank in various categories, examine how they compare to other agencies and see whether they have improved or regressed over time" (GAO, 2012, pp. 5–6; U.S. Office of Personnel Management, 2013, pp. 5, 28; Partnership for Public Service, 2013).

Since its first measurement of DHS employee responses in 2006, the FEVS has consistently found that workforce morale at the department is below that of the government as a whole. Although incremental improvements in the Job Satisfaction Index occurred at DHS and government-wide and the gap between the department and the overall federal government scores narrowed in 2008 and 2010, both have declined since then, and the gap has widened. A similar trend is present in results for the Employee Leadership and Knowledge Management Index (Figure 9.3) (GAO, 2013b, pp. 8–12; U.S. Office of Personnel Management, 2014, pp. 96, 108).

Another constant in the FEVS findings has been the considerable variation in morale among the various DHS components. In the transportation security realm, the Coast Guard has regularly exceeded government-wide averages, but CBP and especially TSA have fallen below the norm. Within TSA, a GAO analysis of 2011 FEVS data indicated that screeners and Air Marshals had significantly lower job satisfaction and employee engagement index scores than other elements within the agency. Given the predominance of screeners

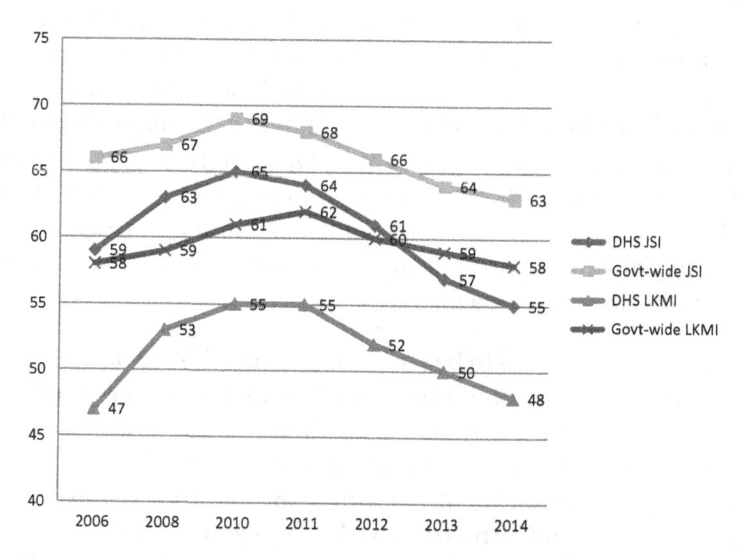

FIGURE 9.3 DHS Employee Job Satisfaction Index (JSI) and Leadership and Knowledge Management Index (LKMI) scores compared with government-wide averages. *(Sources: GAO. December, 2013b. Department of Homeland Security: DHS's efforts to improve employee morale and fill senior leadership vacancies. GAO, Washington, DC, pp. 9, 12; U.S. Office of Personnel Management. 2014. 2014 federal employee viewpoint survey. Washington, DC. pp. 96, 108.)*

Table 9.3 Department of Homeland Security Component Job Satisfaction Index (JSI) and Employee Engagement Index (EEI) Scores, 2013

Component	JSI Score	EEI Score
Government-wide	64	64
DHS-wide	57	56
U.S. Coast Guard	66	70
Customs and Border Protection	58	54
Transportation Security Administration	54	54

(*Source*: GAO. December, 2013b. Department of Homeland Security: DHS's efforts to improve employee morale and fill senior leadership vacancies. GAO, Washington, DC, p. 10.)

within the TSA workforce, this factor has undoubtedly contributed substantially to TSA's relatively poor results.

- Job Satisfaction Index: Federal Security Director staff—67.8; headquarters staff—63.5; Federal Air Marshals—58.4; screeners—53.6.
- Employee Engagement Index: Federal Security Director staff—64.8; headquarters staff—61.5; Federal Air Marshals—52.9; screeners—50.9 (GAO, 2013b, pp. 10, 12) (Table 9.3)

In releasing its 2013 report on "Best Places to Work in the Federal Government," the Partnership for Public Service commented on overall trends within the federal workforce:

> *The 2013 Best Places to Work data present a disturbing picture of federal employees throughout the government who are increasingly dissatisfied with their jobs and workplaces. Government-wide, the federal employee job satisfaction and commitment level dropped for the third year in a row, tumbling 3 points to a score of 57.8 on a scale of 100. This represents the lowest overall Best Places to Work score since the rankings were first launched in 2003, and follows a 3.2-point drop in 2012 and a 1-point decline in 2011. In contrast, private-sector employee satisfaction improved by 0.7 points in 2013 to a score of 70.7, according to the Hay group.... The lower government-wide satisfaction score and the decreases in all 10 workplace issues came during a difficult time for federal employees, who faced a three-year pay freeze, furloughs, hiring slowdowns and across-the-board budget reductions.*

The results for DHS continued to be particularly bad, with the department ranking last in overall score among the 19 largest federal departments and agencies for the second straight year and at or near the bottom on all 10 of the individual workplace issues. In keeping with the FEVS results (not surprising, given that "Best Places to Work" uses FEVS data), the Coast Guard consistently outperformed CBP and TSA on these employee morale measures (Tables 9.4 and 9.5) (Partnership for Public Service, 2013).

In response to a Congressional request, in 2012, the GAO assessed how DHS employee morale compared with that of other federal agencies and the extent to which the department and selected components "have determined the root causes of employee morale,

Table 9.4 Best Places to Work Index Scores (out of 100)

Overall Index Scores	**2013 rank***	**2013**	**2012**	**2011**	**2010**	**2009**	**2007**	**2005**
Private sector	N.A.	70.7	70.0	70.0	70.6	71.2	—	—
Federal government	N.A.	57.8	60.8	64.0	65.0	63.3	61.8	62.1
Department of Homeland Security	19	46.8	52.9	56.6	58.6	56.2	49.8	49.1
DHS Scores in Individual Workplace Categories								
Effective leadership: empowerment	19	33.5	37.0	39.5	42.4	40.4	34.1	31.9
Effective leadership: fairness	19	42.6	44.0	46.2	45.5	42.1	36.4	36.5
Effective leadership: senior leaders	19	35.2	39.3	41.4	43.1	42.7	35.9	34.6
Effective leadership: supervisors	19	54.7	57.4	58.7	59.2	56.2	51.8	51.6
Employee skills-mission match	19	68.8	71.2	73.9	74.1	74.8	71.0	70.9
Pay	18	46.4	53.9	56.8	60.2	56.8	54.5	-----
Strategic management	19	45.0	47.9	50.4	51.2	50.8	47.8	46.3
Teamwork	19	56.8	58.1	59.1	60.2	67.0	63.4	65.5
Training and development	19	48.6	51.7	54.0	55.8	56.2	51.0	50.9
Work–life balance	18	53.3	55.6	56.3	57.7	57.8	51.8	53.1

*Rank out of the 19 largest federal agencies.
(*Source*: Partnership for Public Service. 2013. Best Places to Work 2013. <http://bestplacestowork.org/BPTW/> (accessed 10.26.14.))

and developed action plans to improve morale." In addition to reviewing the FEVS and Best Places to Work data, the GAO conducted interviews with DHS personnel, which helped to identify particular issues that may have contributed to morale problems within the department or its agencies. It concluded:

> *Given the critical nature of DHS's mission to protect the security and economy of our nation, it is important that DHS employees are satisfied with their jobs so that DHS*

Table 9.5 Best Places to Work Index Scores for the Coast Guard, Customs and Border Protection (CBP), and Transportation Security Administration (TSA), 2013 (out of 100)

	Coast Guard		**CBP**		**TSA**	
Overall Index Score	***Rank****	***Score***	***Rank****	***Score***	***Rank****	***Score***
Overall agency	50	68.6	277	45.1	281	43.4
Agency Scores in Individual Workplace Categories						
Effective leadership: empowerment	79	51.8	289	31.0	290	30.9
Effective leadership: fairness	70	59.0	282	43.2	287	38.9
Effective leadership: senior leaders	63	55.2	290	31.6	283	33.6
Effective leadership: supervisors	122	66.4	293	51.4	288	52.9
Employee skills-mission match	83	77.8	286	66.9	270	69.3
Pay	200	51.4	140	55.5	299	29.0
Strategic management	102	55.8	284	42.2	273	44.3
Teamwork	107	68.3	288	55.4	291	54.2
Training and development	147	55.9	276	44.7	200	52.4
Work–life balance	87	63.3	282	50.5	267	52.7

*Rank out of 300 agency subcomponents.
(*Source*: Partnership for Public Service. 2013. Best Places to Work 2013. <http://bestplacestowork.org/BPTW/> (accessed 10.26.14.))

can retain and attract the talent required to complete its work. Employee survey data indicate that when compared to other federal employees, many DHS employees report being dissatisfied and not engaged with their jobs. It is imperative that DHS understand what is driving employee morale problems and address those problems through targeted actions that address employees' underlying concerns. DHS has made efforts to understand morale issues across the department, but those efforts could be improved. Specifically, given the annual employee survey data available through the FEVS, DHS and its components could improve their efforts to determine root causes of morale problems by comparing demographic groups, benchmarking against similar organizations, and linking root cause findings to action plans.... In addition, DHS has established performance measures for its action plans to improve morale, but incorporating attributes such as improved clarity and measurable targets could better position DHS to determine whether its action plans are effective.

GAO, 2012, pp. 3–4, 34–35

GAO did a follow-up evaluation in December 2013 on DHS's response to its 2012 recommendations. Citing continued poor results for DHS in the latest FEVS and Best Places to Work data, it reported that although the department had concurred with those recommendations and had taken steps to address them, full implementation had not yet been achieved.

- DHS created a checklist for its components to use when developing action plans to address morale problems identified in the FEVS, and some agencies (including CBP and TSA) completed a demographic analysis of the FEVS data. However, the agencies were having difficulty in identifying comparable organizations that could serve as suitable benchmarks for evaluating their own results and had not yet completed the analysis of root causes of morale problems.
- DHS officials directed the human resources offices within its component agencies to reevaluate their action plans "to ensure that metrics of success were clear and measurable." However, GAO's review of the 2013 action plans developed by CBP, TSA, the Coast Guard, and U.S. Immigration and Customs Enforcement (ICE) found that of the 53 measures assessed, 16 were not clear, and 35 "lacked measurable targets."
- An analysis of 2012 FEVS results by senior DHS officials "indicated low morale issues may persist because of employee concerns about senior leadership and supervisors, among other things, such as whether their talents are being well-used.... On the basis of the results of this analysis ... the department plans to launch additional employee surveys to probe perspectives on departmental leadership" (GAO, 2013b, pp. 7–13).

COLLECTIVE BARGAINING RIGHTS FOR AIRPORT SCREENERS

Impact on Security Debated

Whether or not to allow TSA's airport screeners (now called Transportation Security Officers [TSOs]) the right to join a union and to engage in collective bargaining with the agency has been a highly contentious matter since the creation of TSA, with the issue serving as the main obstacle to reaching final agreements between congressional Democrats (who generally favored such rights) and President Bush and congressional Republicans (who opposed them) on both the Aviation and Transportation Security Act of 2001 (ATSA), which created TSA, and the Implementing Recommendations of the 9/11 Commission Act of 2007. In its enacted form, ATSA gave the TSA Administrator sole authority in establishing the terms and conditions of employment for the TSA aviation security workforce. Throughout the Bush Administration (2001–2008), collective bargaining for these workers was prohibited, although membership in unions was not. As a result, approximately 13,000 TSOs joined several unions, which provided personal representation but could not engage in collective bargaining for them.

After the 2008 election of Barack Obama (who endorsed collective bargaining rights for TSOs as a candidate) as president, the issue of TSO representation was reviewed by his Administration and TSA. In February 2011, TSA Administrator Pistole issued a decision that authorized an election among TSOs to determine whether a majority wished to have exclusive union representation, and—in the event a majority voted in the affirmative—set forth specific terms "for limited, clearly-defined collective bargaining within a framework consistent with TSA's security mission." The Pistole decision authorized collective bargaining at the national level only and limited it to nonsecurity employment issues, including shift bids, transfers, and awards. Bargaining on security-related topics (including security policies, procedures, or personnel or equipment deployments; pay or other forms of compensation; proficiency testing; job qualifications; and discipline standards) was prohibited, as were strikes or any form of work slowdown (Transportation Security Administration, 2011).

During the summer of 2011, the American Federation of Government Employees (AFGE) won the election among TSOs to represent them and proceeded to negotiate the first collective bargaining agreement with TSA, which was completed in August 2012 and ratified by a vote of 17,326 to 1774 in November of that year. The 3-year agreement created a new performance management system, an increased clothing allowance, and a standardized bidding process for work shifts and vacation time, among other provisions (Davidson, 2012; Sciarrino, 2012).

Many congressional Republicans and others continued to object to collective bargaining for TSOs. For example, Chairman Mike Rogers (R-AL) of the House Homeland Security transportation subcommittee stated he was "concerned that TSA's collective bargaining agreement may impact security operations and further insulate its bloated workforce and bureaucracy from transforming into a smarter and leaner organization." A blogger for the Heritage Foundation based his opposition on worries about the future direction the collective bargaining agreements could take.:

> *First, TSA employees perform a vital function. What happens if they strike? The Obama Administration has prohibited screener strikes, but government unions often strike illegally. Consider the illegal Detroit teachers [2006] or New York City transit [2005] strikes.... Second, though collective bargaining at the TSA is currently limited, it could be expanded to the detriment of passenger safety.... Currently TSA has the flexibility to assign agents where they are most needed. If future negotiations extend to staffing and scheduling decisions—as they do at Customs and Border Protection—that could change, making it hard for the TSA to rapidly adapt to new threats. Third, collective bargaining contracts usually sharply limit performance pay. Right now, the TSA provides merit pay to its top performers.*

> *Historically, unions have opposed merit pay.... Limiting performance bonuses would make it much harder to reward and motivate diligent employees.*
>
> *Davidson, 2012; Sherk, 2012*
>
> On the other hand, Rep. Nita Lowey (D-NY) commented, "By allowing TSOs to bargain collectively, TSA will engage employees, improve morale and increase our national security. This is critical to keeping experienced screeners on the job and protecting the safety of the travelling public." Rep. Bennie Thompson (D-MS), ranking Democrat on the House Homeland Security Committee, stated "As proven by other federal security officers [who have such rights], collective bargaining does not diminish our security—it can actually enhance workforce productivity and TSA's mission." AFGE National President J. David Cox expressed his view that the "union contract will improve their [TSOs'] working lives and bring stability to the workforce. This agreement will mean better working conditions, fair evaluation practices and safer workplaces, and in doing so, it will improve morale. This is important because low morale leads to unsafe levels of attrition in an agency where a stable, professional workforce of career employees is vital to its national security mission" (Davidson, 2012; American Federation of Government Employees, 2012).

■ ■ Critical Thinking ■

Based on the description of checkpoint and checked bag screening operations in the previous chapter and the results from the Best Places to Work index scores on TSA worker responses for individual workplace categories, what do you think are the major sources of lower morale among airport screeners? What can or should be done to address these issues?

Public Opinion

Kraft and Furlong (2007) define public opinion as "what the public thinks about a particular issue or set of issues at any point in time" and elaborate on its influence on public policy:

> *As one would expect in a democracy, public opinion is a major force in policymaking, even if it constitutes an indirect or passive form of action on the public's part. Public opinion influences what elected officials try to do, especially on issues that are highly salient, or of great importance to voters, or on those that elicit strong opinions.... Although public opinion is rarely the determinative influence on policymaking, it sets boundaries for public policy actions. Policymakers cross those boundaries at their own risk. The broad direction of public policies therefore tends to reflect the concerns, fears, and preferences of the U.S. public.*
>
> *Kraft and Furlong, 2007, pp. 51–52*

Although the data are somewhat limited, the available evidence indicates that the American public has been generally supportive of the government's transportation security efforts and, up to a point, tolerant of specific security measures despite inconvenience or loss of civil liberties.

Risk Assessment

With the passage of time since 9/11, public perceptions about the terrorist threat in general and the threat to passenger aviation in particular have generally subsided (although there have been upticks in concern, such as in the immediate aftermath of the December 2009 attempted suicide bombing of Northwest Flight 253).

- CBS News Polls

How likely is it that the United States will face a terror attack in the next few months?

	10/2001	3/2005	8/2006	1/2010	8/2011
Very likely	53%	24%	16%	26%	9%
Somewhat likely	35%	47%	43%	40%	33%
Not very/at all likely	10%	27%	39%	30%	55%

(CBS News, 2011).

- Washington Post/ABC News Polls

Are you personally worried about traveling by commercial airplane because of the risk of terrorism, or do you think the risk is not that great?

	9/13/01	9/8/02	9/7/03	9/7/06	11/21/10
Worried	59%	32%	36%	39%	30%
Not worried	40%	67%	63%	60%	66%

(Washington Post, 2010).

Department of Homeland Security and Transportation Security Administration

In contrast to morale problems within the transportation security workforce, public opinion has been favorably disposed toward both DHS and TSA, especially in recent years. An October 2013 survey by the Pew Research Center found that 66% of Americans held a positive view of DHS, which ranked it behind the Centers for Disease Control and Prevention (75%), NASA (73%), the Defense Department (72%), and the Veterans Administration (68%) but ahead of the Food and Drug Administration (65%), the Environmental Protection Agency (62%), the Department of Health and Human Services (61%), the Justice Department (61%), the Federal Reserve (57%), the National Security Agency (54%), the Department of Education (53%), the Internal Revenue Service (44%), and the U.S. Congress (23%). Furthermore, this result represented a significant improvement over the 43% positive mark for DHS recorded in a similar 2010 survey (Pew Research Center, 2013).

Respondents were asked to rate TSA in a July 2012 Gallup poll, and a 54% majority provided a positive job assessment of the agency (including 13% excellent and 41% good) versus 42% who had a negative view (30% only fair, 12% poor), with 4% not expressing an opinion. When asked to evaluate the effectiveness of "TSA's screening procedures . . . in preventing acts of terrorism on U.S. airplanes," the responses were as follows.

Extremely effective	9%
Very effective	32%
Somewhat effective	44%
Not too effective	8%
Not effective at all	5%
Don't know	2%

(Gallup, 2012).

Transportation Security Measures

Most of the polling about specific transportation security policies has concerned commercial aviation. In an August 2011 CBS News survey, 70% expressed the opinion that "the security measures put into place since 9/11 at airports have made the public safer" (26% "a lot safer," 44% "somewhat safer"), and just 23% indicated such measures had not made the public safer. In the same survey, 59% thought that "airport security personnel who check passengers in the airport are . . . generally doing the right thing" versus 23% who believed the security personnel "are going too far" and 12% who thought they "are not doing enough to check passengers" (CBS News, 2011).

One particular commercial aviation security measure that attracted significant media and public attention was the accelerated deployment of advanced imaging technology equipment (including backscatter x-ray and millimeter wave machines) at the nation's airports in the months after the December 2009 attempted bombing of Flight 253 and the more intrusive physical searches (pat-downs) put into place for those who refused to pass through the AIT equipment. The issue came to a head in late November 2010 after a series of media reports highlighting the privacy impact of these measures. A series of surveys in November and December 2010 found strong public support for the AIT equipment but opinion was divided on use of the enhanced pat-downs.

- November 21, 2010 Washington Post/ABC News poll

  The Transportation Security Administration is increasing the use of so-called "full-body" digital x-ray machines to screen passengers in airport security lines. Supporters say these machines improve the ability to spot hidden weapons and explosives, and reduce the need for physical searches. Opponents say these machines invade privacy by producing x-ray images of passengers' naked body that security officials can see, and don't provide enough added security to justify this. . . . . Do you support or oppose using these scanners in airport security lines?

Strongly support	37%
Somewhat support	27%
Somewhat oppose	14%
Strongly oppose	18%
No opinion	4%

The TSA says it will hand-search people who don't want to be screened electronically, as well as those whose electronic screening raises a question. A TSA screener of the same sex as the passenger checks for hidden objects by placing his or her palms and fingers on the passenger's body, including sensitive areas such as the groin and breast. This replaces earlier hand-screening in which sensitive areas were touched only with the back of the hand. Do you think these new hand-pat measures are justified to try to prevent terrorism, or do you think they go too far in invading personal privacy?

Strongly justified	29%
Somewhat justified	19%
Goes somewhat too far	14%
Goes strongly too far	37%
No opinion	2%

(Washington Post, 2010).

- November 29, 2010 CBS News poll

Some airports are now using "full-body" digital x-ray machines to electronically screen passengers in airport security lines. Do you think these new x-ray machines should or should not be used?

Should	76%
Should not	19%
Don't know	5%

If these new measures ("full-body" digital x-ray screening) are put in place in most airports, how effective do you think these security measures will be in stopping future terrorist attacks on airplanes?

Very effective	36%
Somewhat effective	47%
Not too effective	11%
Not effective at all	4%
Don't know	2%

Under new procedures by TSA, if any passenger refuses to undergo a full-body digital screening, a TSA employee of the same gender will search that passenger by hand, a procedure often referred to as a pat-down. The pat-down could now include the TSA employee touching some sensitive areas of the body. Do you think these pat-downs for people who refuse to undergo full-body screening are too intrusive or not?

Too intrusive	40%
Not too intrusive	57%
Don't know	3%

(CBS News, 2010).

Limited polling on the effectiveness of security measures other than in commercial aviation has indicated less public confidence in those efforts. A survey conducted for the 10th anniversary of 9/11 found that just 45% believed that post-9/11 security measures for bridges and tunnels had "made the public safer" (13% "a lot safer," 32% "somewhat safer"), 26% thought that they had not improved safety, and 29% were unsure. The same survey asked New York City residents whether security efforts in the New York City subway system were sufficient; 57% answered no (CBS News, 2011).

Civil Liberties

The libertarian Reason Foundation sponsored a poll in the fall of 2011 that addressed several issues concerning the balance between security and civil liberties. The results reveal the complexity of this question in the minds of the general public, yielding some seemingly contradictory attitudes.

Thinking about the increased security measures that have been introduced by federal, state, and local governments since 9/11, I am going to read you several statements. For each statement, please tell me if you agree or disagree with that particular statement.

- *We are safer now*

Agree strongly	25%
Agree somewhat	36%
Disagree somewhat	18%
Disagree strongly	18%
Don't know	4%

- *We have less personal freedom now.*

Agree strongly	40%
Agree somewhat	22%
Disagree somewhat	18%
Disagree strongly	18%
Don't know	3%

- *We have less privacy now.*

Agree strongly	55%
Agree somewhat	24%
Disagree somewhat	11%
Disagree strongly	8%
Don't know	2%

- *Today's security measures may be inconvenient, but they are generally worth it.*

Agree strongly	49%
Agree somewhat	32%
Disagree somewhat	8%
Disagree strongly	8%
Don't know	3%

- *We have given up too much freedom and privacy in the name of security.*

Agree strongly	35%
Agree somewhat	20%
Disagree somewhat	23%
Disagree strongly	20%
Don't know	2%

(Reason, 2011).

In the 2011 CBS News survey, 52% believed the U.S. government has struck the right balance between fighting terrorism and maintaining civil liberties, 25% thought it has gone too far in restricting people's liberties, 17% thought it has not gone far enough in fighting terrorism, and 6% were unsure (CBS News, 2011).

Conclusion

It is difficult to properly evaluate the effectiveness of transportation security programs for a number of reasons. The main threat these programs are designed to counter—terrorism—is a rare event, yielding limited information for establishing benchmarks for success (or failure). In addition, there are significant difficulties in quantifying certain aspects of effectiveness (e.g., deterrence), and most of the actual measurements of outcomes are classified because of the understandable desire to avoid disclosure of specific vulnerabilities to the "bad guys."

DHS and its components have coped with these challenges in varying ways. DHS's annual performance measures mostly focus on assessing security outputs (e.g., the percentage of cargo screened) or inspection-based compliance with security standards. The few that do measure effectiveness (e.g., for airline passenger screening) are classified. The Coast Guard has developed risk-based outcome measures, but these are heavily dependent on the reliability of the scenarios and risk estimates used in constructing them.

The DHS OIG and GAO provide the major source of independent assessments of the U.S. government's homeland security activities (including those involved in transportation security). Both have cited ongoing improvements but noted continuing shortcomings in DHS's evaluative efforts, with GAO's 2013 analysis commenting, "DHS continues to miss opportunities to optimize performance across its missions due to a lack of reliable performance information or assessment of existing information; evaluation among possible alternatives; and, as appropriate, adjustment of programs or operations that are not meeting mission needs."

Congressional oversight of homeland security has fallen short of its potential to offer coherent evaluation and direction because of the severe fragmentation of committee responsibilities. Instead, the existing structure has produced an unnecessary drain on resources at DHS and in Congress (because of duplicative reporting requirements), diminished influence of the homeland security committees (evident in the lack of *any* regular authorization bill for DHS or its components), and a lack of responsiveness to emerging security concerns (because of committee jurisdictional overlaps).

Another problem that has plagued DHS since its inception is the lower morale of its workforce, whether measured in absolute terms (on various index scores) or compared with the private sector and other federal agencies. The same holds true for two of DHS's three major divisions that carry out transportation security missions: CBP and TSA (with the Coast Guard the notable exception in registering above-average morale). On the other hand, public opinion has tended to offer positive assessments of the effectiveness of DHS and TSA while expressing some reservations about the impact of security measures on personal liberties.

Discussion Questions

1. Compare and contrast the performance measures used by DHS (in its *Annual Performance Report*), the TSA Office of Investigations, and the U.S. Coast Guard. What are the strengths and weaknesses of each?
2. What are the DHS OIG and GAO, and what role(s) do they play in transportation security policy?
3. What are the consequences of fragmented congressional oversight of homeland security? Cite some specifics involving transportation security.
4. What are the major measures of federal workforce morale? How do DHS and its components fare in these measurements?
5. Describe briefly U.S. public opinion on the threat to aviation security, the performance of DHS and TSA, and the balance between security and civil liberties.

References

American Federation of Government Employees. November, 2012. TSA workers ratify first union contract. <https://www.afge.org/?PressReleaseID(1399> (accessed 10.26.14.)

Annenberg Foundation Trust and Aspen Institute. September, 2013. Task force report on streamlining and consolidating congressional oversight of the U.S. Department of Homeland Security. Washington, DC.

Bipartisan Policy Center. n.d. About the [Homeland Security] Project. <http://bipartisanpolicy.org/projects/homeland-security-project/about> (accessed 10.26.14.)

Bipartisan Policy Center. September, 2011. Tenth anniversary report card: the status of the 9/11 Commission recommendations. Washington, DC.

CBS News. December, 2010. Poll: most support new TSA measures. <http://www.cbsnews.com/8301-503544_162-20024625-503544.hml?tag=mncol;1st;1> (accessed 10.26.14.)

CBS News. September, 2011. Most say U.S. will always face terrorism threat. <http://www.cbsnews.com/news/most-say-us-will-always-face-terrorism-threat/> (accessed 10.26.14.)

Davidson, J., 2012. TSA, union land first labor pact for security officers. Washington Post, B4. August 3, 2012.

Economist. March, 2012. Airport security: Have changes made to airport security since 9/11 done more harm than good? <http://www.economist.com/debate/days/view/820/print> (accessed 10.26.14.)

Gallup. August, 2012. Americans' views of TSA more positive than negative. <http://www.gallup.com/poll/156491/americans-views-tsa-positive-negative.aspx> (accessed 10.26.14.)

GAO. n.d. About GAO. <http://www.gao.gov/about/> (accessed 10.26.14.)

GAO. September, 2012. Department of Homeland Security: Taking further action to better determine causes of morale problems would assist in targeting action plans. Washington, DC.

GAO. February, 2013a. Department of Homeland Security: Progress made and work remaining after nearly 10 years in operation. Washington, DC.

GAO. December, 2013b. Department of Homeland Security: DHS's efforts to improve employee morale and fill senior leadership vacancies. Washington, DC.

Jackson, B.A., LaTourrette, T., Chan, E.W., Lundberg, R., Morral, A.R., Frelinger, D.R., 2012. Efficient aviation security: strengthening the analytic foundation for making air transportation security decisions. Rand Corporation, Santa Monica, CA.

Jenkins, B.M. February, 2014. The RAND Blog. Experts are working to develop evidence-based ways to measure anti-terrorism efforts. <http://www.rand.org/blog/2014/02/experts-are-working-to-develop-evidence-based-ways.html> (accessed 10.26.14.)

Kaiser, F.M., January, 2006. Congressional oversight. Congressional Research Service, Washington, DC.

Kraft, M.E., Furlong, S.R., 2007. Public policy: Politics, analysis, and alternatives, 2nd ed. CQ Press, Washington, DC.

Partnership for Public Service. December, 2013. Best Places to Work 2013. <http://bestplacestowork.org/BPTW/> (accessed 10.26.14.)

Pew Research Center. November, 2013. Homeland Security is viewed favorably by Americans ahead of Jeh Johnson's hearing. <http://www.pewresearch.org/fact-tank/2013/11/13/senate-considers-new-leader> (accessed 10.26.14.)

Reason. September, 2011. 40% of Americans are somewhat confident in the Department of Homeland Security. <http://reason.com/2011/09/06/40-of-americans-somewhat-confi> (accessed 10.26.14.)

Schneier, B., November, 2009. Beyond Security Theater. New Internationalist, pp. 10–13. <https://www.schneier.com/essay-292.html> (accessed 10.26.14).

Sciarrino, R., November, 2012. Airport screeners ratify first-ever union contract with TSA. Star-Ledger. <http://www.nj.com/news/index.ssf/2012/11/airport_screeners_ratify_first.html> (accessed 10.26.14.)

Sherk, J., August, 2012. The Heritage Blog. TSA collective bargaining could endanger Americans. <http://blog.heritage.org/2012/08/02/tsa-collective-bargaining-could-endanger-americans/> (accessed 10.26.14).

The 9/11 Commission. 2004 . The 9/11 Commission Report: Final Report of the National Commission on Terrorist Attacks Upon the United States. Norton, New York.

Transportation Security Administration. February, 2011. TSA Administrator Pistole's decision on collective bargaining. Washington, DC.

Transportation Security Administration. January, 2014. Testimony of Roderick Allen, Assistant Administrator, Office of Inspection, before House of Representatives Committee on Homeland Security Subcommittee on Transportation Security. Washington, DC.

U.S. Department of Homeland Security, Office of Inspector General. n.d. What we do. <http://www.oig.dhs.gov/index.php?option=com_content&view=article&id=94:what-we-do&catid=2&Itemid=63> (accessed 10.26.14.)

U.S. Department of Homeland Security, Office of Inspector General. November, 2012. Major management challenges facing the Department of Homeland Security. Washington, DC.

U.S. Department of Homeland Security. April, 2013a. Annual performance report, fiscal years 2012-2014. Washington, DC.

U.S. Department of Homeland Security, Office of Inspector General. September, 2013b. TSA Office of Inspection's efforts to enhance transportation security. Washington, DC.

U.S. Department of Homeland Security, Office of Inspector General. December, 2013c. Major Management and Performance Challenges Facing the Department of Homeland Security. Washington, DC.

U.S. Department of Homeland Security, Office of Inspector General. September, 2014. Annual review of the United States Coast Guard's mission performance (FY 2013). Washington, DC.

U.S. Office of Personnel Management. 2013. 2013 Federal Employee Viewpoint Survey Results. Washington, DC.

U.S. Office of Personnel Management. 2014. 2014 Federal Employee Viewpoint Survey. Washington, DC.

Washington Post. 2010. Washington Post poll. November 22, 2010. <http://www.washingtonpost.com/wp-srv/politics/polls/postpoll_11222010.html> (accessed 10.26.14.)

10 Transportation Security in Context

CHAPTER OBJECTIVES:

In this chapter, you will learn about efforts to assess and achieve the proper balance between transportation security measures and:

- Economic efficiency
- Personal privacy
- Budgetary constraints

Introduction

The mission statement of the Transportation Security Administration (TSA), which is responsible for many, but far from all, aspects of transportation security in the United States, indicates, "TSA's mission is to maximize transportation security in response to evolving threats while protecting passengers' privacy and facilitating the flow of legal commerce." This simple declaration raises at once the great challenges that have faced transportation security efforts, before and after 9/11, of how to balance security with other key societal priorities (U.S. Department of Homeland Security, 2013a, p. 131.)

■ ■ Critical Thinking ■

Based on your personal experiences as a passenger, what conflicts exist in trying to balance security with personal privacy, convenience, and economic efficiency? In your opinion, how should such conflicts be resolved?

Economics

The large role played by transportation systems in the global economy has made those systems attractive targets for terrorist and other criminal attacks, and the potential economic consequences of such actions have helped to fuel the major expansion of transportation security measures after the 9/11 disaster. For example, the Department of Homeland Security (DHS) has approximated the economic costs of 9/11 to be $375 billion, and independent estimates of potential damage from terrorist attacks on various parts of the U.S. transportation sector range from $1.2 to $1.5 billion for an attack on Seattle's highway system, to $1.1 to $34 billion for an assault on the Los Angeles-Long Beach port, to $214 to $421 billion for an attack on commercial aviation (Farrow and Shapiro, 2009).

However, the security measures themselves are not without costs, in addition to their expense to taxpayers. A 2011 report by the U.S. Travel Association (USTA), composed of

1300 member organizations within the travel industry, focused on the impact on commercial aviation and related industries:

> *As aviation security continues to evolve, the combination of new screening procedures, technologies, regulatory requirements, and evolving threats are putting increased strain on aviation stakeholders and the traveling public. Many are starting to question whether the current system strikes the proper balance between facilitating the movement of goods and people, and providing protection from the continued threat of terrorist attack. According to a number of economic impact studies and consumer surveys, American business and leisure travelers face greater hassles, endure longer travel times, and lose economic opportunities as the result of the current airport security system. In fact, air traveler surveys in 2008 and 2010 show that traveler frustration and the economic consequences of the current system are getting progressively worse.*

The USTA report went on to cite findings from the 2008 survey indicating that 41 million travelers avoided flights because of security-related inconveniences during the period between May 2007 and May 2008, which—according to USTA—translated "into a $26.5 billion loss to the U.S. economy, including $9.4 billion to airlines, $5.6 billion to hotels, $3.1 billion to restaurants, and $4.2 billion in federal, state and local tax revenue" (U.S. Travel Association, 2011, pp. 5–6).

Concern about the economic and other consequences of governmental regulation long predated 9/11, and the expansion of U.S. regulations in the 1960s and 1970s covering such fields as transportation safety, environmental protection, and food safety led to the federal government's "adoption of benefit-cost analysis[1] as a means of assessing regulations and assisting in the centralized control of laws based on agency regulations." Farrow and Shapiro (2009) outline some of the problems faced in applying such an analytical approach to homeland security:

> *Security rules create unique challenges for benefit-cost analyses. While there are numerous problems in calculating the costs of these regulations, the primary challenges are in measuring the benefits and associated probabilities of security rules. Since much of the information required to assess the value of preventing terrorist attacks is not only highly uncertain but also classified, rules on security have generally escaped serious economic analysis.... Minimal components for the benefit-cost analysis of a homeland security regulation are: benefits using estimates of costs avoided; probabilities; and costs to industry, citizens and government to implement a regulation. However, there is no established*

[1]"Benefit–cost analysis" (often called cost–benefit analysis) is defined by OMB as "a systematic quantitative method of assessing the desirability of government projects or policies when it is important to take a long view of future effects and a broad view of possible side-effects." Its major elements include the policy rationale or justification, the explicit assumptions used in arriving at estimates of future benefits and costs, an evaluation of alternatives, and retrospective analyses to determine whether anticipated benefits and costs have been realized (Office of Management and Budget, 1992).

template or model for applying benefit-cost analysis to homeland security issues where the probabilities, and to a lesser extent the avoided costs, are poorly understood.

The process for developing assessments of the costs and benefits of federal regulations was set forth in Executive Order 12866, issued by President Clinton in September 1993, and modified by Executive Order 13563, issued by President Obama on January 18, 2011. Under the current system, an agency intending to issue a regulation must:

- Determine that the regulation's benefits justify its costs (while recognizing that some benefits and costs are difficult to quantify).
- Tailor the regulation to impose the least burden on society, consistent with obtaining its objectives.
- Select the alternative that maximizes net benefits (including economic, environmental, and public health and safety).
- Specify, to the extent feasible, performance objectives rather than specific behaviors or compliance measures that regulated entities must adopt.
- Identify and assess alternatives, including providing economic incentives to encourage the desired behavior.
- Use "the best available techniques to quantify anticipated present and future benefits and costs as accurately as possible. Where appropriate and permitted by law, each agency may consider (and discuss qualitatively) values that are difficult or impossible to quantify, including equity, human dignity, fairness, and distributive impacts" (Federal Register, 2011a, p. 3821).

In its *2008 Report to Congress on the Benefits and Costs of Federal Regulations and Unfunded Mandates on State, Local, and Tribal Entities*, the Office of Management and Budget (OMB) reported that as of that date, 14 "major, economically significant homeland security rules ... have been finalized since the creation of the DHS" with a total cost of between $3.4 billion and $6.9 billion a year. Of these, eight involved transportation security, with estimated annual costs of between $1.8 billion and $4.4 billion. The projected benefits were generally not quantified "because the benefits of homeland security regulation are a function of the likelihood and severity of a hypothetical future terrorist attack, [and] are very difficult to forecast, quantify, and monetize."

1. **Rule:** Required advance electronic presentation of cargo information

   Agency: Customs and Border Protection (CBP)

   Finalized: Fiscal year (FY) 2004

   Benefits: "The rule's primary benefit would be to improve cargo security. Once implemented, this rule will give CBP more time to analyze cargo data, thereby enabling it to target attention on high-risk cargo or carriers. In addition to improving the effectiveness of inspections, improved targeting may act as a deterrent."

   Annual costs: $334 million to $2.094 billion

2. **Rule:** Area Maritime Security

   Agency: Coast Guard

   Finalized: FY 2004

   Benefits: "This final rule, along with the Vessel Security and Facility Security final rules, was published jointly as part of the implementation of the National Maritime Security Initiative. This initiative is designed to reduce the risk and impact of a transportation security incident."

   Annual costs: $66 million

3. **Rule:** Vessel Security

   Agency: Coast Guard

   Finalized: FY 2004

   Benefits: "Reduce the risk and impact of a transportation security incident."

   Annual costs: $188 million

4. **Rule:** Facility Security

   Agency: Coast Guard

   Finalized: FY 2004

   Benefits: "Reduce the risk and impacts of a transportation security incident."

   Annual costs: $743 million

5. **Rule:** Electronic transmission of passenger and crew manifests for vessels and aircraft

   Agency: CBP

   Finalized: FY 2005

   Benefits: "Submission of manifest information is a necessary component of the nation's continuing program of ensuring aviation and vessel safety and protecting national security. The required information also will assist in the efficient inspection and control of passengers and crew members and thus will facilitate the effective enforcement of the customs, immigration, and transportation security laws."

   Annual costs: $127 million

6. **Rule:** Passenger manifest for commercial aircraft and vessels arriving in and departing from the United States

   Agency: CBP

   Finalized: FY 2007

Benefits: "The goal is to prevent high-risk passengers from boarding aircraft bound for or departing from the U.S., and to prevent such passengers and crew from departing on vessels leaving the U.S. DHS performed a break-even analysis, which identified annual risk reductions required for the rule to breakeven for three attack scenarios. DHS also estimated quantified benefits of $14 million per year, primarily due to fewer diverted aircraft."

Annual costs: $94 million to $134 million

7. **Rule:** Documents required for travel within the Western Hemisphere

   Agency: CBP

   Finalized: FY 2007

   Benefits: "The goal of this rule is to increase security in the air environment by requiring a passport at all airports of entry. The rule addresses a vulnerability of the U.S. to entry by terrorists or other persons by false documents or fraud under the previous documentary exemptions for travel within the Western Hemisphere. These vulnerabilities have been noted extensively by Congress and others."

   Annual costs: $131 million to $664 million

8. **Rule:** Transportation Worker Identification Credential (TWIC) implementation in the maritime sector

   Agency: TSA

   Finalized: FY 2007

   Benefits: "The goal of the rule is to increase the security of the maritime transportation sector by reducing the number of high-risk individuals with unescorted access to secure areas in vessels and facilities."

   Annual costs: $88 million to $415 million (Office of Management and Budget, 2008, pp. 4–5, 11–12; 2005, pp. 14–15; 2006, p. 10)

Farrow and Shapiro (2009) observe that the OMB cost calculations are likely to seriously underestimate the total economic impact:

> *One reason is that OMB does not include all regulations in its estimate. The other reason is omissions in the calculations of the costs of individual regulations. There have been far more than fourteen rules issued since 2002 that impact homeland security. OMB has never estimated the cost of rules not deemed "economically significant" but has stated that the rules included in their totals ... likely make up the bulk of regulatory costs.... Forty-nine other final security rules have been promulgated by agencies between 2002 and 2008, in addition to the 14 economically significant. Many*

of these rules are not counted because the promulgating agency estimates that they cost less than $100 million per year. Even if each of these rules only cost $25 million/year, their inclusion would add another billion dollars to the costs.

In the case of transportation security, the principal reason for underestimation of costs is that a significant number of the relevant rules (a total of 17), including a number of major aviation security measures promulgated shortly after 9/11, were initiated before the creation of DHS and thus were not included in the 2008 accounting (Table 10.1).

Table 10.1 Transportation Security Rules Not Included in Office of Management and Budget Cost Estimates, 2002 to 2008

Agency	Rule
Justice	Screening of aliens and other designated individuals seeking flight training*
TSA	Aviation security infrastructure fees*
TSA	Civil aviation security rules*
TSA	Security programs for aircraft with a maximum certified takeoff weight of 12,500 lb or more*
TSA	Transportation of explosives from Canada to the United States via commercial motor vehicle and railroad carrier*
TSA	Aviation security: private charter security rules*
TSA	Threat assessments regarding U.S. citizens who hold or apply for an FAA certificate*
FAA	Aircraft security under general operating and flight rules*
FAA	Flight crew compartment access and door design*
FAA	Enhanced security procedures for operations in certain airports in the Washington, DC Metropolitan Area Special Flight Rules area*
FAA	Security considerations for the flight deck on foreign-operated transport category airplanes*
FAA	Picture identification requirements*
FAA	Ineligibility for an Airman Certificate on security grounds*
DOT	Limitation on the issuance of commercial driver's licenses with a hazardous materials endorsement*
DOT	U.S. locations requirement for dispatching of United States rail operation*
DOT	Hazardous materials: security requirements for offerors and transporters of hazardous materials*
FAA	Screening of aliens and other designated individuals seeking flight training*
Coast Guard	Automatic Identification System carriage requirements
TSA	Threat assessments regarding alien holders of and applicants for FAA certificates
Coast Guard	TWIC implementation in the maritime sector; hazardous materials endorsement for a commercial driver's license
TSA	HAZMAT fee rule: Fees for security threat assessments on Hazmat drivers
Coast Guard	Notification of arrival in U.S. ports, certain dangerous cargoes, and electronic submission
CBP	Documents required for travelers departing from or arriving in the United States at air ports-of-entry from within the Western Hemisphere
CBP	Letters and documents; advanced electronic presentation of cargo data
Coast Guard	Long-range identification and tracking of ships

DOT, Department of Transportation; FAA, Federal Aviation Administration; TSA, Transportation Security Administration; TWIC, Transportation Worker Identification Credential.

*Initiated before formation of the Department of Homeland Security.

(*Source*: Farrow, S., Shapiro, S., 2009. The benefit-cost analysis of security focused regulations. Journal of Homeland Security and Emergency Management, 6(1), Appendix I. <http://www.degruyter.com/view/j/jhsem.2009.6.1/jhsem.2009.6.1.1482/> (accessed 10.26.14.))

BENEFIT–COST ANALYSIS

The Transportation Security Administration's 2011 Final Rule on Air Cargo Screening

In 2011, TSA finalized its regulation to codify a requirement of the Implementing Recommendations of the 9/11 Commission Act that the agency create a system to screen 100% of all cargo transported on passenger aircraft. Specifically, the rule established the Certified Cargo Screening Program (CCSP) under which TSA certifies facilities across the country to screen cargo before transport on passenger aircraft. The Certified Cargo Screening Facilities (CCSF) must implement a TSA-approved security program, including use of a strict chain of custody standards.

Cost estimates: TSA estimated the 10-year costs of the program to be $1.82 billion, of which $73.4 million would be borne by TSA and the remainder by industry ($1.367 billion in direct costs and another $376.5 million because of delays caused by the screening). TSA considered two models for projecting costs:

- A "bottom-up" approach based primarily on the projected industry participation in the CCSP combined with the estimated costs of program compliance
- A "top-down" approach using actual industry costs incurred under the United Kingdom's similar Known Consignor program

The latter was ultimately used because "TSA considers the top-down cost approach more accurate considering the level of uncertainty in TSA's estimate of the number of firms choosing to become CCSFs. Also, the top-down approach is more likely to reflect the efficiencies captured by allowing the market to allocate screening measures."

In arriving at its estimates, TSA projected that 57% of cargo shipped on passenger aircraft would be screened at CCSFs, and 28% would be screened by aircraft operators, in addition to the other 15% assumed to have been screened by the air carriers before the rulemaking. These figures were based, in part, on experience under the CCSP pilot program that began in 2009. Whereas TSA's costs were derived from anticipated implementation and enforcement expenditures, the industry cost was a reflection of the fees under the United Kingdom's Known Consignor program. The delay cost "assumes the 43 percent of cargo (15 percent screened prior to the CCSP and an additional 28 percent under the CCSP) expected to be screened by the aircraft operators will be the only cargo subject to delay" (Table 10.2).

Qualitative benefits: "By screening 100 percent of cargo shipped on passenger aircraft, the passenger airline industry will have more protection against an act of terrorism or other malicious behavior. Second, allowing the screening process to occur throughout the supply chain via the CCSP reduces potential bottlenecks and delays at the airports. Third, the CCSP allows the market to identify the most efficient venue for screening along the supply chain.... Finally, the CCSP enables members to screen valuable cargo earlier in the supply chain and avoid any potentially invasive screening that may occur at the airport operator level."

Benefit estimates: "The main benefit of this regulation, decreased terrorism risk, cannot be quantified given current data limitations. When it is not possible to quantify or monetize the important incremental benefits of a regulation, OMB recommends conducting a threshold, or 'break-even' analysis. According to OMB, such an analysis answers the question, 'How small could the value of the non-quantified benefits be (or how large would the value of the non-quantified costs need to be) before the rule would yield net benefits?' Consequently ... TSA performed a series of break-even analyses. In these ... TSA compared the annualized costs of the rule's requirements to the expected benefits of preventing certain potential terrorist attacks...."

For example, TSA considered the direct costs of a scenario where the explosive device placed in cargo shipped on a passenger plane destroys a standard narrow body aircraft during flight. This incident is assumed to result in the loss of the lives of all passengers and crewmembers on board, along with the total destruction of the aircraft.... Assuming that the aircraft is destroyed and minimal impact damage is done [and utilizing standard valuations for passenger and crew size, Value of a Statistical Life—which is a statistical model of an individual's willingness to pay to avoid a fatality and is set at $6 million per lost life here, and the replacement cost for the aircraft], TSA estimates the total direct monetary consequences of the attack ... at $732.5 million. Dividing the $732.5 million in estimated direct consequences, by the $178.1 million (the annualized cost of the rule discounted at seven percent), shows that in order for the rule to break even, it will need to reduce the existing or baseline frequency of a terrorist attack by one attack every 4.1 years.... The estimate of the economic impacts of the attack scenarios used in these break-even analyses is limited to direct costs only (value of casualties and lost aircraft). This analysis does not consider any indirect or macroeconomic consequences these terrorist attacks might cause. Consequently ... [it provides] a lower-bound estimate of the economic impact of these attacks" (Federal Register, 2011b, pp. 51848, 51861–51862, 51865–51866).

Starting with the 2009 annual report on benefits and costs of federal regulations, the OMB began to include quantified benefit estimates for certain rules, along with projected costs. However, only one transportation security measure has been listed since then, a CBP rule from FY 2008 on "Documents required for travelers entering the United States at sea and land ports-of-entry from within the Western Hemisphere." The benefits were not

Table 10.2 10-Year Total Cost Summary of the Certified Cargo Screening Program (in $ millions)

Year	TSA Cost	Industry Cost	Delay Cost	Total Cost
1	32.7	109.7	30.1	172.5
2	5.4	115.0	31.6	152.0
3	4.9	120.5	33.1	158.5
4	4.1	126.3	34.7	165.1
5	4.1	132.3	36.4	172.9
6	4.5	138.7	38.2	181.4
7	4.3	145.3	40.1	189.7
8	4.3	152.3	42.0	198.6
9	4.6	159.6	44.0	208.2
10	4.4	167.3	46.2	217.9
Total	73.4	1367.0	376.5	1816.8
Low estimate*	55.0	1139.2	296.5	1490.7
High estimate*	91.7	1594.8	463.3	2149.9

*The low and high estimates represent variance around the Transportation Security Administration's (TSA's) primary estimate to allow for uncertainties with the inputs used to estimate the total cost of the rule.

(*Source*: Federal Register. August 18, 2011b. Transportation Security Administration: Air cargo screening; final rule. Washington, DC, p. 51862.)

estimated, and the costs were projected to be between $268 million and $284 million a year (Office of Management and Budget, 2009, p. 15).

Before 2007, DHS's benefit–cost analyses generally did not include quantified benefits and focused largely on direct implementation expenses. Since then, DHS and its components have often used "break-even" analysis in which benefits are calculated based on a comparison of losses (typically from a terrorist attack) expected to occur with and without the proposed regulation, with the net loss reduction representing the benefit. This, in turn, is set against the anticipated costs, with the break-even point representing the level of loss reduction needed to equal the cost and the regulation deemed cost effective when the benefits exceed this point. Farrow and Shapiro (2009) note that these "break-even" analyses "are, without question, improvements" over the previous efforts, but "remain incomplete substitutes for a true benefit-cost analysis … due to the implicit assumptions [used] … but also due to further complications, such as the potential displacement of attacks from one site to another or the existence of budget limitations."

The 2012 Rand report *Efficient Aviation Security* examined "key uncertainties and knowledge gaps in aviation security," with the goal of obtaining a clearer understanding "of what security measures truly cost and what we get when we buy those measures [in order to] get closer to the efficient security we must aspire to in a world of finite resources and many varied policy areas that demand funding and attention." Although the analysis is confined to the aviation sector, many of the findings are applicable to all areas of transportation security:

- *While it is broadly accepted that security measures have intangible costs—and that those costs affect the utility of the aviation system—it is less clear how to appropriately capture them in security analysis. Building from accepted cost-benefit methodologies … even approximate estimates for such effects can be used when different security measures are compared or—as has been the strategy in aviation—when increasing numbers of security measures are added on top of one another as threats change over time.*
- *If the intangible costs of security translate into reduced passenger demand, the benefits of security in reducing attack risk are quickly overwhelmed by the losses stemming from the reduced value of the aviation system. Even a slight reduction in passenger demand can greatly reduce or even negate the net benefit of a security investment. This essentially raises the bar for the performance of security measures: Not only do they need to be effective in reducing the risk of attack, they must do so without sacrificing too much of the value of the system they seek to protect. Recognizing the strong influence of the indirect costs of security emphasizes the importance of designing security approaches that avoid such costs, by assembling systems of security measures that minimize the effect on passengers and other users' experience.*
- *The problem of uncertainty regarding the full costs of security measures in many ways mirrors a difficulty on the "other side of the ledger" for making security decisions: uncertainty regarding the risk from terrorism complicating assessment of the benefits of new security measures. While it is well established that there is some risk of terrorism that security measures seek to address, the magnitude of that risk—whether the*

expected annual losses from attacks are (at least in monetary terms) in the millions, billions, or even approaching trillions of dollars—is uncertain. If the true risk is low, then the potential benefits of improved security will be low—since they would be reducing a comparatively smaller risk. If the true risk is high, then even small percentage reductions in risk could amount to very substantial benefits. To address this uncertainty, rather than seeking to calculate a single benefit value and assess new security measures against it, analysts have instead used ranges of terrorism risk values.... Though such analyses do not provide single answers regarding the cost-benefit balance of specific security measures, they can be useful for framing choices (pp. xvii–xviii, 47–48, and 135).

Privacy

The 9/11 Commission noted that the increased security measures being recommended and implemented in response to the 9/11 attacks would have a substantial impact on civil liberties and that steps should be taken to safeguard those liberties:

The terrorists have used our open society against us. In wartime, government calls for greater powers, and then the need for those powers recedes after the war ends. This struggle will go on. Therefore, while protecting our homeland, Americans should be mindful of threats to vital personal and civil liberties. This balancing is no easy task, but we must constantly strive to keep it right. This shift of power and authority to the government calls for an enhanced system of checks and balances to protect the precious liberties that are vital to our way of life.... The burden of proof for retaining a particular governmental power should be on the executive [branch], to explain (a) that the power actually materially enhances security and (b) that there is adequate supervision for the executive's use of the powers to ensure protection of civil liberties.... We must find ways of reconciling security with liberty, since the success of one helps protect the other. The choice between security and liberty is a false choice, as nothing is more likely to endanger America's liberties than the success of a terrorist attack at home. Our history has shown us that insecurity threatens liberty. Yet, if our liberties are curtailed, we lose the values that we are struggling to defend.

The 9/11 Commission Report, 2004, pp. 394–395

Since the issuance of the 9/11 Commission final report, new institutions have been created within the federal government that are designed to address privacy and civil liberties issues arising out of specific homeland security policies.

Privacy and Civil Liberties Oversight Board

The 9/11 Commission recommended "there should be a board within the executive branch to oversee adherence to the [privacy and civil liberties] guidelines we recommend [on information sharing within the government and other homeland security measures] and the commitment the government makes to defend our civil liberties." The Intelligence Reform and Terrorism Prevention Act of 2004, which was developed in response to the 9/11 Commission report, established a Privacy and Civil Liberties Oversight Board within the Executive Office of the President, which was to consist of five members appointed by the president, with the chair and vice chair subject to Senate confirmation. The statute directed the Board to "ensure that concerns with respect to privacy and civil liberties are appropriately considered in the implementation of laws, regulations, and executive branch policies related to efforts to protect the Nation against terrorism." However, its authority to obtain relevant information from federal and nonfederal entities was somewhat limited; for example, the Board was not granted subpoena power to compel the production of relevant information. In addition, implementation of the new panel proceeded slowly, with President Bush not submitting his nominees until June 2005, the Senate confirming the chair and vice chair in February 2006, and the first meeting held the following month. As observed in a report by the Congressional Research Service, in its early stages, "the PCLOB [Privacy and Civil Liberties Oversight Board], to some, appeared to be a presidential appendage, devoid of the capability to exercise independent judgment and assessment or to provide findings and recommendations" (Hatch, 2012, pp. 1–5).

Such dissatisfaction led to a reconstitution of the PCLOB as a bipartisan, independent agency in the Implementing Recommendations of the 9/11 Commission Act of 2007 (9/11 Act). That legislation made all five Board members (four part-time members and a full-time chairman) subject to Senate confirmation and specified two primary purposes: to review and analyze actions the executive branch takes to protect the nation from terrorism, ensuring the need for such actions is balanced with the need to protect privacy and civil liberties, and to ensure that liberty concerns are appropriately considered in the development and implementation of laws, regulations, and policies related to efforts to protect the nation against terrorism.

To implement these objectives, the 9/11 Act directed the Board to:

- Provide advice to the president and federal agencies on policy development and implementation.
- Oversee and continually review federal agency implementation of regulations, policies, and procedures.
- Work with federal agency privacy officers, and when appropriate, coordinate their activities on relevant interagency matters.
- Submit semiannual reports to Congress and the president.
- Inform the public by releasing its reports in unclassified form to the greatest extent possible and holding public hearings.

Finally, the 9/11 Act authorized the Board to access all relevant agency documents (including classified information); interview any executive branch officer or employee; request information or assistance from state, local, and tribal governments; and request in writing that the attorney general subpoena any nonexecutive branch entity to produce relevant information. The Board was not given enforcement power and may not order government agencies to alter their practices (Privacy and Civil Liberties Oversight Board, n.d.; Stanley, 2013b).

Again, activation of the PCLOB took considerable time. In February 2008, President Bush submitted nominations for three positions on the Board but refused to submit the two candidates put forth by the Senate Democratic leadership; in response, the Senate took no further action on the nominees. President Obama did not submit any nominations to the Board until December 2010, when he transmitted two names to the Senate. Taking note of this state of affairs, the National Security Preparedness Group's 2011 report on the status of the 9/11 Commission's recommendations stated, "If we were issuing grades, the implementation of this recommendation [to establish a privacy and civil liberties oversight board] would receive a failing mark. A robust and visible Board can help reassure Americans that these [homeland security] programs are designed and executed with the preservation of our core values in mind. Board review can also give national security officials an extra degree of assurance that their efforts will not be perceived later as violating civil liberties" (Stanley, 2013a; Bipartisan Policy Center, 2011, p. 16).

In December 2011, the president submitted names for the three remaining positions on the PCLOB, and the Senate confirmed four of the nominees (all but the chair) in August 2012. Finally, the nominee for chairman was confirmed in May 2013. In that same month, President Obama issued an executive order directing DHS to consult with the PCLOB in evaluating interagency cybersecurity information sharing. Since that time, much of the Board's activity has been directed at investigating the National Security Agency's telephone metadata program,[2] and thus far it has had relatively little involvement in transportation security-related issues (Stanley, 2013a; Privacy and Civil Liberties Oversight Board, 2013, pp. 9–14).

Department of Homeland Security Privacy Office

The DHS's Privacy Office was the first statutorily mandated privacy office in any federal agency, being required under the terms of the Homeland Security Act of 2002 that established DHS. Its mission is "to protect all individuals by embedding and enforcing privacy protections and transparency in all DHS activities ... [to] work with every [DHS] component and program to ensure that privacy considerations are addressed when planning or updating any program, system or initiative ... [and] to ensure that technologies used at the Department sustain, and do not erode, privacy protections." The Office operates under the DHS's Fair Information Practice Principles (FIPP), which include transparency,

[2]Under this program, the NSA "collect[s] nearly all call detail records generated by certain telephone companies in the United States," including information that appears on one's telephone bill (e.g., date and time of call, its duration, and the participating phone numbers). It does not include the content of the conversation. In its January 2014 report, the PCLOB found that there was no proper legal basis for the program and recommended that it be ended (PCLOB, 2014, pp. 8, 10, 168).

individual participation, purpose specification, data minimization, use limitation, data quality and integrity, security, and accountability and auditing (U.S. Department of Homeland Security, n.d.a).

The Privacy Office focuses on:

- Requiring compliance with federal privacy and disclosure laws and policies in all DHS programs, systems, and operations
- Centralizing Freedom of Information Act (FOIA) and Privacy Act operations to provide policy and programmatic oversight, to support operational implementation within the DHS components, and to ensure the consistent handling of disclosure requests
- Providing leadership and guidance to promote a culture of privacy and adherence to the FIPPs across the Department
- Advancing privacy protections throughout the federal government through active participation in interagency forums
- Conducting outreach to the Department's international partners to promote understanding of the U.S. privacy framework generally and the Department's role in protecting individual privacy
- Ensuring transparency to the public through published materials, reports, formal notices, public workshops, and meetings

From July 2012 through June 2013, the DHS Privacy Office issued three major policy documents (outlining principles and procedures for DHS research projects, Privacy Officer investigations, and DHS applications for FOIA exemptions), approved 87 new or updated Privacy Impact Assessments[3] and 24 System of Records Notices (SORN),[4] reviewed 241 intelligence products and 519 Intelligence Information Reports "to ensure that only the minimum amount of personally identifiable information necessary to the intelligence value of the product is included," and received 811 FOIA requests and processed 746 of these. In addition, the Office conducted a series of Privacy Compliance Reviews of DHS programs, the most important of which in the transportation security field were those concerning CBP's Automated Targeting System and the 2011 United States–European Union (EU) Passenger Name Record (PNR) Agreement (U.S. Department of Homeland Security, 2013c, pp. 1–4).

[3]The E-Government Act of 2002 sought to address the potential compromise of personal information stemming from advances in information technology and required federal agencies to conduct Privacy Impact Assessments of the privacy impact of new or substantially revised government information technology systems (U.S. Department of Homeland Security, n.d.b, "Privacy Office - Privacy Impact Assessments (PIA)." <http://www.dhs.gov/privacy-office-privacy-impact-assessments-pia> accessed 06.18.14.).

[4]The Privacy Act of 1974 requires federal agencies to publish a notice in the *Federal Register* of all of its systems of records, which are groups "of any records under the control of any agency from which information can be retrieved by the name of the individual or by some identifying number, symbol, or other identifier assigned to the individual" (U.S. Department of Homeland Security, n.d.c, "System of Record Notices (SORNs)." <http://www.dhs.gov/system-records-notices-sorns> accessed 06.18.14.).

Automated Targeting System Privacy Impact

In a notice of proposed rulemaking published in the May 23, 2012, *Federal Register*, DHS announced its intention to continue to exempt the Automated Targeting System (ATS) system of records from certain requirements of the Privacy Act. The Privacy Act allows federal agencies to provide such exemptions but requires them to issue a notice "to make clear to the public the reasons why a particular exemption is claimed." The May 2012 notice was necessary because of an alteration of the ATS program. In the notice, DHS presented its justifications:

> *These exemptions are needed to protect information relating to DHS activities from disclosure to subjects or others related to these activities. Specifically, the exemptions are required to preclude subjects of these activities from frustrating these processes; to avoid disclosure of activity techniques; to protect the identities and physical safety of confidential informants and law enforcement personnel; to ensure DHS' ability to obtain information from third parties and other sources; to protect the privacy of third parties; and to safeguard officially classified and/or controlled information. Disclosure of information to the subject of the inquiry could also permit the subject to avoid detection or apprehension. The exemptions proposed here are standard law enforcement and national security exemptions exercised by a large number of federal law enforcement and intelligence agencies.*
>
> *Federal Register, 2012, pp. 30433–30435*

The DHS Privacy Impact Assessment for the proposed exemptions was issued by CBP on June 1, 2012. In it, the ATS was described in some detail, and potential privacy concerns, as well as mitigating measures, were highlighted:

1. *ATS aggregates data from many systems, which may exceed the minimal amount necessary to achieve its mission. Mitigation: To mitigate the risks posed in the collection of large amounts of data, CBP has imposed strict controls to maximize the security of the information that is being stored. Officers rely on data to make accurate determinations and are trained to identify inaccurate information. Data are kept in secure areas protected by armed guards. Access to ATS records is limited to those individuals who have a need to know the information for the performance of their official duties and who have appropriate clearances or permissions.*
2. *Information about two different individuals with similar names and dates of birth could be mischaracterized as the same individual, thus attributing the wrong information to the wrong individual. Mitigation: DHS personnel are required to review and cross reference the records in ATS to improve the level of confidence and reliability in derogatory information before any action is taken against an individual.*
3. *One potential risk to individuals from the use of ATS is that a traveler, conveyance, or cargo in which an individual has an interest may be referred to secondary inspection even though the traveler, conveyance, or cargo does not present any risk of harm to*

the United States and has not committed or been associated with any violation of U.S. law. Mitigation: Referral to secondary inspection, as necessary, permits an officer to intercede and resolve mis-identifications, and to clarify information associated with an individual's travel document records. Determinations in secondary regarding admissibility are made by a CBP officer or supervisor. Secondary processing is a necessary component of CBP's admissibility determination for each person arriving in the United States when admissibility cannot be determined at primary inspection.....

4. *Data may be retained for too long. Mitigation: ATS retains data according to the SORN requirements of the system from which the data was obtained. PNR is retained for five years in an active state and ten years in a dormant state. However, users will only be able to use PII [personally identifiable information] for six months. After six months PII will be masked and require each user [of ATS] to obtain supervisory approval before unmasking the PII....*
5. *ATS may retain data longer than the source system. Mitigation: In general, ATS has implemented controls that delete data in ATS if such data are deleted in a source system. For data that has been identified by a CBP officer in ATS as having law enforcement relevance, the record may be maintained longer than allowed for in the source system....*
6. *Information may be shared under inappropriate circumstances. Mitigation: Risks related to sharing of information outside DHS, including any potential risk of further dissemination of information by the external agency to a third agency, are mitigated through arrangements governing access to ATS by external parties and sharing of ATS information with external parties.... The arrangements generally require the external party accessing or receiving the information to employ measures relating to security, privacy, and safeguarding of information that are equivalent to measures employed by DHS. As a general matter, the arrangements also stipulate that any further dissemination of ATS information is subject to prior authorization by CBP....*
7. *Individuals may not get the level of redress they desire. Mitigation: As set forth in the SORN published in connection with this PIA, DHS has exempted portions of ATS from the access, amendment, and certain accounting provisions of the Privacy Act.... DHS and CBP, however, will consider each request for access to records maintained in ATS to determine whether or not information may be released. Also, individuals may, pursuant to FOIA, seek access to information for which ATS is the source system or which originates from another government source system and as a matter of CBP policy, redress may also be requested [from the CBP INFO Center].... However, individuals, regardless of nationality, country of origin or place of residence who believe their PNR has been used in an inappropriate manner may seek redress, including but not limited to, through the DHS Traveler Redress Inquiry Program.*

U.S. Department of Homeland Security, 2012, pp. 17–18, 21, 25–26, 28, 31–32

The Electronic Privacy Information Center (EPIC), which is "an independent non-profit research center [that] works to protect privacy, freedom of expression, [and] democratic values" and has been actively involved in raising privacy-related concerns about a number of transportation security policies (including Advanced Imaging Technology screening systems, Pre✓, and behavioral profiling, among others), filed its comments on the DHS notice on June 21, 2012:

> *EPIC submits these comments to address the substantial privacy and security raised by the [ATS] database, to urge that CBP cease retaining personal information on American citizens in the ATS, and to demand that CBP significantly narrow the Privacy Act exemptions for the system if the proposal goes forward.... The ATS database contains an excess of personally identifiable information ranging from names, addresses, nationalities, and Social Security Numbers to "information that could directly indicate the racial or ethnic origin, political opinions, religious or philosophical beliefs, trade union membership, health, or sex life" of individuals. On the basis of this information, and not any actual conduct, a federal agency of the United States makes determinations about the rights and opportunities of U.S. citizens. Since 2006, EPIC has consistently recommended that CBP suspend ATS, or in the alternative, fully apply all Privacy Act safeguards to any person subject to ATS.*
>
> *Electronic Privacy Information Center, n.d.a; Electronic Privacy Information Center, 2012, pp. 1–2*

The EPIC comments also addressed the DHS Privacy Impact Assessment:

> *The ATS Privacy Impact Assessment does nothing to ameliorate concerns about the impact of the Automated Targeting System. In fact, the Privacy Impact Assessment makes clear that the program should not continue. The assessment sets out the various privacy risks associated with "access to datasets used and stored in ATS," yet does nothing to solve them. For example, ATS is accessible over web-based "DHS infrastructure or remotely through secure-encrypted mobile devices with one-factor authentication...." CBP insufficiently safeguards against unauthorized access by solely requiring one-factor authentication to this information. Government databases are frequently hacked and compromised. One-factor authentication increases the likelihood that ATS can be compromised.*
>
> *Electronic Privacy Information Center, 2012, pp. 8–9*

United States–European Union Passenger Name Record Agreement

Under the Aviation and Transportation Security Act of 2001 (ATSA) and the ensuing regulations designed to implement it, each air carrier operating passenger flights to and from the United States was required to transfer personal data contained in the carrier's Passenger Name Records[5] to U.S. security officials. In June 2002 the European Commission informed

[5]A PNR "is a record of travel information created by commercial air carriers that could include each passenger's name, destination, and method of payment, flight details, and a summary of communications with airline representatives" (U.S. Department of Homeland Security, 2013b, p. 4).

the U.S. government that these requirements may conflict with European Union (EU) and its member states' legislation on data protection, which impose conditions on the transfer of personal information. Consequently, negotiations were begun between the United States and the EU aimed at reaching an agreement on sharing air passenger data while providing an adequate level of privacy protection (European Commission, 2013a, p. 2).

In 2004, CBP issued a set of undertakings indicating how the agency would process and transfer PNR data on flights between the EU and the United States, which was followed by a finding by the European Commission that these actions would adequately meet the EU's privacy concerns, and subsequently by a formal agreement between DHS and the European Commission authorizing the transfer of PNR data. However, in May 2006, the European Court of Justice found that the 2004 agreement was entered into without appropriate EU legal authority and was therefore invalid. In July 2007, DHS and the EU signed a new agreement on PNR transfer, which provisionally went into effect upon signature of the two parties, but this accord was never ratified by all EU member states, and the newly empowered European Parliament informed the European Commission that it would not ratify the 2007 agreement, directing the Commission to negotiate a new agreement. The resulting negotiations produced the 2011 U.S.-EU PNR Agreement, which was signed in December 2011 and ratified by the European Parliament in April 2012 (U.S. Department of Homeland Security, 2013b, pp. 8–10).

Under the terms of the agreement:

- PNR data may only be used for the prevention, detection, investigation, and prosecution of terrorism and certain transnational crimes (e.g., drug trafficking and human trafficking).
- PNR data may be retained for 15 years for terrorist-related offenses, and 10 years for other transnational crimes. However, after the first 6 months, the data are to be "de-personalized" and after 5 years are to be moved to a "dormant" database, with additional access controls.
- PNR data is to be transmitted under a clear set of rules, with the "push method" (under which airlines submit the data to CBP rather than having CBP directly access airline reservation systems) the preferred approach.
- In addition to the "de-personalization" of PNR data (which involves the masking of all elements that specifically identify the individual, .e.g., name and contact information), personal data are protected by providing passengers with a means to access their PNR, request correction of any inaccurate data, and seek administrative and judicial redress. In addition, stricter rules are applied to prevent the loss or unauthorized disclosure of personal data, with oversight by independent bodies, including the DHS Privacy Office, the DHS Inspector General, and the Government Accountability Office (GAO).
- Data sharing is to be limited to only those countries that offer a high level of data protection and is to be done only on a case-by-case basis.
- The agreement is for 7 years, with provisions for automatic renewal and for termination by either party at any point (European Commission, 2013b, pp. 1–3).

The agreement also provided for a review of its implementation by both parties 1 year after it entered into force. On the U.S. side, the DHS Privacy Office issued its assessment in July 2013 and concluded that DHS practices were "generally compliant with the 2011 Agreement," with one exception wherein "the notification of EU Member States was not taking place after sharing [of PNR data] with one of our international partners." Similarly, in its November 2013 compliance report, the EU Commission indicated, "DHS implements the Agreement in accordance with the terms of the Agreement. DHS respects its obligations as regards the access rights of passengers and has a regular oversight mechanism in place to guard against unlawful discrimination" (U.S. Department of Homeland Security, 2013b, p. 4; European Commission, 2013a, p. 20).

Advanced Imaging Technology Lawsuit

Another example of efforts to reconcile security and civil liberties concerns was in the lawsuit filed in 2010 by EPIC that sought to suspend the deployment of Advanced Imaging Technology (AIT) equipment to screen airline passengers. EPIC contended that the security measure violated the Fourth Amendment's protection against unreasonable searches and is "unlawful, invasive, and ineffective," citing the invasive nature of the devices and TSA's disregard of public opinion concerning their usage (Electronic Privacy Information Center, n.d.b).

In a July 15, 2011, decision, the U.S. Court of Appeals for the District of Columbia rejected EPIC's claims of constitutional and statutory violations, making the following finding with respect to the application of the Fourth Amendment in this case:

> *As other circuits have held, and as the Supreme Court has strongly suggested, screening passengers at airports is an "administrative search" because the primary goal is not to determine whether any passenger has committed a crime but rather to protect the public from a terrorist attack.... Whether an administrative search is "unreasonable" within the condemnation of the Fourth Amendment is "determined by assessing, on the one hand, the degree to which it intrudes upon an individual's privacy and, on the other, the degree to which it is needed for the promotion of legitimate governmental interests." That balance clearly favors the government here. The need to search airline passengers "to ensure public safety" can be particularly acute, and crucially, an AIT scanner, unlike a magnetometer, is capable of detecting, and therefore of deterring, attempts to carry aboard airplanes explosives in liquid or powder form. On the other side of the balance, we must acknowledge the steps TSA has already taken to protect passenger privacy, in particular distorting the image created using AIT and deleting it as soon as the passenger has been cleared. More telling, any passenger may opt-out of AIT screening in favor of a patdown, which allows him to decide which of the two options for detecting a concealed, nonmetallic weapon or explosive is least invasive.*
>
> *U.S. District Court of Appeals for the District of Columbia Circuit, 2011, pp. 16–17*

Critical Thinking Question

Consider the preceding three instances where privacy concerns were weighed against security considerations (involving ATS, PNR data, and AIT) and the different methods of resolution used in each case (regulatory, international negotiations, and judicial, respectively). What are the strengths and weaknesses of each approach? In your opinion, was each resolution appropriate? Why or why not?

DIGGING DEEPER
SECURITY OF TRANSPORTATION SYSTEMS' CRITICAL CYBER INFRASTRUCTURE

An Emerging Priority

Nowhere is the nexus of transportation security, economics, and privacy more apparent than in the field of cybersecurity,[6] where cyber attacks pose simultaneous threats to commerce, personal information, and security. Protection of information systems against attacks was recognized as an important objective well before 9/11. For example, the GAO designated protection of federal information systems as a high-risk area beginning in 1997, and securing cyberspace[7] was made a part of the evolving homeland security structure by the Bush Administration via the issuance of the National Strategy to Secure Cyberspace in 2003 and the Comprehensive National Cybersecurity Initiative in 2008. However, "threats to [cyber] systems supporting critical infrastructure and federal information systems are evolving and growing. Advanced persistent threats—where adversaries that possess sophisticated levels of expertise and significant resources ... pursue [their] objectives repeatedly over an extended period of time—pose increasing risks." In response, in 2009, President Obama stated that the cyber threat was "one of the most serious economic and national security challenges we face as a nation ... America's economic prosperity in the 21st century will depend on cybersecurity" (Figure 10.1) (GAO, 2013, pp. 3, 21–23).

In November 2011, DHS issued its "Blueprint for a Secure Cyber Future," which sought to provide a strategic framework for the creation of "a safe, secure, and resilient cyber environment," with goals for protecting critical information infrastructure (reduce exposure to cyber risk, ensure priority response and recovery, maintain shared situational awareness, and increase resilience) and strengthening the cyber environment (empower individuals and organizations to operate securely, make and use more trustworthy cyber protocols and architectures, build collaborative communities, and establish transparent processes) (U.S. Department of Homeland Security, 2011, p. iii).

---

[6] Cybersecurity is "the activity or process, ability or capability, or state whereby information and communications systems and the information contained therein are protected from and/or defended against damage, unauthorized use or modification, or exploitation" (National Initiative for Cybersecurity Careers and Studies, n.d.).

[7] Cyberspace is "the independent network of information technology infrastructures that includes the Internet, telecommunications networks, computer systems, and embedded processors and controllers" (National Initiative for Cybersecurity Careers and Studies, n.d.).

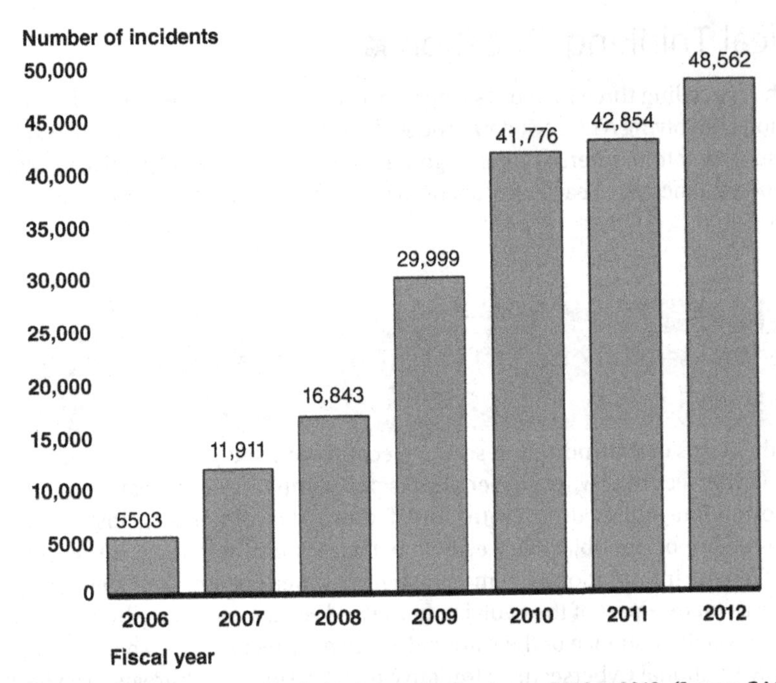

FIGURE 10.1 Cybersecurity incidents reported by federal agencies, fiscal years 2006 to 2012. *(Source: GAO. February, 2013. Cybersecurity: National strategy, roles, and responsibilities need to be better defined and more effectively implemented. Washington, DC. February 14, 2013, p.8.)*

The National Cyber Security Division is the DHS entity primarily responsible for coordinating federal cybersecurity preparedness and response, with the U.S. Computer Emergency Readiness Team (US-CERT) its operational arm. US-CERT "provides response support and defense against cyber attacks for the federal civil executive branch and provides information sharing and collaboration with federal, state, and local government; industry; the research community; and international partners." DHS has also deployed the EINSTEIN 2 system (which is "an automated cyber surveillance system that monitors federal internet traffic for malicious intrusions and provides near real-time identification of malicious activity) at 15 federal departments and agencies. In 2011, US-CERT responded to more than 106,000 incident reports and released more than 5,000 cyber security alerts and other information products (U.S. Department of Homeland Security, 2010, pp. 185–186; n.d.d).

The Obama Administration requested $1.25 billion in funding for DHS cybersecurity activities in its FY 2015 budget. Included was $377.7 million for the more advanced EINSTEIN3 Accelerated program, and additional funding for the TSA Cybersecurity Assessment and Risk Management Approach (CARMA) initiative, which "aims to assist the pipeline industry in developing a prioritized list of risks to business functions from cyber-attacks and a prioritized list of activities to address those risks" (U.S. Department of Homeland Security, 2014, p. 14; Transportation Security Administration, 2014, p. 12).

In a 2013 assessment, the GAO cited progress but identified a number of problem areas in federal cybersecurity efforts.

- *Risk management:* Only eight of 22 major federal agencies reported compliance with risk management requirements for cyber systems under the Federal Information Security Management Act, and DHS had not yet identified cybersecurity guidance applicable to nonfederal critical infrastructure.
- *Detection, response and mitigation of cyber incidents:* "Difficulties in sharing and accessing classified information and the lack of a centralized information-sharing system continue to hinder progress. According to DHS, a secure environment for sharing cybersecurity information ... is not expected to be fully operational until fiscal year 2018."
- *Workforce training:* As of November 2011, only three of eight federal agencies reviewed by GAO had developed a department-wide training program for their cybersecurity workforce (GAO, 2013, Highlights).

One of the most significant cybersecurity efforts in the transportation field was the development of the 2012 "Roadmap to Secure Control Systems in the Transportation Sector." With the assistance of governmental and private sector stakeholders, DHS's National Cyber Security Division facilitated the development of the document, which is intended to serve as an aid for transportation entities in improving the security of their industrial control systems (ICSs)[8] (Foreword, p. 2).

The report outlines the particular vulnerabilities of modern transportation systems to cyber threats:

> *New generation aircraft and legacy aircraft are designed or retrofitted with technologies such as Ethernet IP-enabled networks, wireless connectivity (e.g., Bluetooth) capabilities, and GPSs. Similarly, trains are now supplied with onboard IT systems that provide and receive real-time updates on track conditions, train position, train separation, car status, and other operational data. While such technologies are designed to provide faster and more reliable communications, these wireless communication advances result in aircraft and trains no longer functioning as closed systems, thus increasing the e-enabled threats and risks to these transportation mediums. Many pipelines are now supplied with SCADA systems, RTUs, and automated pressure regulators and control valves. If this pipeline infrastructure is intentionally attacked, many control valves and pressure regulators could simultaneously be affected.... Today's control systems in the Highway and Maritime modes are often not only automated but also highly integrated.... If an individual system or device was deliberately attacked, the potential to affect multiple control systems would be a distinct reality.*
>
> *Roadmap to Secure Control Systems in the Transportation Sector, 2012, p. 37*

To address such threats, the "Roadmap" presents a voluntary approach that:

- Defines a consensus-based strategy that addresses the specific cybersecurity needs of transportation asset owners and operators

[8] For purposes of the "Roadmap," ICSs include all process control systems, functional and operational systems, safety systems tied to operational systems, supervisory control and data acquisition systems, distributed control systems, programmable logic controllers, and general-purpose process controllers but do not include business systems and information technology systems.

- Proposes a comprehensive plan for improving the security, reliability, functionality, and oversight of transportation ICSs
- Proposes methods and activities that encourage participation and compliance by all stakeholders
- Guides modal cybersecurity efforts
- Presents a vision—along with a supporting framework of goals, objectives, and milestones and metrics—for continuous improvement of the cybersecurity of ICSs in the Transportation Sector (p. 3)

How would you achieve the proper balance in federal cybersecurity efforts between having a program strong enough to afford businesses and individuals a reasonable expectation of security against cyber attacks yet not unduly impeding commerce or raising civil liberties concerns about government access to personal information?

Budgetary Constraints

In May 2005 testimony to Congress, GAO raised questions about the future funding of homeland security programs:

> *Where the money will come from is unclear. In our 2002 statement on national preparedness, we highlighted the need to examine the sustainability of increased funding ... for homeland security efforts.... The current economic environment makes this a difficult time for private industry and state and local governments to make security investments and sustain increased security costs. According to industry representatives and experts we contacted, most of the transportation industry operates on a very thin profit margin, making it difficult to pay for additional security measures.*
>
> *GAO, 2005, pp. 23–24*

However, over the short term, GAO's concerns about funding constraints with respect to U.S. transportation security programs largely failed to materialize, and funding continued to rise, albeit at a slower pace, through FY 2009. Budgetary pressures have increasingly manifested themselves since then, with actual reductions in overall transportation security appropriations in FYs 2011 to 2013.

Fiscal Year 2014 Department of Homeland Security Appropriations

As has become the rule in recent years, DHS joined the rest of the federal government in not receiving its spending authority for FY 2014 before the start of that fiscal year (October 1, 2014). Indeed, congressional action on all appropriations m easures was even more fraught with partisan division and delay than in preceding years. A Congressional Research Service report described the train of events:

> *From October 1, 2013, through October 16, 2013, the federal government (including DHS) operated under an emergency shutdown furlough due to the expiration*

> *of annual appropriations for FY 2014. More than 31,000 DHS employees were furloughed. Tens of thousands of others that were excepted from furlough,*[9] *and those whose salaries were paid through annual appropriations, worked without pay until the lapse was resolved by passage of a short-term continuing resolution. From October 17, 2013, to January 17, 2014, the federal government operated under the terms of two consecutive continuing resolutions: PL 113-46, which lasted until its successor was enacted on January 15, 2014; and PL 113-73, which lasted until the Omnibus Appropriations Act, 2014 (PL 113-76) was enacted on January 17, 2014.*
>
> *Painter, 2014, Summary*

Funding for DHS was included in Division F of PL 113-76, which provided $39.27 billion to the department in adjusted net discretionary budget authority for FY 2014, approximately $922 million above its FY 2013 appropriation level after taking into account the impact of sequestration. However, a change made by the Budget Control Act of 2011 that facilitated inclusion of disaster funding in regular appropriations bills, rather than in emergency or supplemental measures, has made year-to-year comparisons of DHS funding even more problematic than previously. Leaving aside disaster relief, the level of annual appropriations for DHS continued to decline in FY 2014, as it has done each year since its FY 2010 peak (Painter, 2014, Summary, pp. 5–6).

A closer look at major transportation security programs within DHS reveals some departures from the overall trend, with TSA's major aviation security accounts (aviation security, the Aviation Security Capital Fund, and Federal Air Marshals) that had continued to rise through FY 2012 dropping back to their FY 2008 level in FY 2013 and only recovering to slightly above FY 2009 funding in FY 2014. (Much of the increase in TSA appropriations for aviation security in the FY 2010 to 2012 period was in response to the December 2009 attempted bombing of Northwest Flight 253.) The Coast Guard's ports, waterways, and coastal security program has remained at roughly the same appropriation level going back to FY 2004, but CBP's container security activities have experienced greater year-to-year fluctuations in resources but with the recent trend (after FY 2010) being downward. Finally, TSA's land (or surface) transportation security programs have continued to operate at a very limited funding level, which has been decreasing since FY 2012 (Table 10.3).

The budgetary constraints reflected in the diminished resources made available for transportation security programs in FYs 2013 and 2014 are also evident in the president's proposed FY 2015 budget, which recommends a $1.05 billion reduction in DHS's net discretionary budget authority compared with FY 2014. Among the proposed reductions in transportation security programs are the following:

- $100 million through TSA "risk-based security efficiencies"

[9]Furloughs are "the placing of an employee in a temporary non-duty, non-pay status because of lack of work or funds, or other non-disciplinary reasons." Excepted employees are those "who are exempt from a furlough by law because they are (1) performing emergency work involving the safety of human life or the protection of property, (2) involved in the orderly suspension of agency operations, or (3) supporting the discharge of the President's constitutional duties to nominate and appoint officers of the Government" (White House, U.S. Office of Government Ethics, 2011).

Table 10.3 Select Department of Homeland Security Program Funding Levels, Fiscal Years 2013 and 2014 (in $ thousands)

	FY 2013: Final†	FY 2014: Enacted‡
Transportation Security Administration		
Aviation security	4,766,114	4,982,735
Avsec Capital Fund	250,000	250,000
Federal Air Marshals	874,557	818,607
Aviation subtotal	*5,890,671*	*6,051,342*
Surface transportation security	122,015	108,618
Intelligence and Vetting§	267,537	237,489
Transportation security support	908,417	962,061
Other fees	5,117	5,000
Total budget authority	7,193,757	7,364,510
Less prior year rescissions	25,035	59,209
Net budget authority	7,168,722	7,305,301
U.S. Coast Guard		
Ports, waterways, and coastal security	1,800,274	1,777,419
Customs and Border Protection (CBP)		
CBP container security¶	231,889	220,377

*Totals may not add because of rounding.
†FY 2013: Final includes across-the-board rescissions made pursuant to PL 113-6 and sequestration under PL 112-25 (except for CBP container security, which reflects enacted level only).
‡FY 2014: Enacted based on funding provided by PL 113-76.
§Formerly Transportation Threat Assessment and Credentialing.
¶Includes International Cargo Screening (including CSI), Customs-Trade Partnership Against Terrorism, and inspection and detection technology
(*Sources*: U.S. Department of Homeland Security. March, 2014. Budget-in-brief: Fiscal year 2015. Washington, DC, p. 72; U.S. Coast Guard. 2013 Performance Highlights, 2015 Budget in Brief, 2014, p. 24; U.S House of Representatives, Committee on Appropriations, Department of Homeland Security Appropriations Bill, 2014 (H. Report 113-91) and 2015 (H. Report 113-481).)

- $20 million through cutting TSA "Playbook" operations at selected airports
- $4.9 million in the TSA Federal Flight Deck Officer program through "efficiencies gained from the implementation of an Inactive Reserve Force, the consolidation of requalification facilities and the elimination of unfilled program management vacancies"
- $19.5 million in the Federal Air Marshal Service
- $10.9 million from elimination of four TSA Visible Intermodal Prevention and Response (VIPR) teams, decreasing the number of such teams from 37 to 33
- $1.4 million from elimination of four Coast Guard Vessel Board and Search Teams used for ports, waterways, and coastal security enforcement activities (U.S. Department of Homeland Security, 2014, pp. 1, 73–74, 84)

Passenger Aviation Security Fee

The impasse that led to the federal government shutdown during the first half of October 2013 was partially resolved via the enactment of the Bipartisan Budget Act (BBA) of 2013

(PL 113-67), which was negotiated by the chairs of the House and Senate Budget Committees, Representative Paul Ryan (R-WI) and Senator Patty Murray (D-WA) respectively. Under its terms, $63 billion was provided to restore a portion of sequester cuts, offset by reductions elsewhere in the budget, and the overall discretionary spending caps for FYs 2014 and 2015 were raised. This facilitated subsequent action on FY 2014 appropriations, including the restoration of funding for transportation security programs, in the Omnibus Appropriations Act of 2013 (U.S. House of Representatives, 2013, p. 1).

Among the BBA's provisions designed to offset the increases in discretionary spending was an increase in the passenger aviation security fee from $2.50 per enplanement (with a $5.00 maximum for a one-way trip) to a single $5.60 charge per one-way trip, effective July 1, 2014. It is projected that, over the next 10 years, the increased fee will generate $16.9 billion in additional collections. The BBA directed that $12.63 billion of this total is to be deposited in the general funds of the Treasury and subject to the regular appropriations process, with the remainder used as offsetting fees for aviation security costs. Additionally, effective October 1, 2014, the BBA repealed the Aviation Security Infrastructure Fee paid by the airlines, resulting in a $3.89 billion drop in offsetting fees for aviation security between FYs 2014 and 2023. Thus, the net effect of the new law is to make an additional $387.4 million available in offsetting fees specifically allocated for aviation security over the next 10 years ($23.86 million collected for this purpose under the new system versus $23.47 million collected under the old fee structure) while providing $12.6 billion in additional collections that may, or may not, be used for that purpose. (the president's FY 2015 budget proposal assumes the latter will be used for deficit reduction) (Federal Register, 2014, pp. 35463–35464, 35467–35470; Office of Management and Budget, 2014, p. 194).

The Administration's FY 2015 budget also proposed further increases in the aviation passenger security fee: to $6.00 per one-way trip in FY 2015, $6.50 in FY 2016, $7.00 in FY 2017, and $7.50 in FY 2018. These increases would result in an additional $11.3 billion in revenue from FY 2016 to FY 2024, which the Administration proposes to divide into $5.9 billion in offsetting fees to pay for aviation security and $5.4 billion to be used for nonsecurity purposes. Additionally, the president's budget calls for reinstatement of the Aviation Security Infrastructure Fee on the airlines, which would generate an additional $4.2 billion in offsetting collections over 10 years (Office of Management and Budget, 2014, p. 194).

In a May 2014 statement, Airlines for America (the principal trade association for U.S. airlines) expressed its concerns over increased passenger fees, which provide "no incremental [security] benefit for air travelers," and their potential economic impact:

> *The passenger airline industry remains a low-margin business, significantly lagging Standard and Poor's average net profitability—7.9 percent net margin for passenger airlines in 2013 compared to the S&P 500 average net margin of 10.4 percent. The airlines' first quarter 2014 net margin was a paltry 1.1 percent. One culprit is that airlines remain highly susceptible to volatile jet fuel prices. Jet fuel prices in 2013 exceeded $50 billion for the third straight year despite improved fuel efficiency.... [An increase in security fees] ignores the fact that air travel is often discretionary; higher costs count*

when consumers make the decision to fly or stay home, or to ship an item. The elasticity in demand for air travel is well documented. In 2012, GAO found that a one percent increase in the cost of an airline ticket (including taxes and fees) would result in a 1.12 percent reduction in the quantity of tickets sold. That unmistakably implies that further increases in government-imposed taxes and fees will dampen demand, reduce airline revenue and diminish overall U.S. economic activity.

Airlines for America, 2014

■ ■ Critical Thinking ■

What was the outcome of the FY 2015 appropriations process with respect to TSA, Coast Guard ports, waterways and coastal security, and CBP container security programs? Did these results follow the FY 2012 to 2013 patterns discussed earlier? Why or why not? Consult Congress.gov, the House and Senate Appropriations Subcommittees on Homeland Security, the Coast Guard's 2016 Budget in Brief, and other sources as necessary to research your answer.

Long-Term Sustainability

Funding for transportation and homeland security does not operate within a vacuum, and future prospects for these programs will be strongly affected by the overall budget situation in the United States and other countries. In the United States, the Budget Control Act of 2011 and other budgetary and appropriations decisions taken to reduce the size of the federal deficit—which had mushroomed as a result of the Great Recession of 2008 to 2009 and its aftermath—were successful in significantly reducing the deficit, in part by curbing spending. Transportation security was not immune to this trend, although the 2009 failed terrorist attempt to blow up the Northwest aircraft produced a temporary boost for aviation security. Despite this recent history, the Congressional Budget Office (CBO) has raised serious concerns about the budgetary outlook over the coming decade:

CBO now estimates that if the current laws that govern federal taxes and spending do not change, the budget deficit in fiscal year 2014 will be $492 billion. Relative to the size of the economy, that deficit—at 2.8 percent of gross domestic product (GDP)—will be nearly a third less than the $680 billion shortfall in fiscal year 2013, which was equal to 4.1 percent of GDP. This will be the fifth consecutive year in which the deficit has declined as a share of GDP since peaking at 9.8 percent in 2009. But if current laws do not change, the period of shrinking deficits will soon come to an end. Between 2015 and 2024, annual budget shortfalls are projected to rise substantially—from a low of $469 billion in 2015 to about $1 trillion from 2022 to 2024—mainly because of the aging population, an expansion of federal subsidies for health insurance, and growing interest payments on federal debt. CBO expects that cumulative deficits during that decade will equal $7.6 trillion if current laws remain unchanged. As a share of GDP, deficits are projected to rise from 2.6 percent in 2015 to about 4 percent near the

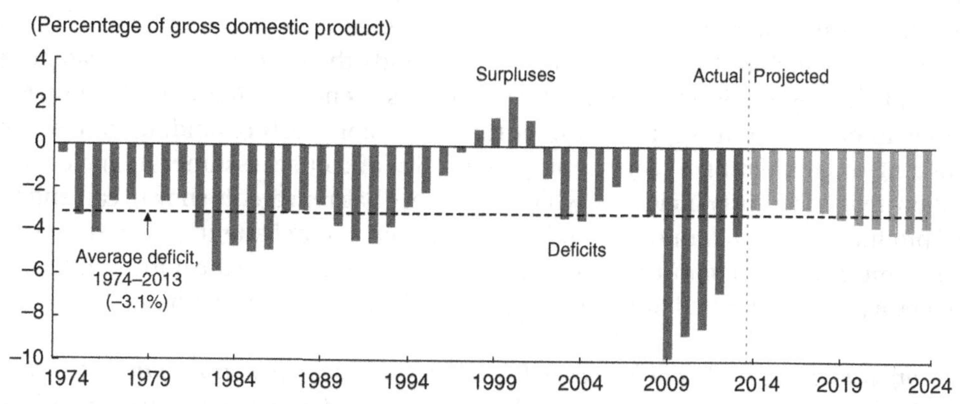

FIGURE 10.2 Total deficits or surpluses (percentage of gross domestic product). *(Source: Congressional Budget Office. April, 2014.* Updated budget projections: 2014 to 2024. *Washington, DC, p. 2.)*

> *end of the 10-year period.... Such high and rising debt would have serious negative consequences. Federal spending on interest payments would increase considerably when interest rates rose to more typical levels. Moreover, because federal borrowing would eventually raise the cost of investment by businesses and other entities, the capital stock would be smaller, and productivity and wages lower, than if federal borrowing was more limited* (Figure 10.2).

Furthermore, it should be noted that even in this "baseline" scenario of unchanged policies and rising deficits, the category of discretionary spending, under which most transportation security funding falls, is projected to grow very slowly, by just 1% from 2014 to 2017 and then by 2% annually thereafter (Congressional Budget Office, 2014, pp. 1, 5).

Other long-term constraints will also complicate future funding of transportation security programs. A 2012 Rand report focused on aviation security in the United States, but the same observations apply, to varying degrees, to the other modes and other nations, as well:

> *Looking to the future of aviation security in the United States, the resource constraints that are almost certain to affect most policy areas will be a challenge. Such constraints will be even more difficult to navigate as the lifespan of technologies and systems used now is exhausted and decisions to recapitalize, replace, or improve them must be made over the short-, medium-, and long-term policy horizons. Major investments have been made in imaging technologies, for example, whose operational lifetime is finite—meaning that even as resources may be declining, there will be requirements to spend just to maintain the status quo, much less expand or reform the aviation security system.*
>
> *Jackson et al., 2012, p. 137*

Looking ahead, it seems likely that transportation security policymakers will respond to resource constraints by using some or all of the methods used in the FY 2015 Obama budget proposal: efficiencies claimed through expanded targeting of programs (i.e., risk management), which will allow reductions in labor and other resource costs; greater usage of fees and offsetting collections from transportation system users (although, at best, these are likely to be utilized to supplant, rather than augment, existing funding sources); and downsizing or elimination of certain programs. This, of course, assumes no major transportation-related security incidents, which, as history has repeatedly shown, can and likely will produce major and unforeseeable shifts in governmental attention and resources.

The same 2012 Rand report highlighted the challenges, but also the opportunities, presented by a period of limited or no growth in transportation security funding:

> *For organizations and people charged with protecting citizens from harm, the potential for cuts in resources is always difficult to consider and implement. In addition to the highly charged politics surrounding homeland security measures, there will always be an understandable trepidation to make cuts out of fear that imprudent action will undermine effective security efforts. But if a sufficient analytical basis for assessing security measures and strategies is available, that trepidation can be made into an opportunity. Constraints force choices, which in turn force evaluation to help ensure that we are not spending limited national resources in ways that are not achieving what they are intended to achieve. In aviation security, where the total cost of the national effort has expanded significantly since 9/11, such evaluation could pay dividends not just in reduced national expenditures, but also by helping to identify ways to get comparable or better security for less cost—more efficient aviation security—that could make our homeland security efforts more sustainable and make the country better off in the long run.*
>
> *Jackson et al., 2012, p. 137*

Conclusion

Transportation security must be a priority for national governments. Harm to passengers and disruption of commerce can impose immense physical, psychological, and economic costs, and preventing such losses will continue to be an important objective. However, neither before nor after 9/11 has it been a transcendent goal whose attainment outweighs all other societal imperatives, including economic efficiency, civil liberties, and fiscal responsibility. The 9/11 attacks certainly produced a period—not only in the United States but also in much of the rest of the world—when these other priorities were subordinated to homeland security, but as time has passed and absent other calamities of similar magnitude, "these other claims have predictably and necessarily, reasserted themselves" (Johnstone, 2006, p. 108).

Attempts to weigh the economic costs and benefits of transportation security measures have proven problematic because of the uncertainties involved in calculating each of them. The chief benefit (of preventing a successful terrorist attack) is very difficult to quantify, given limitations in estimations of the probability of an attack absent the security measures as well as the magnitude of the consequences of an averted attack. On the other side, although the government's implementation expenses can be fixed with some precision, the indirect costs to industry (e.g., shipping delays, loss of passengers) and individuals (inconvenience in various forms) are much harder to determine. As a result, these economic impact analyses have been of limited value thus far in identifying an optimal balance between security and commerce.

The need for more invasive prescreening and screening technologies and systems, as well as for other expanded security activities, led the 9/11 Commission to call for "an enhanced system of checks and balances to protect the precious liberties that are vital to our way of life." The centerpiece of the Commission's recommendations in this regard was its proposal to create an independent board within the executive branch to oversee governmental compliance with privacy and civil liberties protections in carrying out its homeland security responsibilities. For a variety of reasons—many of them political—this body (called the Privacy and Civil Liberties Oversight Board) has been slow in activation, only reaching full operational status in May 2013, almost 9 years after the original 9/11 Commission recommendation. To date, it has played little role in addressing privacy-related disputes involving transportation security measures.

DHS's Privacy Office, which was created at the same time as the department, has sought to ensure compliance by DHS components with federal privacy laws and policies by issuing policy guidance; reviewing and approving Privacy Impact Assessments prepared by DHS entities for their programs; and conducting privacy compliance reviews of major DHS initiatives, including CBP's Automatic Targeting System and the U.S.-EU PNR Agreement. Despite these efforts, civil liberties groups have continued to raise strong objections to certain transportation security measures, especially those involved in the prescreening and screening of aviation passengers.

The increasing technological sophistication and interconnection of transportation control systems have amplified the vulnerability of such systems to cyber attacks. However, with most of such systems under private ownership and the U.S. government possessing limited authority and resources to enforce security requirements, most transportation cybersecurity efforts have been confined to dissemination of best practices.

Transportation security activities are, of course, dependent on the resources made available to carry them out. Prospects for future funding of these programs in the U.S. will likely be impacted by several limiting factors, including:

- Continuing profitability concerns in the transportation industry
- Ongoing partisan divisions in the Congress, leading to stalemates and delays in the appropriations process
- Rising federal budget deficits after FY 2015, which will produce growing pressures for cuts in federal spending
- The need to replace aging technologies and detection systems (e.g., the imaging systems used at airport checkpoints or the radiation portal monitors deployed at seaports)

Discussion Questions

1. Describe the process for assessing costs and benefits of proposed regulations in the United States and the difficulties in applying this process to transportation security programs.
2. What were the projected costs and "qualitative" benefits in TSA's 2011 rulemaking on air cargo screening? What was the "break-even" analysis?
3. What are the Privacy and Civil Liberties Oversight Board and the DHS Privacy Office? Which of these has been more active with respect to transportation security programs?
4. Describe the major privacy concerns posed by the ATS and how these have been addressed by DHS.
5. What are some of the cyber threats to transportation systems? (You may refer to threats not included in the text.)
6. Name factors that may limit future funding of transportation security programs and describe how these limitations may be dealt with by policymakers.

References

Airlines for America. May, 2014. Keynote address to the Federal Bar Association Transportation Security Administration annual legal conference. <http://www.airlines.org/Pages/Keynote-Address-to-the-Federal-Bar-Association-Transportation-Security-Administration-Annual-Legal-Conference.aspx> (accessed 10.27.14.)

Bipartisan Policy Center, Homeland Security Preparedness Group. September, 2011. Tenth anniversary report card: The status of the 9/11 Commission recommendations. Washington, DC.

Congressional Budget Office. April, 2014. Updated budget projections: 2014 to 2024. Washington, DC.

Electronic Privacy Information Center. n.d.a. About EPIC. <http://epic.org/epic/about.html> (accessed 10.27.14.)

Electronic Privacy Information Center. n.d.b. Whole body imaging technology and body scanners. <http://epic.org/privacy/airtravel/backscatter/> (accessed 10.27.14.)

Electronic Privacy Information Center. June, 2012. Comments of the Electronic Privacy Information Center on the U.S. Customs and Border Protection of the Department of Homeland Security Automated Targeting System notice of Privacy Act system of records and proposed rule: Privacy Act of 1974 exemptions. Washington, DC.

European Commission. November, 2013a. Joint review of the implementation of the Agreement between the European Union and the United States of America on the processing and transfer of passenger name records to the United States Department of Homeland Security. Brussels.

European Commission. November, 2013b. Frequently asked questions: The EU-US agreement on the transfer of Passenger Name Record (PNR) data. Brussels.

Farrow, S., Shapiro, S., April, 2009. The benefit-cost analysis of security focused regulations. Journal of Homeland Security and Emergency Management 6 (1). <http://www.degruyter.com/view/j/jhsem.2009.6.1/jhsem.2009.6.1.1482/> (accessed 10.27.14.)

Federal Register. January, 2011a. Presidential documents: Executive Order 13563, improving regulation and regulatory review. Washington, DC.

Federal Register. August, 2011b. Transportation Security Administration: Air cargo screening; final rule. Washington, DC.

Federal Register. May, 2012. Privacy Act of 1974: Implementation of exemptions; Automated Targeting System. Washington, DC.

Federal Register. June, 2014. Adjustment of Passenger Civil Aviation Security Service Fee. Washington, DC.

GAO. May, 2005. Maritime security: enhancements made, but implementation and sustainability remain key challengers. Washington, DC.

GAO. February, 2013. Cybersecurity: national strategy, roles, and responsibilities need to be better defined and more effectively implemented. Washington, DC.

Hatch, G., August, 2012. Privacy and Civil Liberties Oversight Board: New independent agency status. Congressional Research Service, Washington, DC.

Jackson, B.A., LaTourrette, T., Chan, E.W., Lundberg, R., Morral, A.R., Frelinger, D.R., 2012. Efficient aviation security: Strengthening the analytic foundation for making air transportation security decisions. Rand Corporation, Santa Monica, CA.

Johnstone, R.W., 2006. 9/11 and the Future of Transportation Security. Praeger, Westport, CT.

National Initiative for Cybersecurity Careers and Studies. n.d. Explore Terms: A Glossary of Common Cybersecurity Terminology. <http://niccs.us-cert.gov/glossary> (accessed 10.27.14.)

Office of Management and Budget. October, 1992. Circular no. A-94 revised: guidelines and discount rates for benefit-cost analysis of federal programs. <http://www.whitehouse.gov/omb/circulars_a094> (accessed 10.27.14.)

Office of Management and Budget. 2005. Validating regulatory analysis: 2005 report to Congress on the costs and benefits of federal regulations and unfunded mandates on state, local, and tribal entities. Washington, DC.

Office of Management and Budget. 2006. 2006 report to Congress on the costs and benefits of federal regulations and unfunded mandates on state, local, and tribal entities. Washington, DC.

Office of Management and Budget. 2008. 2008 report to Congress on the benefits and costs of federal regulations and unfunded mandates on state, local, and tribal entities. Washington, DC.

Office of Management and Budget. 2009. 2009 report to Congress on the benefits and costs of federal regulations and unfunded mandates on state, local, and tribal entities. Washington, DC.

Office of Management and Budget. March, 2014. Budget of the U.S. Government, fiscal year 2015: analytical perspectives. Washington, DC.

Painter, W.L., March, 2014. Department of Homeland Security appropriations: FY2014 overview and summary. Congressional Research Service, Washington, DC.

Privacy and Civil Liberties Oversight Board. n.d. About the PCLOB. <http://www.pclob.gov/about-us> (accessed 10.27.14.)

Privacy and Civil Liberties Oversight Board. November, 2013. Semi-annual report March 2013- September 2013. Washington, DC.

Privacy and Civil Liberties Oversight Board. 2014. Report on the Telephone Records Program conducted Under Section 215 of the USA PATRIOT Act and on the Operations of the Foreign Intelligence Surveillance Court. January, 2014, pp. 8, 10, 168.

Roadmap to Secure Control Systems in the Transportation Sector. August, 2012. Washington, DC.

Stanley, J., February, 2013a. Small but significant privacy oversight institution almost a reality after pathetic story of delay. [Web log post]. <https://www.aclu.org/print/blog/national-security-technology-and-liberty/> (accessed 10.27.14.)

Stanley, J., November, 2013b. What powers does the civil liberties oversight board have? [Web log post]. <https://www.aclu.org/print/blog/national-security-technology-and-liberty/> (accessed 10.27.14.)

The 9/11 Commission. 2004. The 9/11 Commission Report: Final Report of the National Commission on Terrorist Attacks Upon the United States. Norton, New York.

Transportation Security Administration. March, 2014. Transportation Security Administration, surface transportation security fiscal year 2015 congressional [budget] justification. Washington, DC.

U.S. Department of Homeland Security. n.d.a. About the privacy office. <http://www.dhs.gov/about-privacy-office> (accessed 10.27.14.)

U.S. Department of Homeland Security, Privacy Office. n.d.b. Privacy Office—Privacy Impact Assessments (PIA). <http://www.dhs.gov/privacy-office-privacy-impact-assessments-pia> (accessed 10.27.14.)

U.S. Department of Homeland Security. n.d.c. System of Records Notices (SORNs). <http://www.dhs.gov/system-records-notices-sorns (accessed 10.27.14.)

U.S. Department of Homeland Security. n.d.d. Cybersecurity results. <http://www.dhs.gov/cybersecurity-results> (accessed 10.27.14.)

U.S. Department of Homeland Security. 2010. Transportation Systems Sector-specific Plan: an annex to the National Infrastructure Protection Plan. Washington, DC.

U.S. Department of Homeland Security. November, 2011. Blueprint for a secure cyber future: the cybersecurity strategy for the homeland security enterprise. Washington, DC.

U.S. Department of Homeland Security. June, 2012. Privacy Impact Assessment for the Automated Targeting System. Washington, DC.

U.S. Department of Homeland Security. April, 2013a. Budget-in-brief: Fiscal year 2014. Washington, DC.

U.S. Department of Homeland Security, Privacy Office. July, 2013b. A report on the use and transfer of passenger name records between the European Union and the United States. Washington, DC.

U.S. Department of Homeland Security, Privacy Office. November, 2013c. 2013 report to Congress. Washington, DC.

U.S. Department of Homeland Security. March, 2014. Budget-in-brief: Fiscal year 2015. Washington, DC.

U.S. District Court of Appeals for the District of Columbia Circuit. July, 2011. Electronic Privacy Information Center, et al., v. United States Department of Homeland Security, et al. Washington, DC.

U.S. House of Representatives, Committee on the Budget. December, 2013. Summary of the Bipartisan Budget Act of 2013. Washington, DC.

U.S. Travel Association. 2011. A better way: building a world class system for aviation security. Washington, DC.

White House, U.S. Office of Government Ethics. 2011. Plan for Shutting Down Operations in the Event of a Lapse of Appropriations. Washington, DC.

Conclusion

Despite the surge in funding and attention, pursuit of optimal transportation security policies in the United States since 9/11 has been complicated for a number of reasons:

- At the very outset, the newly created Transportation Security Administration (TSA) was focused on meeting a series of specific, congressionally established mandates, primarily pertaining to commercial aviation security screening, rather than on developing a comprehensive approach to transportation security. Furthermore, this legacy—in the form of the screener workforce and the technologies they operate—continues to dominate the budget and personnel of the agency.
- One year after TSA was created, the Homeland Security Act of 2002 was adopted, and a new Department of Homeland Security (DHS) was formed out of a merger of 22 existing federal departments and agencies, including the components made responsible for various aspects of transportation security: TSA, the U.S. Coast Guard, and the Customs Service (now Customs and Border Protection [CBP]). These agencies brought with them their own institutional history and approaches and, in the case of the Coast Guard and Customs, retained significant non–homeland security responsibilities that vie for their budgets and policy focus.
- When DHS was first established, departmental leadership did *not* form a distinct policy office, only doing so—under some external prodding—in 2005. Since then, that office has evolved into its current role as the lead in intradepartmental policy development, coordination, and integration while also overseeing DHS's international engagement efforts and informing the department's budget process through program and acquisitions guidance and strategy development. Yet it has been consistently underfunded in the congressional appropriations process (Carafano and Zuckerman, 2012). The Office of Policy's final funding level in fiscal year (FY) 2013 was $41.6 million, supporting 194 positions. In FY 2014, this was trimmed to $36.5 million, for 173 positions, and the president's FY 2015 budget proposed only a partial restoration ($38.5 million for 180 positions) (U.S. Department of Homeland Security, 2014a).
- Congress has failed to offer adequate direction and prioritization (having never enacted a comprehensive authorization bill for DHS since the department's creation) or oversight (which has been severely fragmented, with more than 100 committees and subcommittees asserting some form of jurisdiction over DHS in 2011–2012). In addition, partisan gridlock has seriously hampered the appropriations process in recent years.
- Individual security incidents continue to play a leading role in shaping policy development, for good and ill. For example, the Christmas 2009 attempted bombing

of Northwest Flight 253 produced an $800 million infusion of resources into the already heavily funded airport checkpoint screening technologies (as well as canine explosives teams and Federal Air Marshals). This resulted in more rapid deployment of Advanced Imaging Technology equipment with greater capabilities to detect explosives but also to the wasting of millions of dollars on the purchase and installation of the "backscatter" version of this equipment, which was subsequently decommissioned because of privacy concerns.

Despite these impediments, progress has been made in most fields of transportation security policy since 2001. International frameworks for aviation, ship and port, and supply chain security have been updated and strengthened. Advances have been made in detection technologies, risk management principles have been implemented in an increasing array of programs (e.g., TSA's Pre✓ and CBP's Automated Targeting System [ATS]), better and more widely disseminated intelligence and information has been developed, more industry "best practices" have been implemented, and more vulnerability assessments and security training exercises have been conducted. Last, a substantial number of strategies and plans have been issued, even if many are at a rudimentary stage of development. As the Government Accountability Office (GAO) and the DHS Office of Inspector General regularly document, there is considerable room for improvement in almost every one of these undertakings, but a key point is that, after more than a decade, foundations for transportation security policies are being established, and after these foundations are "put down on paper" (or deployed in the field), they can be examined, questioned, and improved.

However, it is the view of this author that these efforts at policy improvement will continue to fall short of their potential, as well as what should ultimately be required of them, without proper attention to fundamental questions about balance, funding, and organization, attention that can only come from policymakers at the highest levels of the executive and legislative branches (Johnstone, 2006). (Balancing security needs with other national priorities, including economic efficiency and privacy, and addressing future funding constraints were considered in Chapter 10, and the issue of roles and responsibilities was presented in Chapter 4.)

Venturing even more deeply into the realm of personal opinion, the following recommendations are offered as steps toward answering these questions:

1. *Congress needs to pass an authorization bill for DHS and its components (including TSA, the Coast Guard, and CBP).* To quote again from a 2005 House Committee report, "The complexity of the [Homeland Security] Department's missions, coupled with the enormity of its management and operational challenges, requires the close and continuing oversight that an annual Congressional re-authorization provides.… DHS should be subject to an annual authorization process through which the evolving needs of the Department can be met, and through which Congressional direction, oversight, and prioritization can take place. An annual authorization will help the Department improve the overall management and integration of its various legacy agencies, to guide resource allocation and prioritization, to set clear and

achievable benchmarks for progress and success, and to enhance the Department's implementation of its critical mission" (see Chapter 5). In their current gridlocked state, passing such an authorization bill—an achievement that has eluded previous Congresses operating in somewhat less divided times—would appear to be beyond the reach of the U.S. House and Senate for the foreseeable future. Nonetheless, it remains an important objective in improving homeland security (and transportation security) policy, and Congress should be held to account for its continuing absence.

2. *A prestigious independent advisory panel should be created to help reconcile the security needs addressed by the ATS (and similar systems) with privacy concerns.* On 9/11, the only security layer that "worked" was the Computer-Assisted Passenger Prescreening System (CAPPS), which targeted for security attention passengers whose profiles (mainly derived from ticketing information) indicated that they may pose security threats. On September 11, 2001, the CAPPS system directly selected seven of the 19 hijackers, with two more added at the discretion of an airline ticket counter customer representative because of suspicious behavior and a 10th (the ringleader Atta himself) targeted by the system's random selection feature. Because the consequences of CAPPS designation were at that time limited to screening of selectees' checked baggage, the success of the system in identifying potential threats proved irrelevant in dealing with the 9/11 plot (see Chapter 2). However, its performance illustrated the potential of prescreening in helping to target security resources toward the greatest security threats. Deployment of the CAPPS system—with its usage of personal information from all passengers, including the vast majority who had done nothing wrong and posed no threat to civil aviation—understandably raised significant worries about privacy. When the Gore Commission (the White House Commission on Aviation Safety and Security), which was the body that initially recommended the adoption of what became the CAPPS system, faced this question in 1997, it established a highly regarded Civil Liberties Advisory Board (chaired by Floyd Abrams, a noted expert on Constitutional law) to meet with the Commission "to discuss civil liberties concerns pertaining to profiling." The Advisory Board subsequently submitted its recommendations for addressing such issues, and the Commission included many of these in its own recommendations in the belief that "civil liberties that are so fundamentally American should not, and need not, be compromised by a profiling system" (White House Commission on Aviation Safety and Security, 1997). Although the privacy protections incorporated into the Gore Commission recommendations did not at the time of their presentation, or later, fully resolve concerns held by civil liberties advocates, they did help win general acceptance of the CAPPS system and facilitated its deployment. In large part because of its inability to successfully address privacy issues highlighted by Congress and others, TSA has made little advance in further development of CAPPS or any successor system, and most of the impetus for refinements in profiling (or targeting) has come through CBP's ATS. These augmentations in ATS capability have continued to raise questions about the impact on civil liberties (see Chapter 10). Profiling systems

will be key in developing transportation security measures that are more efficient, effective, and sustainable than those currently in use (by targeting security measures on the greatest potential threats while easing the security burden on non-threats), and thus further development of ATS (and CAPPS) would be desirable. However, if these policies are to win widespread acceptance (and thus remain sustainable over the long run), concerns about civil liberties must be taken fully into account at each stage of such development. Although the DHS Privacy Office and the Privacy and Civil Liberties Oversight Board can and should play a role in representing those interests, at present, neither of these possesses the reputation, the clout, or the independence to adequately validate that effort. Therefore, DHS should follow the precedent established by the Gore Commission and establish an independent advisory board, composed of well-known and well-regarded civil liberties and constitutional law experts, to review and make recommendations on privacy protections for ATS and other targeting systems using personal information.

3. *The executive branch should explicitly delineate the roles of federal and nonfederal entities involved in transportation security, and should detail how transportation security policies are to be prioritized, funded, and sustained over time.* In 2004, the 9/11 Commission (2004) wrote:

   The U.S. government should identify and evaluate the transportation assets that need to be protected, set risk-based priorities for defending them, select the most practical and cost-effective ways of doing so, and then develop a plan, budget, and funding to implement the effort. The plan should assign roles and missions to the relevant authorities (federal, state, regional, and local) and to private stakeholders.

Over the past 10 years, much has been accomplished within DHS and elsewhere in the nation in identifying transportation assets, developing plans for protecting them, and seeking funding for the specific security measures adopted (see Chapters 3, 5, 6, 7, & 8). Furthermore, arrangements have been made for the coordination of intra- and interdepartmental and governmental and private sector transportation security activities when roles or interests overlap (see Chapter 4). However, the comprehensive budget plan and clear assignment of roles and responsibilities envisioned by the 9/11 Commission remain largely unrealized objectives. For example, although the 9/11 Act of 2007 mandated a Quadrennial Homeland Security Review (QHSR) that was to contain, among other things, an identification of "the budget plan required to provide sufficient resources to successfully execute the full range of missions called for in the national homeland security strategy," neither the first such review (issued in 2010) nor the second one (issued in June 2014) did so. In addressing the budget issue, the latter simply stated:

The out-year funding assumptions applied for this quadrennial review are based on the economic and policy assumptions underpinning the President's 2015 Budget submission to Congress. Since the last Quadrennial Homeland Security Review, economic conditions have had wide-ranging impacts across homeland security partners and stakeholders, affecting both daily operations and current investments to meet longer-term needs and challenges.... Going forward, the budgets of many homeland security partners are assumed to maintain parity with inflation or modestly decline in real terms. We also assume that state budgets will be constrained by reductions in federal grants, which are projected to remain below their 2007 historic high (as a percentage of gross domestic product). International partners will likely face similar constraints. Economic pressures on families, nonprofits, and the private sector may also adversely affect local investment in the security and resilience of our communities.

U.S. Department of Homeland Security, 2014b

These observations are all reasonable (and indeed likely) but appear to fall well short of offering the type of budgetary *guidance* (for agencies and Congress) in setting funding priorities advocated by the 9/11 Commission. Such guidance will be all the more necessary as the era of constrained resources identified in the 2014 QHSR comes to pass.

4. *DHS should step up efforts to improve its benefit–cost analyses and performance metrics.* Attempts to enhance transportation security are crucially dependent on the quality of information used in risk management, benefit–cost analysis, and performance measurement. Although the first of these—calculations of threat, vulnerability, and consequences—has appropriately received considerable attention, both internally within DHS and externally (from such sources as GAO), less focus has been directed at the other two, which are also vital in order for a proper determination to be made of the cost effectiveness of proposed security measures and the efficacy of such measures after they have been deployed. The benefit–cost analyses for transportation security issued by the Office of Management and Budget or included in DHS rulemaking documents have been limited, both quantitatively and qualitatively, with many important programs omitted entirely by the former and the latter beset by a number of uncertainties in its attempts to estimate both costs and benefits (see Chapter 10). Assessments of the actual performance of specific transportation security measures are also plagued by significant problems. Most of DHS's performance indicators do not truly measure performance or security outcomes but rather "symptoms" of performance, such as tallying outputs (e.g., percentage of cargo screened) or reporting on inspection-based compliance with "leading security indicators." The Coast Guard uses a different approach, using risk-based outcome measures, but these have been developed for internal use only (and thus not part of department-wide performance evaluations) and the reliability of the factors used in arriving at the results cannot be determined through publicly available information (see Chapter 9). A variety of approaches can be taken to upgrade DHS's economic and performance metrics. The department has established an annual process for performance measure improvement, but this is largely confined to internal assessments within the department. Providing a greater role for outside experts (including the National Research Council, which has been involved in evaluating

DHS risk management efforts, and "think tanks" such as the Rand Corporation) would appear to offer a number of advantages in drawing on a variety of perspectives and expertise and in offering a means for external validation of the measurements. Similarly, expanded use of independent, external reviews of DHS benefit–cost analyses should lead not only to improved metrics but also to greater confidence in their findings. Congress could aid these efforts by tasking GAO to undertake a comprehensive evaluation of DHS performance measures (including those employed by the Coast Guard) and benefit–cost analyses and compare these with the largely nonquantitative assessments conducted by that office.

In a relatively short period of time, transportation security has moved from being a minor aspect of transportation policy, which was primarily concerned with mobility and safety, to an independent discipline, with its own organizations, policies, and programs. The rapid expansion and evolution of this enterprise has produced significant accomplishments in increasing protections for transportation systems and their passengers and cargo but at a price in financial costs (both direct and indirect) and inconvenience (including infringements on civil liberties). Unlike the early years of this expansion, when resources and policy attention rose dramatically, transportation security programs now face an era of budgetary constraints and diminished focus, requiring policymakers (including those in Congress) to pay greater attention to setting priorities and determining the most cost-effective security measures. However, as in the past, all of this is subject to the occurrence (or absence) of major security incidents that can produce substantial and hurried shifts in funding and policies.

References

Carafano, J., Zuckerman, J., 2012. DHS Office of Policy: Misguided reorganization threatens homeland security strategic planning. Heritage Foundation, Washington, DC.

Johnstone, R.W., 2006. 9/11 and the Future of Transportation Security. Praeger, Westport, CT, pp. 107–113.

The 9/11 Commission. 2004. The 9/11 Commission Report: Final Report of the National Commission on Terrorist Attacks Upon the United States. Norton, New York, p. 391.

U.S. Department of Homeland Security. 2014a. Congressional Budget Justification FY 2015. Office of the Secretary and Executive Management, Washington, DC, p. 18.

U.S. Department of Homeland Security. 2014b. The 2014 Quadrennial Homeland Security Review, Washington, DC, pp. 11–12, 26–27

White House Commission on Aviation Safety and Security. 1997. Final Report to President Clinton, February 12, 1997.<http://fas.org/irp/threat/212fin~1.html> (accessed 10.27.14.)

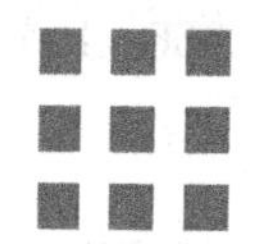

Index

J

K

L

M

O

T

Printed in the United States
By Bookmasters